GROUP SUPERMATRICES IN FINITE ELEMENT ANALYSIS

ELLIS HORWOOD SERIES IN CIVIL ENGINEERING
Series Editors
Structures: Professor H.R. EVANS, Department of Civil Engineering, University College, Cardiff
Hydraulic Engineering and Hydrology: Dr R.H.J. SELLIN, Department of Civil Engineering, University of Bristol
Geotechnics: Professor D. MUIR WOOD, Cormack Professor of Civil Engineering, University of Glasgow

Addis, W.	STRUCTURAL ENGINEERING: The Nature of Theory and Design
Allwood, R	TECHNIQUES AND APPLICATIONS OF EXPERT SYSTEMS IN THE CONSTRUCTION INDUSTRY
Bhatt, P.	PROGRAMMING THE MATRIX ANALYSIS OF SKELETAL STRUCTURES
Blockley, D.I.	THE NATURE OF STRUCTURAL DESIGN AND SAFETY
Britto, A.M. & Gunn, M.J.	CRITICAL STATE SOIL MECHANICS VIA FINITE ELEMENTS
Bljuger, F.	DESIGN OF PRECAST CONCRETE STRUCTURES
Carmichael, D.G.	STRUCTURAL MODELLING AND OPTIMIZATION
Carmichael, D.G.	CONSTRUCTION ENGINEERING NETWORKS
Carmichael, D.G.	ENGINEERING QUEUES IN CONSTRUCTION AND MINING
Cyras, A.A.	MATHEMATICAL MODELS FOR THE ANALYSIS AND OPTIMISATION OF ELASTOPLASTIC SYSTEMS
Dowling, A.P. & Ffowcs-Williams, J.E.	SOUND AND SOURCES OF SOUND
Edwards, A.D. & Baker, G.	PRESTRESSED CONCRETE
Fox, J.	TRANSIENT FLOW IN PIPES, OPEN CHANNELS AND SEWERS
Graves-Smith, T.R.	LINEAR ANALYSIS OF FRAMEWORKS
Gronow, J.R., Schofield, A.N. & Jain, R.K.	LAND DISPOSAL OF HAZARDOUS WASTE
Hendry, A.W., Sinha, B.A. & Davies, S.R.	LOADBEARING BRICKWORK DESIGN, 2nd Edition
Heyman, J.	THE MASONRY ARCH
Holmes, M. & Martin, L.H.	ANALYSIS AND DESIGN OF STRUCTURAL CONNECTIONS: Reinforced Concrete and Steel
Iyengar, N.G.R.	STRUCTURAL STABILITY OF COLUMNS AND PLATES
James, A.N.	SOLUBLE MATERIALS IN CIVIL ENGINEERING
Jordaan, I.J.	PROBABILITY FOR ENGINEERING DECISIONS: A Bayesian Approach
Kwiecinski, M.	PLASTIC DESIGN OF REINFORCED SLAB-BEAM STRUCTURES
Lencastre, A.	HANDBOOK OF HYDRAULIC ENGINEERING
Leroueil, S., Magnan, J.P. & Tavenas, F.	EMBANKMENTS ON SOFT CLAYS
MacLeod, I.	ANALYTICAL MODELLING OF STRUCTURAL SYSTEMS
Megaw, T.M. & Bartlett, J.	TUNNELS: Planning, Design, Construction
Melchers, R.E.	STRUCTURAL RELIABILITY ANALYSIS AND PREDICTION
Mrazik, A., Skaloud, M. & Tochacek, M.	PLASTIC DESIGN OF STEEL STRUCTURES
Pavlovic, M.	APPLIED STRUCTURAL CONTINUUM MECHANICS: Elasticity
Pavlovic, M.	APPLIED STRUCTURAL CONTINUUM MECHANICS: Plates
Pavlovic, M.	APPLIED STRUCTURAL CONTINUUM MECHANICS: Shells
Shaw, E.M.	ENGINEERING HYDROLOGY TECHNIQUES IN PRACTICE
Spillers, W.R.	INTRODUCTION TO STRUCTURES
Vilnay, O.	CABLE NETS AND TENSEGRIC SHELLS: Analysis and Design Applications
Zlokovic, G.	GROUP THEORY AND G-VECTOR SPACES IN STRUCTURAL ANALYSIS: Vibration, Stability and Statics

GROUP SUPERMATRICES IN FINITE ELEMENT ANALYSIS

GEORGE M. ZLOKOVIĆ, D.Sc., C.Eng.
Professor of Structure Systems and Space Structures
Faculty of Architecture
University of Belgrade, Yugoslavia

ELLIS HORWOOD
NEW YORK LONDON TORONTO SYDNEY TOKYO SINGAPORE

First published in 1992 by
ELLIS HORWOOD LIMITED
Market Cross House, Cooper Street,
Chichester, West Sussex, PO19 1EB, England

A division of
Simon & Schuster International Group
A Paramount Communications Company

© Ellis Horwood Limited, 1992

All rights reserved. No part of this publication may be reproduced, stored in a retrieval system, or transmitted, in any form, or by any means, electronic, mechanical, photocopying, recording or otherwise, without the prior permission, in writing, of the publisher

Printed and bound in Great Britain
by Hartnolls, Bodmin

British Library Cataloguing in Publication Data

A catalogue record for this book is available from the the British Library

0–13–365917–8

Library of Congress Cataloging-in-Publication Data

Available from the publisher

Table of contents

Preface 7

Introduction 9

1 Results of groups, vector spaces and representation theory 13
   1.1 Introduction 13
   1.2 Groups 14
   1.3 Vector spaces 20
   1.4 Representation theory 33

2 G-vector analysis 47
   2.1 Nodal configurations with patterns based on the square 48
   2.2 Nodal configuration with a pattern based on the cube 58
   2.3 Nodal configuration with a pattern based on the right parallelepiped 62
   2.4 Nodal configurations with patterns based on the regular icosahedron 67
   2.5 Algorithmic scheme of G-vector analysis 84

3 Group supermatrix transformations 90
   3.1 Definition of group supermatrices and formulation of their relations 90
   3.2 Group supermatrices of the group C_2 94
   3.3 Group supermatrices of the group C_3 104
   3.4 Group supermatrices of the group C_4 107
   3.5 Group supermatrices of the group C_{2v} 111
   3.6 Group supermatrices of the group C_{3v} 121
   3.7 Group supermatrices of the group C_{4v} 126
   3.8 Group supermatrices of the group D_{2h} 130

4 Formulation of shape functions in G-invariant subspaces 146
   4.1 Group supermatrix procedure for derivation of element shape functions
       in G-invariant subspaces 146
   4.2 Four-node rectangular element 153

Table of contents

    4.3 Eight-node rectangular element 159
    4.4 Twelve-node rectangular element 170
    4.5 Sixteen-node rectangular element 185
    4.6 Eight-node rectangular hexahedral element 199
    4.7 Twenty-node rectangular hexahedral element 205
    4.8 Thirty-two-node rectangular hexahedral element 224
    4.9 Sixty-four-node rectangular hexahedral element 241

5 Stiffness equations in G-invariant subspaces 254
    5.1 Group supermatrix procedure for derivation of stiffness equations in G-invariant subspaces 254
    5.2 Stiffness equations in G-invariant subspaces for one-dimensional elements 257
    5.3 Stiffness equations in G-invariant subspaces for the beam element 266
    5.4 Stiffness equations in G-invariant subspaces for the rectangular element for planar analysis 273
    5.5 Stiffness equations in G-invariant subspaces for the rectangular element for plate flexure 290

6 Group supermatrices in formulation and assembly of stiffness equations 326
    6.1 Group supermatrix procedure in the direct stiffness method 326
    6.2 Linear beam element assembly 330
    6.3 Girder grillage 339

Appendix — Character tables 357

Bibliography 370

Index 374

Preface

The object of this work is to introduce and develop the new concept of group supermatrices in the formulations of shape functions and stiffness equations of finite elements possessing symmetry properties, as well as in the formulation and assembly of system stiffness equations.

In comparison with the conventional approach, formulations by group supermatrices in finite element analysis provide substantial qualitative and quantitative advantages realized by the concept of shape functions in G-invariant subspaces and obtained in the form of drastic reductions of the amounts of derivation and computation.

In order to make this present volume self-contained, results of groups, vector spaces and representation theory are given in the form of extracts from my earlier book (Zloković, 1989) published by Ellis Horwood in 1989.

The algorithmic procedure, called G-vector analysis and developed in the above volume, is formulated for configurations with simple and complex symmetry. Beside elements with nodal patterns based on the square, the cube and the right parallelepiped, in this present volume configurations are analysed with nodal patterns having the very complex symmetry of the icosahedral group.

Definition of group supermatrices and formulation of their relations means that the system of equations with the group supermatrix in normal form is turned by the transformation supermatrix into the system of equations with the group supermatrix in diagonal form, which corresponds to decomposition of a G-vector space into G-invariant subspaces. Group supermatrices are derived for particular groups and respective elements with nodal patterns, providing maximum utilization of their symmetry properties.

Element shape functions in G-invariant subspaces are derived by the group supermatrix procedure, which provides decomposition of the function of the displacement field into symmetry-adapted functions in G-invariant subspaces.

One of the original features of the procedure is that the terms of the polynomial of the displacement function are obtained as a product of the column and row vectors of Cartesian sets and products as they stand in addition to the character table of the group pertaining to the symmetry types of irreducible group representations. The terms are obtained in the unique order that produces the system of equations with the group supermatrix in diagonal form, providing the decomposition of the space of shape functions into G-invariant subspaces.

By this procedure shape functions are derived in G-invariant subspaces for rectangular elements with 4, 8, 12 and 16 nodes, and for rectangular hexahedral elements with 8, 20, 32 and 64 nodes. Alternatively, these shape functions are also derived from the known shape functions of serendipity elements.

Element stiffness equations in G-invariant subspaces are derived by means of the element shape functions in the subspaces. This produces the element stiffness equation with the group supermatrix in diagonal form, which is turned by the group supermatrix transformation into the element stiffness equation with the group supermatrix in normal form. This equation is turned by another transformation, concerning positive directions of generalized displacements, into the conventional stiffness equation.

Stiffness equations formulated by the group supermatrix procedure are derived for one-dimensional elements, the beam element, the rectangular element for planar analysis and the rectangular elements for plate flexure. In comparison with conventional methods the group supermatrix procedure provides substantial qualitative and quantitative advantages throughout the entire analysis.

The group supermatrix procedure in the direct stiffness method, developed in my book (1989) and coordinated with formulations in this present volume, includes the process of direct superposition that is applicable in finite element analysis. In this procedure the initial configuration, with generalized displacements and forces, has the symmetry type of the first irreducible representation of the group that describes the symmetry properties of the configuration. With this and other requirements direct superposition of element stiffness matrices compiles the system stiffness group supermatrix in normal form. This supermatrix is turned by the group supermatrix transformation into the stiffness group supermatrix in diagonal form. The system stiffness equation with this supermatrix contains systems stiffness equations in G-invariant subspaces, which provides solutions for each subspace separately. By applying the transformation supermatrix to these solutions the final results are obtained as linear combinations of the results from the subspaces, with dimensions of the matrices in the subspaces which are a fraction of the dimension of the matrix in conventional analysis.

The group supermatrix procedure in finite element analysis has shown that new great possibilities for maximum utilization of symmetry properties of finite elements and their assembly lie in the choice of finite elements (and their nodal patterns) with complex symmetry, which enables decomposition of the system into a great number of subspaces. Then, theoretical formulation, derivation of equations and computation are a small fraction of the amount of these operations in conventional procedures with customary utilization of symmetry.

Finally, I should like to express my gratitude to the publishers Ellis Horwood and Simon and Schuster.

George Zlokovic´

Introduction

When a finite element with its nodal pattern possesses symmetry properties which are described by a group, the group supermatrix procedure can provide formulations with maximum utilization of symmetry. The concept of group supermatrices is developed for formulations of element shape functions, element stiffness equations and assembly of system stiffness equations. In comparison with the conventional use of symmetry, formulations by group supermatrices in finite element analysis have many advantages that result from decomposition of the G-vector space of the system into G-invariant subspaces.

Results of groups, vector spaces and representation theory, presented in Chapter 1, provide the basis for formulation of group supermatrices in G-vector spaces, with the object of achieving maximum utilization of symmetry properties of finite elements and their assembly. Basic results of groups are used to formulate symmetry groups and group representations. Extracts from n-dimensional vector spaces, linear operators, basis transformations and unitary spaces are utilized in formulations of group algebra and G-vector spaces. Results of group representations, group algebra and G-vector spaces are employed in formulations of G-vector analysis and the group supermatrix procedure.

Chapter 2 contains a short presentation of the G-vector analysis which was developed by Zlokovic (1973, 1974, 1976, 1977, 1989) for various fields of the theory of structures. When an element (or a nodal configuration) possesses symmetry properties that can be described by a group, G-vector analysis provides a decomposition of the G-vector space of the element (or the nodal configuration) into G-invariant subspaces. Then, in comparison with conventional utilization of symmetry, it is possible to reduce, to a large extent, the work for derivation of equations and computation.

Beside elements with nodal patterns based on the square, the cube and the right parallelepiped, configurations with patterns based on the regular icosahedron are also analysed. G-vector analysis is systemized in the form of an algorithmic scheme which may serve as a basis for computer programs.

Chapter 3, on group supermatrix transformations, deals with definition of group supermatrices and formulation of their relations. Following their general definition, group supermatrices are derived for particular groups and respective elements with nodes which are permuted within a single nodal set by the action of group elements.

For the most convenient formulation of group supermatrices it is necessary to determine the optimum nodal numbering, which is accomplished by preserving the structure of the matrix of characters (in the character table of the group) for derivation of basis vectors of G-invariant subspaces. The nodal numbering is established by assuming the optimum position for the first node of the nodal pattern and by applying to this node group elements in the order provided by the sequence of conjugate elements in the character table of the group.

Applications of group elements to particular nodes produce respective permutations of the nodes, forming a permutation table, which defines the group supermatrix in normal form when the nodal numbers are replaced by matrices. In order to provide derivations of formulations in supermatrix form, the character table of the group is transformed into supermatrix form. By utilizing such a table the relation of idempotents of the centre of group algebra and the class sums (sums of classes of conjugate elements) is established in supermatrix form too.

Applying the idempotents to the nodal functions derives the basis vectors of G-invariant subspaces of the G-vector space of the element. The basis transformation supermatrix, relating basis vectors to nodal functions, is orthonormal. Conjugating by this supermatrix the group supermatrix in normal form produces the group supermatrix in diagonal form. Thus, a system of equations with the group supermatrix in normal form is transformed into the system of equations with the group supermatrix in diagonal form, where its matrices, symmetry-adapted variables and free members pertain to respective G-invariant subspaces.

Group supermatrices in normal and diagonal forms are derived for basic patterns pertaining to symmetry groups C_2, C_3, C_4, C_{2v}, C_{3v}, C_{4v} and D_{2h}.

In Chapter 4 element shape functions in G-invariant subspaces are formulated by means of the group supermatrix procedure which contains the following.

(1) Introduction of the unique group supermatrix nodal numbering.
(2) Derivation of basis vectors of G-invariant subspaces.
(3) Formulation of relations of basis vectors and nodal functions as systems of equations with supermatrices in diagonal form.
(4) Derivation of the polynomial function of the displacement field decomposed into G-invariant subspaces.
(5) Determination of relations of generalized displacements to coefficients of the displacement polynomial in G-invariant subspaces.
(6) Determination of the coefficients of the displacement polynomial of G-invariant subspaces.
(7) Formulation of displacement fields in G-invariant subspaces.
(8) Derivation of element shape functions in G-invariant subspaces.

In this procedure the terms of the polynomial of the displacement field are formulated by the product of the column and row vectors of Cartesian sets and products given in addition to the character table of the group, where they pertain to the symmetry types of irreducible group representations. The order of the terms obtained in this way is the unique order which enables formulation of the systems of equations with group supermatrices in diagonal form, providing decomposition of the G-vector space of shape functions into G-invariant subspaces.

Element shape functions in G-invariant subspaces can be also derived by group supermatrix transformation of the known shape functions of serendipity elements. These functions are formulated according to the group supermatrix procedure and transformed by the transformation supermatrix into shape functions in G-invariant subspaces.

The group supermatrix procedure was applied for derivation of shape functions in G-invariant subspaces for rectangular elements with 4, 8, 12 and 16 nodes, and for rectangular hexahedral elements with 8, 20, 32 and 64 nodes. For most of them the shape functions at the nodes are derived from shape functions in G-invariant subspaces. Shape functions in G-invariant subspaces were also derived from the known shape functions of serendipity elements.

Chapter 5 deals with derivation of stiffness equations in G-invariant subspaces. Element shape functions in G-invariant subspaces, formulated in Chapter 4, with additional generalization concerning nodal degrees of freedom, are used for derivation of stiffness matrices in the subspaces. Thus, for variious elements, the element stiffness matrix with the group supermatrix in diagonal form is obtained, where matrices, symmetry-adapted variables and free members pertain to G-invariant subspaces. By the group supermatrix transformation this stiffness equation is turned into the element stiffness equation with the group supermatrix in normal form. This equation is turned into the conventional stiffness equation by the basis transformation that transforms the set of positive directions of generalized displacements of the group supermatrix procedure into the conventional set of positive directions. All these transformations can be performed in reverse order.

Stiffness equations in G-invariant subspaces are derived for one-dimensional elements with two and four nodes, the two-node beam element, the four-node 8-degrees-of-freedom rectangular element for planar analysis, the four-node 12-d, -o, -f. rectangular plate bending element, as well as its 16-d, -o, -f. version (to the equation with the supermatrix in diagonal form for determination of the coefficients of the displacement polynomial). In comparison with conventional methods, the above derivations of stiffness equations are accomplished with substantial qualitative and quantitative advantages throughout the entire analysis.

In Chapter 6, on group supermatrices in formulation and assembly of stiffness equations, the group supermatrix procedure in the direct stiffness method is presented as a process of direct superposing which can be used to assemble stiffness equations in finite element analysis.

In comparison with the conventional direct stiffness method, in the group supermatrix procedure the conventions concerning the nodal numbering, positions of origins and directions of axes of the local and global coordinate systems and the sets of positive directions of generalized displacements are changed in order to suit the symmetry type of the first irreducible representation of the group. Then it is possible to formulate system stiffness equations with group supermatrices in normal and diagonal forms and to obtain system stiffness equations in G-invariant subspaces.

The group supermatrix procedure in the direct stiffness method contains the following.

(1) Determination of the symmetry group of the structure.
(2) Determination of the numbering of the nodes and nodal variables.
(3) Determination of the basis vectors of G-invariant subspaces.
(4) Utilization of group supermatrices of the chosen group.
(5) Establishing the correspondence of positive directions of generalized nodal displacements with respective positive directions in the first subspace.
(6) Formulation of element stiffness matrices.
(7) Formulation of system stiffness equations with group supermatrices in normal and diagonal forms.

Thus, by direct superposition of element stiffness matrices that correspond to the configuration suiting the first symmetry type of the group, the system stiffness group supermatrix in normal form is compiled. This supermatrix is transformed by the group supermatrix transformation into the system stiffness group supermatrix in diagonal form.

In this way the solutions are found from system stiffness equations for each G-invariant subspace separately, while the final results are obtained as linear combinations of solutions from the subspaces, which is achieved by a supermatrix transformation.

In comparison with conventional analyses drastic reduction in the amount of computation is realized because in the group supermatrix procedure dimensions of matrices in the subspaces are a fraction of the dimension of the matrix in the conventional analysis.

It can be concluded that the group supermatrix procedure in finite element analysis provides the greatest qualitative and quantitative advantages in cases of finite elements with complex symmetry, where decomposition into a great number of subspaces can be realized.

1

Results of groups, vector spaces and representation theory

1.1 INTRODUCTION

The object of this chapter is to present some results of groups, vector spaces and representation theory which are necessary for the formulation of G-vector spaces, used for maximum utilization of symmetry properties of a space object or a nodal configuration.

This matter, with derivations of formulations and proofs of theorems, is given in Zloković (1989). In order to make this present volume self-contained, results of groups, vector spaces and representation theory are considered here, but for the most part without derivations, proofs and illustrative descriptions.

The object in section 1.2 is to present some basic results of abstract groups which are necessary for formulation of symmetry groups and their classification, as well as for development of group representations. This concise extract of abstract groups is mostly based on the approach of Mathiak and Stingl (1968), while other sources (Zloković, 1969b; Hammermesh, 1962; Kurosh, 1967) were utilized to some extent.

Vector spaces, presented in section 1.3, contain some extracts from n-dimensional vector spaces, linear operators, basis transformations and unitary spaces, which will be utilized in group algebra and G-vector spaces in section 1.4. The approach is mainly based on the work of Mathiak and Stingl (1968) and, to a certain extent, on the work of Hammermesh (1962) and Schonland (1965).

Representation theory in section 1.4 contains some extracts from group representations, group algebra, idempotents of the centre of group algebra and G-vector spaces, which will be utilized in G-vector analysis (Chapter 2) and in the group supermatrix procedure (Chapter 3), as well as in formulation of element shape functions in G-invariant subspaces (Chapter 4), in derivations of stiffness equations of elements in G-invariant subspaces (Chapter 5) and in the direct stiffness method (Chapter 6). The approach in section 1.4 is mostly based on the work of Mathiak and Stingl (1968) and to some extent on the work of Schonland (1965).

1.2 GROUPS

The analysis of the symmetry of a space object is usually based on the geometrical framework formed by the points or nodes of the system. The space object may contain identical nodes occupying physically identical positions in the nodal framework. A rotation about an axis of the nodal framework may rearrange the identical and physically equivalent nodes, i.e. it may bring the nodal framework to a new position which coincides exactly with the original position and is physically indistinguishable from it. The same may be accomplished by a reflection of the nodal framework as a whole in a plane, or by a combination of rotations and reflections. Thus, a symmetry operation has no effect on any physical property of the structure or its initial state with respect to deformation or strain, which is a basic fact of all symmetry considerations.

1.2.1 Abstract group theory

Definition
A non-empty set of elements $\alpha, \beta, \gamma, \ldots$ is said to form a **group** G with respect to a binary operation (multiplication, combination etc.), if for arbitrary elements of G the following axioms are satisfied:

(I) *Closure*. The product (combination) of any two elements of the group is a unique element which also belongs to the group: $\alpha\beta = \gamma$.
(II) *Associative law*. When three or more elements of G are multiplied, the sequence of multiplications (combinations) is irrelevant: $\alpha(\beta\gamma) = (\alpha\beta)\gamma$.
(III) *Identity*. Among the elements of G there is an element ϵ which leaves the elements unchanged on multiplication (combination). $\epsilon\alpha = \alpha\epsilon = \alpha$.
(IV) *Inverses*. Each element α of G has an inverse α^{-1} in the group: $\alpha\alpha^{-1} = \alpha^{-1}\alpha = \epsilon$.

Consequences
(1) Equations $\alpha\sigma = \beta$ or $\sigma\alpha = \beta$, when $\alpha, \beta, \sigma \in G$, are solvable in a unique way.
(2) Equation $\alpha\sigma = \alpha\tau$ or $\sigma\alpha = \tau\alpha$ allow α to be cancelled: $\sigma = \tau$.
(3) Parentheses in products can be omitted.
(4) The group has exactly one identity element ϵ with the property $\epsilon\alpha = \alpha\epsilon = \alpha$.
(5) With every α in G exists only one inverse element such that $\alpha^{-1}\alpha = \alpha\alpha^{-1} = \epsilon$, i.e. $(\alpha^{-1})^{-1} = \alpha$.
(6) $(\alpha\beta)^{-1} = \beta^{-1}\alpha^{-1}$.
(7) $(\alpha^n)^m = \alpha^{nm}$, $\alpha^n\alpha^m = \alpha^{n+m}$ for integers m, n.
(8) If α and β commute, i.e. $\alpha\beta = \beta\alpha$, then $(\alpha\beta)^n = \alpha^n\beta^n$ for natural numbers n.

Abeliean groups satisfy $\alpha\beta = \beta\alpha$ for every α and every β in G. The order of a group is the number of its elements and it is denoted by h.

The **group table** (multiplication table) is a tabular survey of all possible products. If $\alpha\beta = \tau$, this means that β and α are taken from the top and the side respectively, while τ lies at the intersection of the column β and the row α:

	ε	α	β	...	σ	...	ω
ε	ε	α	β	...	σ	...	ω
α	α		τ				
β	β						
⋮	⋮						
σ	σ						
⋮	⋮						
ω	ω						

Some properties of the group table (which follow from the group axioms) are as follows:

(1) The first row and the first column coincide with the top and the side respectively if the group table begins with identity, which is usual.
(2) The group tables of Abelian groups are symmetric with respect to the principal diagonal.
(3) The positions of products with the identity element are symmetric with respect to the principal diagonal.
(4) Every group element appears in each row and in each column only once.

Two groups are **isomorphic** if their elements can be put into one-to-one correspondence which is preserved under multiplication (combination). Isomorphic groups have the same structure and one common group table.

Generators

The smallest positive integer n such that $\alpha^n = \epsilon$ is the order of the element α. A cyclic group Z consists of powers of one element only. Because $\alpha^r \alpha^s = \alpha^{r+s} = \alpha^s \alpha^r$, every cyclic group is Abelian. When the order of a cyclic group equals n, its elements are $\epsilon = \alpha^0, \alpha^1, \alpha^2, \ldots, \alpha^{n-1}$.

If M is a non-empty subset of elements of the group G, while every group element σ is a product of a finite number of elements of M and their inverses, M is a **system of generators** of the group G. In cyclic groups M consists of only one element.

A permutation of objects $1, 2, \ldots, n$ may be described by

$$\begin{bmatrix} 1 & 2 & \ldots & n \\ \alpha(1) & \alpha(2) & \ldots & \alpha(n) \end{bmatrix},$$

which means that 1 transforms into $\alpha(1)$, 2 into $\alpha(2), \ldots, n$ into $\alpha(n)$. Objects in any permutation appear only once. The collection of all possible permutations of n objects forms the **permutation group** S_n.

Subgroups

A subset U of a group G is a **subgroup** of G if U is itself a group with respect to the multiplication (combination) of G. The order of a finite group G is a multiple of the order of every one of its subgroups.

The set of all elements α in G such that $\alpha\sigma = \sigma\alpha$ for all σ in G is called the **centre** of G. The centre is always an Abelian subgroup of G, while the centre of an Abelian group is the group itself because all its elements commute.

Classes

An element α of a group G is said to be **conjugate** to the element β in G if there is an element σ such that $\alpha = \sigma^{-1}\beta\sigma$. The consequences are as follows.

(1) α is a self-conjugate ($\alpha = \epsilon^{-1}\alpha\epsilon$).
(2) If α is conjugate to β, then β is also conjugate to α.
(3) If α is conjugate to β, and β conjugate to γ, then α is also conjugate to γ.

The collection of all elements formed by evaluating $\sigma^{-1}\alpha\sigma$ for all σ in the group is called the **class** of α. The elements of a finite group can be partitioned into nonoverlapping classes. Two elements belonging to different classes are never conjugate to each other, so that every element of G appears in one and only one class. The consequences are as follows.

(1) The identity element always forms a class by itself.
(2) Every element of the centre is in its class alone.
(3) If G is an Abelian group, every element is in a class by itself.
(4) Any two conjugate elements have the same order.
(5) The number of elements of a class is a divisor of the group order.

1.2.2 Symmetry groups

The symmetry of a space object can be described by a set of all transformations that preserve distances between all pairs of points of the space object and bring it into coincidence with itself. A group where all elements are symmetry operations which leave invariant at least one point of the space object is referred to as a symmetry group.

A **symmetry operation** of a space object is a mapping which moves every point P of the space object into some other point $P' = \alpha(P)$ of the space object, while all distances between points remain unchanged. In this way the space object is brought to a new position which coincides exactly with the original position and is physically indistinguishable from it.

Symmetry operations of symmetry groups are

(1) reflection in a plane of symmetry (symbol: σ),
(2) rotation about an axis of symmetry (symbol: C_n) and
(3) rotary reflection, i.e. rotation with reflection in the plane perpendicular to the rotation axis (symbol: S_n).

The identity operation (denoted by E) may be identified with a rotation through the angle 0° or 360°.

The **symmetry element** is a collection of points which are not moved by a symmetry operation (reflection planes or rotation axes). A rotary reflection leaves unmoved only one point in space (the point of intersection of the rotation axis and the reflection plane).

When all symmetry operations of a space object leave only one point unmoved, this point is called the centre of symmetry. The inversion (denoted by i) is a reflection in the centre of symmetry, i.e. it is a special case of rotary reflection where the angle of rotation is 180°: $i = \sigma_h C_2$.

The order of a rotation axis is n if its symmetry element C_n is a rotation through the angle $360°/n$. Powers of these elements, denoted by C_n^2, C_n^3, ... symbolize rotations through angles $2 \times (360°/n)$, $3 \times (360°/n)$, ...

Every symmetry operation has its **inverse symmetry operation** which produces an opposite effect, i.e. after performing both operations a space object is brought back to its original position. It can be stated that

(1) the inverse operation of a reflection is the reflection itself ($\sigma^2 = E$),
(2) the inverse operation of the inversion is the inversion itself ($i^2 = E$);
(3) the inverse operations of rotations and rotary reflections are the same rotations and rotary reflections, but with the opposite direction of rotation.

Classification of symmetry groups

Group theory is not concerned with the nature of group elements, since the structure of a group is defined by the group table alone. Symmetry properties of a space object may be described by a group if its axes and planes of symmetry allow application of a set of symmetry operations that form a group. Two space objects, although different in form and complexity, may have the same system of axes and planes of symmetry and correspond to the same group.

A symmetry operation which rotates or reflects a space object into itself must leave the centre of symmetry of the space object unmoved, while all axes and planes of symmetry must intersect at at least one common point. This accounts for the term **point groups** in crystallography. The number of possible systems of axes and planes of symmetry is restricted by geometrical considerations and it is possible to list all symmetry groups which exist.

The classification of symmetry groups takes into account symmetry elements (axes and planes). Symmetry elements can be associated with certain groups as follows.

(1) Simple groups C_s and C_i possessing one plane of symmetry or one centre of symmetry,
(2) Groups C_n, S_{2n}, C_{nh}, C_{nv}, D_n, D_{nh} and D_{nd} possessing a single n-fold axis of symmetry and
(3) Groups T, T_d, T_h, O, O_h, I and I_h of regular polyhedra with more than one main n-fold axis of symmetry.

Table 1.1 contains a more detailed classification arranged in order of increasing complexity.

Table 1.1. Classification of symmetry groups

Groups	Symmetry elements			Group order	Remarks
	Rotation axes	Rotation–reflection axes	Reflection planes		
C_n $n = 1, 2, \ldots$	C_n $n = 1, 2, \ldots$			n	Cyclic groups
C_i, S_{2k} $k = 2, 3, \ldots$		$S_n, n = 2k$ $k = 1, 2, \ldots$		n	Cyclic groups $C_i = S_2$ (elements E, i)
S_{2k-1} $k = 1, 2, \ldots$		$S_n, n = 2k - 1$ $k = 1, 2, \ldots$		$2n$	Cyclic groups
C_s, C_{nh} $n = 2, 3, \ldots$	C_n $n = 2, 3, \ldots$		σ_h	$2n$	Abelian groups; for odd, $n = 2k - 1$, $k = 1, 2, \ldots C_{nh}$ is a cyclic group; for $n = 2k$, $k = 1, 2, \ldots$ is $C_{nh} = [C_n, i]$
C_{nv} $n = 2, 3, \ldots$	C_n $n = 2, 3, \ldots$		σ_v	$2n$	Non-Abelian groups if $n > 2$
D_n $n = 2, 3, \ldots$	$C_n, n \geq 2$ C_2, horizontal axis			$2n$	Non-Abelian groups if $n > 2$
D_{nh} $n = 2, 3, \ldots$	$C_n, n \geq 2$ C_2, horizontal axis		σ_h	$4n$	Elements: nC_n (E included), nC_2, nS_n ($S_1 = \sigma_h$ included), $n\sigma_v$ (with $\sigma_v C_2 \sigma_h$); another system of generators $D_{nh} = [C_n, \sigma_h, \sigma_v]$
D_{nd} $n = 2, 3, \ldots$	$C_n, n \geq 2$ C_2, horizontal axis		$n\sigma_d$	$4n$	Elements: nC_n (E included), nC_2, $n\sigma_d$, nS_{2n}; another system of generators $D_{nd} = [S_{2n}, \sigma_d]$

Table 1.1. (continued)

Groups	Symmetry elements			Group order	Remarks
	Rotation axes	Rotation-reflection axes	Reflection planes		
T	$3C_2, 4C_3$			12	Tetrahedral groups
T_d	$3C_2, 4C_3$	$3S_4$	6σ	24	
T_h	$3C_2, 4C_3$	$4S_6, i$	3σ	24	
O	$3C_2, 6C_2', 4C_3, 3C_4$			24	Octahedral groups
O_h	$3C_2, 6C_2', 4C_3, 3C_4$	$3C_4, 4S_6, i$	9σ	48	
I	$15C_2, 10C_3, 6C_5$			60	Icosahedral groups
I_h	$15C_2, 10C_3, 6C_5$	$10S_6, 6S_{10}, i$	15σ	120	

Isomorphic groups

$C_2 \cong C_i \cong S_2 \cong C_s \cong Z_2$, $C_{2h} \cong C_{2n} \cong D_2 \cong V_4$, $C_{4n} \cong D_4 \cong D_{2d}$, $C_{6v} \cong D_6 \cong D_{3h}$, $T_d \cong O$, $C_n \cong S_n \cong Z_n$ for even $n \geqslant 2$, $C_{nh} \cong S_{2n} \cong Z_{2n}$ for odd $n \geqslant 3$, $C_{nv} \cong D_n$ for $n \geqslant 2$, $D_{nd} \cong D_{nh}$ for odd $n \geqslant 3$.

Direct products

$C_{nh} = C_n \times C_s$ for $n \geqslant 2$, $C_{nh} = C_n \times C_i$ for even $n \geqslant 2$, $D_{nh} = D_n \times C_s$ for $n \geqslant 2$, $D_{nh} = D_n \times C_i$ for even $n \geqslant 2$, $D_{nd} = D_n \times C_i$ for odd $n \geqslant 3$, $T_h = T \times C_i$, $O_h = O \times C_i$, $I_h = I \times C_i$.

Determination of the group of a space object

After all symmetry elements of a space object have been found, its symmetry group may be determined by applying the procedure in Table 1.2.

Table 1.2. Determination of the group of a space object

	Axes	Groups
1	No C_n or S_n axes	$C_1 = E$ $C_s = [\sigma]$ $C_i = [i]$
2	One C_n axis $(n \geqslant 2)$, no S_{2n} axis	$C_n = [C_n]$ $C_{nh} = [C_n, \sigma_h]$ $C_{nv} = [C_n, \sigma_v]$
3	One C_n-axis $(n \geqslant 2)$, No S_{2n} axis, horizontal C_2 axes	$D_n = [C_n, C_2]$ $D_{nh} = [C_n, C_2, \sigma_h]$
5	Several C_n axes, $n = 2, 3, \ldots$	T, T_h, T_d O, O_h I, I_h

1.3 VECTOR SPACES

This section contains some extracts from real n-dimensional vector spaces, linear operators, basis transformations and unitary spaces, which will be utilized in group algebra and G-vector spaces in section 1.4.

1.3.1 n-dimensional vector spaces

Definition
An n-dimensional vector space V is a set of vectors $A, B, \ldots, X, \ldots$ which satisfy the following axioms.

(1) V is an Abelian additive group:
     (a)   any two vectors A, B of V determine a unique vector $A + B$ of V;
     (b) $A + (B + C) = (A + B) + C$;
     (c)   there is a unique null vector 0 satisfying $0 + A = A + 0 = A$ for all A;
     (d)   any vector A of V determines a unique vector $-A$ of V satisfying $(-A) + A = 0$;
     (e) $A + B = B + A$.
(2)   Multiplication of vectors A of V by real numbers $a, b, \ldots, x, \ldots$ (in this relation they are named scalars), is defined in the following way:

(a) Any vector A of V and any scalar a determine a unique vector aA of V;
(b) $a(bA) = (ab)A$;
(c) $a(A + B) = aA + aB$;
(d) $(a + b)A = aA + bA$;
(e) $1A = A$.

(3) There are n vectors $B_1, B_2, \ldots, B_n$ of V such that any vector A of V can be expressed by

$$A = a_1 B_1 + a_2 B_2 + \ldots + a_n B_n,$$

where the scalars $a_1, a_2, \ldots, a_n$ are determined by A in a unique way (A is a linear combination of $B_1, B_2, \ldots, B_n$ with coefficients $a_1, a_2, \ldots, a_n$).

Summary of elements and operations of an n-dimensional vector space:

elements	$A, B, \ldots$
neutral element	0 such that $0 + A = A$
inverse element	$-A$ such that $-A + A = 0$
addition	$A + B = C$ (a unique vector sum)
multiplication	aA (a is a scalar)
associative law	$A + (B + C) = (A + B) + C$
distributive law	$a(A + B) = aA + aB$
commutative law	$A + B = B + A$.

Basis of a vector space

A basis of a vector space V is a set of n non-zero vectors $B_1, B_2, \ldots, B_n$ by which any vector A of V can be expressed as their linear combination in a unique way

$$A = a_1 B_1 + a_2 B_2 + \ldots + a_n B_n,$$

as was already stated by axiom (3). Because of this property the set $B_1, B_2, \ldots, B_n$ is called a generator system of V.

The m non-zero vectors $A_1, A_2, \ldots, A_m$ of V are called linearly independent when

$$a_1 A_1 + a_2 A_2 + \ldots + a_m A_m = 0$$

if and only if $a_1 = a_2 = \ldots = a_m = o$.

Since $oA = 0$ for any vector A of V, one may write $oB_1 + oB_2 + \ldots + oB_n = 0$. The vector 0, like any vector of V, can be linearly combined in a unique way:

$$b_1 B_1 + b_2 B_2 + \ldots + b_n B_n = 0.$$

The vectors $B_1, B_2, \ldots, B_n$ are basis vectors and cannot be zero vectors, so that $b_1 = b_2 = \ldots = b_n = o$ must always hold. Therefore, the basis vectors are always linearly independent.

If the number of vectors in a given set is smaller than n, not all vectors of V can be expressed as linear combinations of them, and if their number is larger than n, all vectors of V can be expressed as linear combinations of them, but in more than one way. It is usual to say that an n-dimensional space V is spanned by a set of n linearly independent vectors. The dimension of V is denoted by Dim $V = n$.

When in V a basis $B_1, B_2, \ldots, B_n$ is established, any vector A is expressed in one unique way by

$$A = a_1 B_1 + a_2 B_2 + \ldots + a_n B_n,$$

i.e. A is determined by scalars $a_1, a_2, \ldots, a_n$. This set of scalars is called the coordinates of A with respect to the basis $B_1, B_2, \ldots, B_n$, and it is written as a column or a row. Two vectors A, B of V are equal if they coincide in all coordinates. If A has coordinates $a_1, a_2, \ldots, a_n$ and B has coordinates $b_1, b_2, \ldots, b_n$, then the coordinates of $A + B$ are $a_1 + b_1, a_2 + b_2, \ldots, a_n + b_n$. The coordinates of aA (a is a scalar) are $aa_1, aa_2, \ldots, aa_n$. The coordinates of the basis vectors $B_1, B_2, \ldots, B_n$ are

$$(1, 0 \ldots, 0), \quad (0, 1, 0, \ldots, 0), \ldots, (0, \ldots, 0, 1),$$

i.e. unit columns are associated with the coordinates columns.

The coordinates of any vector X of V can be expressed as linear combinations of unit columns

$$X = \begin{bmatrix} x_1 \\ x_2 \\ \vdots \\ x_n \end{bmatrix} = \begin{bmatrix} x_1 \\ 0 \\ \vdots \\ 0 \end{bmatrix} + \begin{bmatrix} x_2 \\ \vdots \\ 0 \end{bmatrix} + \ldots + \begin{bmatrix} 0 \\ 0 \\ \vdots \\ x_n \end{bmatrix} = x_1 \begin{bmatrix} 1 \\ 0 \\ \vdots \\ 0 \end{bmatrix} + x_2 \begin{bmatrix} 0 \\ 1 \\ \vdots \\ 0 \end{bmatrix} + \ldots + x_n \begin{bmatrix} 0 \\ 0 \\ \vdots \\ 1 \end{bmatrix}.$$

Unit vectors $E_1, E_2, \ldots, E_n$ of coordinate axes can be represented as directed line segments extending from the origin and having the magnitude (length) 1. Any vector X of V can be expressed as a linear combination of $E_1, E_2, \ldots, E_n$:

$$X = x_1 E_1 + x_2 E_2 + \ldots + x_n E_n.$$

Subspaces

Analogous to subgroups of groups, the notion of a subspace of an n-dimensional vector space can be introduced.

Definition

A non-empty subset U of a vector space V (Dim $V = n$) is a k-dimensional subspace of V ($0 \leq k \leq n$) when U itself is a vector space with respect to all combinations of V.

This is valid if for any vector $A, B, \ldots$ of U the following are also vectors of U:

(1) $A + B$,
(2) aA (for any scalar a),
(3) $(-1)A = -A$,
(4) $oA = 0$.

The dimension k of U is the maximum number of linearly independent vectors of U ($0 \leq k \leq n$).

1.3.2 Linear operators

Definition
An **operator** σ of an n-dimensional vector space V is a mapping which associates with every vector A of V another vector σA of V.

The operator σ is a **linear operator** of V, when, in addition to that, the following linearity properties are valid for all vectors A, B and all scalars a:

(1) $\sigma(A + B) = \sigma A + \sigma B$ and
(2) $\sigma(aA) = a(\sigma A)$.

The elements and operations of an n-dimensional vector space with linear operators are as follows:

elements	σA (image vector of A),
neutral elements	o such that $o + \sigma A = \sigma A$,
inverse element	$-\sigma A$ such that $-\sigma A + \sigma A = 0$,
addition	$\sigma_1 A + \sigma_2 B$ (a unique solution for given σ_1, σ_2, A, B),
multiplication	$a(\sigma A)$ (a is a scalar),
associative law	$\sigma_1 A + (\sigma_2 B + \sigma_3 C) = (\sigma_1 A + \sigma_2 B) + \sigma_3 C$,
distributive law	$\sigma(A + B) = \sigma A + \sigma B$,
commutative law	$\sigma_1 A + \sigma_2 B = \sigma_2 B + \sigma_1 A$.

The elements and operations of linear operators are as follows:

elements	$\alpha, \beta, \ldots, \sigma, \tau, \ldots,$
neutral elements	
addition	o such that $o + \sigma = \sigma$,
operator multiplication	ϵ such that $\epsilon\sigma = \sigma\epsilon = \sigma$,
scalar multiplication	1 such that $1\sigma = \sigma$,
inverse element	
addition:	$-\sigma$ such that $(-\sigma) + \sigma = o$,
multiplication	σ^{-1} such that $\sigma\sigma^{-1} = \sigma^{-1}\sigma = \epsilon$
	(σ^{-1} exists if σ is invertible),
addition	$\sigma_1 + \sigma_2 = \sigma_3$,
multiplication	
scalars multiplication	$a\sigma$,
operator multiplication	$\sigma_1 \sigma_2 = \sigma_3$,
	(application of σ_2 first, then σ_1),
associative laws	
addition	$\sigma_1 + (\sigma_2 + \sigma_3) = (\sigma_1 + \sigma_2) + \sigma_3$,
multiplication	$\sigma_1(\sigma_2 \sigma_3) = (\sigma_1 \sigma_2)\sigma_3$, $a(b\sigma) = (ab)\sigma$,
distributive laws	$\sigma_1(\sigma_2 + \sigma_3) = \sigma_1 \sigma_2 + \sigma_1 \sigma_3$, $(a + b)\sigma = a\sigma + b\sigma$,
	$a(\sigma_1 + \sigma_2) = a\sigma_1 + a\sigma_2$,

commutative laws
- addition $\quad\sigma_1 + \sigma_2 = \sigma_2 + \sigma_1,$
- multiplication $\quad\sigma_1 \sigma_2 \neq \sigma_2 \sigma_1$ (in general).

If U is a subspace of V, the set σU of vectors which are the images of the vectors of U produced by the mapping σ again forms a subspace of V.

Regular linear operators

Two different vectors A and B may have, in general, identical images $\sigma A = \sigma B$. For example, different vectors (x, y, a) and (x, y, b) have the same image $(x, y, 0)$ when the operator representing the vertical projection on the xy plane is applied. With symmetry operators this is not the case. Linear operators that always transform all vectors into different images are called **regular** (nonsingular or invertible) **operators**.

By application of a regular linear operator σ of V, every vector of V appears as an image vector. If σ is a regular linear operator, so is its inversion σ^{-1}, which relates every image vector σA to its original vector A.

Since all group axioms are satisfied for regular linear operators with respect to multiplication, the set of all regular linear operators of an n-dimensional vector space V forms a group with respect to combinations one after another, i.e. the linear group of V.

Linear operators represented by matrices

Transformation of a vector by a linear operator

When $X(x_1, \ldots, x_n)$ is a vector in an n-dimensional vector space V with an established basis $B_1, \ldots, B_n$ and σ a regular linear operator

$$\sigma X = \sigma(x_1 B_1 + \ldots + x_n B_n) = x_1(\sigma B_1) + \ldots + x_n(\sigma B_n) = \sum_{k=1}^{n} x_k(\sigma B_k).$$

Like every vector, the vector σB_k $(k = 1, \ldots, n)$ can be expressed by means of the basis vectors $B_1, \ldots, B_n$

$$\sigma B_1 = \sum_{i=1}^{n} a_{i1} B_i$$

$$\vdots$$

$$\sigma B_n = \sum_{i=1}^{n} a_{in} B_i \quad \text{or}$$

$$\sigma B_k = \sum_{i=1}^{n} a_{ik} B_i \quad (k = 1, \ldots, n),$$

so that

Vector spaces

$$\sigma X = \sum_{k=1}^{n} x_k \left(\sum_{i=1}^{n} a_{ik} B_i \right) = \sum_{i=1}^{n} \left(\sum_{k=1}^{n} x_k a_{ik} \right) B_i.$$

Since expressing this by means of the basis vectors is unique for coordinates x'_i of the image vector σX, one obtains

$$x'_1 = a_{11}x_1 + a_{12}x_2 + \ldots + a_{1n}x_n$$
$$x'_2 = a_{21}x_1 + a_{22}x_2 + \ldots + a_{2n}x_n$$
$$\vdots$$
$$x'_n = a_{n1}x_1 + a_{n2}x_2 + \ldots + a_{nn}x_n \quad \text{or}$$

$$x'_i = \sum_{k=1}^{n} a_{ik} x_k \quad (i = 1, \ldots, n).$$

Therefore, every linear operator of an n-dimensional vector space V has a corresponding set of n linear transformation equations for the coordinates of an arbitrary vector X, i.e. linear equations for x'_i are linear combinations of x_k. Conversely, every set of such linear transformation equations has a corresponding linear operator σ, which is exactly that which turns a vector X with coordinates $x_1, \ldots, x_n$ into the vector σX with coordinates $x'_1, \ldots, x'_n$.

Correspondence of operations

When linear operators α, β with respect to an established basis are represented by matrices (a_{ik}), (b_{ik}), all operations with operators must correspond to respective operations with matrices, as follows.

(1) $(\alpha + \beta)X = \alpha X + \beta X$ $\quad \sum_{k=1}^{n} a_{ik} x_k + \sum_{k=1}^{n} b_{ik} x_k = \sum_{k=1}^{n} (a_{ik} + b_{ik}) x_k$

$(i = 1, \ldots, n)$.

(2) $(a\alpha)X = a(\alpha X)$ $\quad a \left(\sum_{k=1}^{n} a_{ik} x_k \right) = \sum_{k=1}^{n} (a a_{ik}) x_k \quad (i = 1, \ldots, n)$.

(3) When a linear operator σ is regular (nonsingular or invertible), then the corresponding matrix A must be regular (nonsingular or invertible) too. An operator σ is invertible if its system of transformation equations

$$x'_1 = a_{11}x_1 + \ldots + a_{1n}x_n$$
$$\vdots$$
$$x'_n = a_{n1}x_1 + \ldots + a_{nn}x_n$$

is solvable for $x'_1, \ldots, x'_n$ in a unique way (the original vector is transformed into a unique image vector). Since σ^{-1} is also a linear operator, the coordinates $x_1, \ldots, x_n$ are linear combinations of $x'_1, \ldots, x'_n$:

$$x_1 = a'_{11}x'_1 + \ldots + a'_{1n}x'_n$$
$$\vdots$$
$$x_n = a'_{n1}x'_1 + \ldots + a'_{nn}x'_n.$$

The coefficients a'_{ik} ($i, k = 1, \ldots, n$) are determined by a_{ik} in a unique way and they form the inverse matrix A^{-1}.

(4) The product of two linear operators is an application of one operator and then the other. This is done when the transformation equations

$$x'_j = \sum_{k=1}^n b_{jk}x_k \quad \text{and} \quad x''_i = \sum_{j=1}^n a_{ij}x'_j$$

represented by matrices $A = (a_{ij})$ and $B = (b_{jk})$ are substituted into $AB = C$, so that the equations for x''_i can be expressed by x_k:

$$x''_i = \sum_{j=1}^n a_{ij} \left(\sum_{k=1}^n b_{jk}x_k \right) = \sum_{k=1}^n \left(\sum_{j=1}^n a_{ij}b_{jk} \right) x_k.$$

Therefore, the transformation of x_k into x''_i is represented by the matrix C, the elements of which are

$$c_{ik} = \sum_{j=1}^n a_{ij}b_{jk} = a_{i1}b_{1k} + a_{i2}b_{2k} + \ldots + a_{in}b_{nk} \qquad (i, k = 1, 2, \ldots, n).$$

This coincides with the matrix multiplication rule.

Since the correspondence with respect to all operations of linear operators and matrices is established, all group axioms are satisfied for invertible matrices. Therefore, invertible $n \times n$ matrices, like regular linear operators of an n-dimensional vector space, form a group with respect to matrix multiplication.

In the classification of the properties of matrices and their operations when they represent linear operators, matrices $A, B, \ldots$ will be called elements (not to be confused with a_{ik} which are elements of a matrix).

The matrix A with elements a_{ij} ($i, j = 1, \ldots, n$) is sometimes denoted by (a_{ij}).

The elements and operations of matrices that represent linear operators are as follows:

elements $A, B, \ldots, X, Y, \ldots,$
neutral elements
   addition O such that $O + A = A$,
   matrix multiplication E such that $EA = AE = A$,
   scalar multiplication   1 such that $1A = A$,
inverse elements
   addition $-A$ such that $(-A) + A = O$,
   multiplication A^{-1} such that $A^{-1}A = AA^{-1} = E$
                                 (A must be invertible),

addition	$A + B = (a_{ij}) + (b_{ij}) = (a_{ij} + b_{ij})$	$(i, j = 1, \ldots, n)$,
multiplication		
by scalar	$aA = (aa_{ik})$ $(i, k = 1, \ldots, n)$,	
matrix multiplication	$AB = C$,	

$$c_{ik} = \sum_{j=1}^{n} a_{ij} b_{jk} \qquad (i, k = 1, \ldots, n),$$

associative laws
  addition $A + (B + C) = (A + B) + C$,
  multiplication $A(BC) = (AB)C$, $a(bA) = (ab)A$,
distributive laws $A(B + C) = AB + AC$, $(a + b)A = aA + bA$,
                            $a(A + B) = aA + aB$,

commutative law
  addition $A + B = B + A$,
  multiplication $AB \neq BA$ (in general).

Conjugate matrices
In a matrix group, analogous to groups, classes of conjugate elements can be formed. Invertible matrices A and B are conjugate matrices if an invertible matrix C exists so that $A = C^{-1}BC$.

The trace of a matrix equals the sum of its diagonal elements

$$\mathrm{Tr}(a_{ik}) = \sum_{i=1}^{n} a_{ii}.$$

Traces of conjugate matrices are equal.

When a matrix representing a linear operator is considered, it is assumed that a fixed basis is established. In different bases an operator is represented by different matrices.

1.3.3 Basis transformations

Relation between two bases of an n-dimensional vector space
When $B_1, \ldots, B_n$ is a basis of an n-dimensional vector space V and $\bar{B}_1, \ldots, \bar{B}_n$ another basis of the same space, then it is always possible to combine linearly the vectors $\bar{B}_1, \ldots, \bar{B}_n$ by means of $B_1, \ldots, B_n$

$$\bar{B}_1 = c_{11} B_1 + c_{21} B_2 + \ldots + c_{n1} B_n$$
$$\bar{B}_2 = c_{12} B_1 + c_{22} B_2 + \ldots + c_{n2} B_n$$
$$\vdots$$
$$\bar{B}_n = c_{1n} B_1 + c_{2n} B_2 + \ldots + c_{nn} B_n \quad \text{or}$$

$$\bar{B}_r = \sum_{i=1}^{n} c_{ir} B_i \qquad (r = 1, \ldots, n),$$

or in matrix form (with indices i and r of rows and columns respectively)

$$[\bar{B}_1 \ \bar{B}_2 \ \ldots \ \bar{B}_n] = [B_1 \ B_2 \ \ldots \ B_n] \begin{bmatrix} c_{11} & c_{12} & \ldots & c_{1n} \\ c_{21} & c_{22} & \ldots & c_{2n} \\ \vdots & \vdots & & \vdots \\ c_{n1} & c_{n2} & \ldots & c_{nn} \end{bmatrix} \quad \text{or}$$

$\bar{B} = BC$, with $C = (c_{ir})$.

Conversely, $B_1, \ldots, B_n$ can be expressed by means of $\bar{B}_1, \ldots, \bar{B}_n$

$$B_1 = \bar{c}_{11}\bar{B}_1 + \bar{c}_{21}\bar{B}_2 + \ldots + \bar{c}_{n1}\bar{B}_n$$
$$B_2 = \bar{c}_{12}\bar{B}_1 + \bar{c}_{22}\bar{B}_2 + \ldots + \bar{c}_{n2}\bar{B}_n$$
$$\vdots$$
$$B_n = \bar{c}_{1n}\bar{B}_1 + \bar{c}_{2n}\bar{B}_2 + \ldots + \bar{c}_{nn}\bar{B}_n \quad \text{or}$$

$$B_k = \sum_{r=1}^{n} \bar{c}_{rk}\bar{B}_r \quad (k = 1, \ldots, n),$$

or in matrix form (with r and k as indices of rows and columns respectively)

$$[B_1 \ B_2 \ \ldots \ B_n] = [\bar{B}_1 \ \bar{B}_2 \ \ldots \ \bar{B}_n] \begin{bmatrix} \bar{c}_{11} & \bar{c}_{12} & \ldots & \bar{c}_{1n} \\ \bar{c}_{21} & \bar{c}_{22} & \ldots & \bar{c}_{2n} \\ \vdots & \vdots & & \vdots \\ \bar{c}_{n1} & \bar{c}_{n2} & \ldots & \bar{c}_{nn} \end{bmatrix} \quad \text{or}$$

$B = \bar{B}\bar{C}$, with $\bar{C} = (\bar{c}_{ir})$.

By substituting

$$\bar{B}_r = \sum_{i=1}^{n} c_{ir} B_i \quad (r = 1, \ldots, n)$$

into

$$B_k = \sum_{r=1}^{n} \bar{c}_{rk} \bar{B}_r \quad (k = 1, \ldots, n)$$

one finds that

$$B_k = \sum_{r=1}^{n} \bar{c}_{rk} \sum_{i=1}^{n} c_{ir} B_i = \sum_{i=1}^{n} \left(\sum_{r=1}^{n} c_{ir} \bar{c}_{rk} \right) B_i.$$

This equation is true if the sum

$$\sum_{r=1}^{n} c_{ir}\bar{c}_{rk}$$

equals zero when $i \neq k$ and if it equals unity when $i = k$, i.e.

$$\sum_{r=1}^{n} c_{ir}\bar{c}_{rk} = \delta_{ik}$$

where (δ_{ik}) is the unit matrix E. In matrix notation this is $C\bar{C}=E$, where $\bar{C}=C^{-1}=(\bar{c}_{rk})$ is the inverse matrix of $C=(c_{ik})$. Therefore, the matrices of basis transformations are always invertible.

Transformation of vector coordinates by change of the basis

The relation between coordinates $x_1, \ldots, x_n$ of a vector X with respect to a basis $B_1, \ldots, B_n$, and the coordinates $\bar{x}_1, \ldots, \bar{x}_n$ of the same vector with respect to another basis $\bar{B}_1, \ldots, \bar{B}_n$, can be expressed by

$$X = \sum_{k=1}^{n} \bar{x}_k \bar{B}_k = \sum_{k=1}^{n} \bar{x}_k \left(\sum_{i=1}^{n} c_{ik} B_i \right) = \sum_{i=1}^{n} \left(\sum_{k=1}^{n} c_{ik} \bar{x}_k \right) B_i.$$

The coordinates $\bar{x}_1, \ldots, \bar{x}_n$ are transformed into coordinates $x_1, \ldots, x_n$ by means of

$$x_i = \sum_{k=1}^{n} c_{ik} \bar{x}_k.$$

Transformation of a matrix by change of the basis

For basis vectors, the matrix of the linear operator σ, vectors X and σX, with respect to the initial and to the new basis respectively, the notation in Table 1.3 will be applied.

Table 1.3. Notation for transformation of a matrix

	Initial basis	New basis
Basis vectors	$B_1, \ldots, B_n$	$\bar{B}_1, \ldots, \bar{B}_n$
Matrix of σ	$A = (a_{ik})$	$\bar{A} = (\bar{a}_{rs})$
Coordinates of X	x_k	$\bar{x}_s$
Coordinates of σX	$x'_i = \sum_k a_{ik} x_k$	$\bar{x}'_r = \sum_s \bar{a}_{rs} \bar{x}_s$

The coordinates of X are $x_k = \sum_s c_{ks} \bar{x}_s$, the rth coordinate of the vector σX with respect to $\bar{B}_r$ is

$$\bar{x}'_r = \sum_s \left(\sum_{i,k} \bar{c}_{ri} a_{ik} c_{ks} \right) \bar{x}_s,$$

while

$$\bar{a}_{rs} = \sum_{i,k} \bar{c}_{ri} a_{ik} c_{ks}.$$

Conclusions
(1) When A is the matrix of a linear operator with respect to a basis of an n-dimensional vector space and C the matrix of the transformation of the initial basis to a new basis, the matrix of the linear operator with respect to the new basis is $C^{-1}AC$.
(2) Since traces of conjugates $C^{-1}AC$ coincide, the trace of the matrix of a linear operator of an n-dimensional vector space has a fixed value regardless of the choice of the basis.
(3) Since $\mathrm{Det}(C^{-1}AC) = (\mathrm{Det}\, C)^{-1}\, \mathrm{Det}\, A \cdot \mathrm{Det}\, C = \mathrm{Det}\, A$, the value of the determinant $\mathrm{Det}\, A$ of the matrix of a linear operator is the same in all bases.

1.3.4 Unitary spaces

Definitions

When a complex number (X, Y) is associated in a unique way with any pair of vectors X, Y in an n-dimensional vector space V, this complex number is called the **scalar product** if the following axioms are satisfied.

(1) Commutative law: $(X, Y) = (Y, X)^*$ (where the asterisk (*) denotes the complex conjugate).
(2) Associative law: $(aX, Y) = a(X, Y)$.
(3) Distributrive law: $(X_1 + X_2, Y) = (X_1, Y) + (X_2, Y)$.
(4) $(X, X) \geqslant 0$ with $(X, X) = 0$ only if $X = 0$.

A vector space V in which a scalar product is defined is called a **unitary space**. Any function satisfying the above axioms can define a scalar product in V.

The scalar product (X, Y) of any pair of vectors X, Y of V can be also defined as a real number, so that Axiom (1) becomes $(X, Y) = (Y, X)$, while the other axioms remain the same.

The consequence of the axioms is the following equation (1), while equations (2), (3) and (4) define length, normalization and orthogonality.

(1) $(0, X) = (X, 0) = 0$.
(2) $|X| = \sqrt{X, X}$ (definition of the length of X).
(3) $\dfrac{X}{|X|} = \sqrt{\dfrac{X}{|X|}, \dfrac{X}{|X|}} = \dfrac{1}{|X|} \sqrt{(X, X)} = 1$ (normalization of X).
(4) $(X, Y) = 0$ (orthogonality relation of X, Y).

The scalar product (X, Y) can be expressed as a function of coordinates x_i, y_i in an established basis. If the basis vectors are $B_1, \ldots, B_n$ (they need not be orthogonal), the scalar product is determined by the numbers m_{ij}

$$m_{ij} = (B_i, B_j).$$

The Axiom (1) shows that $(X, Y) = (Y, X)^*$, so that

$$m_{ij} = m_{ji}^*.$$

The numbers m_{ij} defined by $m_{ij} = (B_i, B_j)$, form the metric matrix M, which, because $m_{ij} = m_{ji}^*$, has the property

$$M = M^+$$

(where ($^+$) denotes 'conjugate transpose'). The elements of M^+ are m_{ji}^*, while M^+ is the **adjoint** of the matrix M. A matrix identical with its conjugate transpose is a **self-adjoint matrix**, and M is such a matrix.

Orthonormal bases

Two vectors X, Y are orthogonal if $(X, Y) = 0$. Two subspaces U_1, U_2 of V are orthogonal if all vectors X_1 of U_1 and all vectors X_2 of U_2 satisfy the orthogonality relation $(X_1, X_2) = 0$, i.e. every vector of U_1 is orthogonal to every vector of U_2.

The orthogonal non-zero vectors $A_1, \ldots, A_k$ are linearly independent, since from

$$a_1 A_1 + \ldots + a_i A_i + \ldots + a_k A_k = 0$$

after forming the scalar product by A_i $(i = 1, \ldots, k)$ it follows that

$$a_1 (A_1, A_i) + \ldots + a_i (A_i, A_i) + \ldots + a_k (A_k, A_i) = 0.$$

Since A_i are non-zero vectors, $a_i = 0$ must hold for each i, which proves the linear independence of $A_1, \ldots, A_k$.

A basis consiting of normalized orthogonal vectors is an orthonormal basis. Every basis of V can be transformed into an orthonormal basis of V, i.e. by using a set of m $(1 \leq m \leq n)$ linearly independent vectors it is possible to combine linearly m orthogonal and normalized vectors.

It is evident that for all orthonormal vectors the following are valid.

(1) $(B_i, B_k) = \delta_{ik}$.

(2) $(X, Y) = \left(\sum_i x_i B_i, \sum_k y_k B_k \right) = \sum_i x_i \left(B_i, \sum_k y_k B_k \right) =$

$$= \sum_{i,k} x_i y_k (B_i, B_k) = \sum_{i,k} x_i y_k \delta_{ik} = \sum_i x_i y_i.$$

(3) $|X| = \sqrt{(X, X)} = \sqrt{(x_1^2 + \ldots + x_n^2)}$.

The change from an orthonormal basis $B_1, \ldots, B_n$ to another orthonormal basis $\bar{B}_1, \ldots, \bar{B}_n$ is affected by a matrix C with the property $C^{-1} = C^T$, i.e. its inverse equals its transpose (in the real space).

When the scalar product is defined as a complex number (X, Y) associated with any vector X, Y of V, the Axiom (1) gives $(X, Y) = (Y, X)^*$ and in an orthonormal basis the metric matrix M (defined in this section) reduces to the unit matrix, since $(B_i, B_j) = \delta_{ij}$. Then the scalar product is

$$(X, Y) = \sum_i x_i^* y_i = X^+ Y.$$

When changing from the orthonormal basis B_i to another basis $\bar{B}_i$, the new basis will be orthonormal if

$$\delta_{ij} = (\bar{B}_i, \bar{B}_j) = (a_{ik}^* B_k, a_{jl} B_l) = a_{ik}^* a_{jl} (B_k, B_l) = a_{ik}^* a_{jl} \delta_{kl} = a_{ik}^* a_{jk}$$

since

$$\bar{B}_i = \sum_{j=1}^n a_{ij} B_j \quad (i = 1, \ldots, n).$$

In matrix form this is

$$AA^+ = A^+ A = E \quad \text{and} \quad A^+ = A^{-1}.$$

A matrix A which satisfies $A^+ = A^{-1}$ is unitary.

Function spaces

A function space is a type of a vector space where 'vectors' are functions which have some properties in common.

The elements and operations of a function space ϕ are analogous to the elements and operations of an n-dimensional vector space and are as follows:

elements	ψ
neutral element	0 function such that $0 + \psi = \psi$,
inverse element	$-\psi$ such that $-\psi + \psi = 0$,
addition	$\psi_1 + \psi_2$,
multiplication	$a\psi$ (a is a scalar),
associative law	$\psi_1 + (\psi_2 + \psi_3) = (\psi_1 + \psi_2) + \psi_3$,
distributive law	$a(\psi_1 + \psi_2) = a\psi_1 + a\psi_2$,
commutative law	$\psi_1 + \psi_2 = \psi_2 + \psi_1$.

An axiom of function spaces is also that that are n functions $\varphi_1, \varphi_2, \ldots, \varphi_n$ of ϕ such that any function of ϕ can be expressed by

$$\psi = a_1 \varphi_1 + a_2 \varphi_2 + \ldots + a_n \varphi_n,$$

where the scalars $a_1, a_2, \ldots, a_n$ are determined by ψ in a unique way.

A basis of an n-dimensional function space ϕ is a set of n functions $\varphi_1, \varphi_2, \ldots, \varphi_n$ by which any function of ϕ can be expressed as their linear combination in a unique way

$$\psi = \sum_{i=1}^n x_i \varphi_i.$$

A scalar product (ψ_1, ψ_2) of two functions ψ_1, ψ_2 is defined by

$$(\psi_1, \psi_2) = \int \psi_1 \psi_2 \, d\tau,$$

where $\int \ldots d\tau$ indicates integration with respect to all variables of which ψ_1, ψ_2 are functions over a suitably defined range.

The functions $\psi_1, \psi_2, \ldots, \psi_n$ are orthogonal if

$$(\psi_i, \psi_j) = \delta_{ij}.$$

The scalar product of any pair ψ_1, ψ_2 of ϕ is then

$$(\psi_1, \psi_2) = \left(\sum_{i=1}^{n} x_i \varphi_i, \sum_{k=1}^{n} y_k \varphi_k \right) = \sum_{i,k=1}^{n} x_i y_k (\varphi_i, \varphi_k) = \sum_{i,k=1}^{n} x_i y_i \delta_{ik}$$

$$= \sum_{i=1}^{n} x_i y_i.$$

The length of

$$\psi = \sum_{i=1}^{n} x_i \varphi_i$$

is

$$|\psi| = \sqrt{(\psi, \psi)} = \sqrt{\sum_{i=1}^{n} x_i^2}.$$

The normalization of ψ is $\psi' = 1/|\psi|$, while $|\psi'| = 1$.

1.4 REPRESENTATION THEORY

When in an n-dimensional vector space V a basis is established, elements of a group G can be mapped on a set of $n \times n$ matrices; this provides an n-dimensional representation of the group G. A change of the basis of V produces a new set of matrices that is another representation of G which is equivalent to the original representation. It is possible to find a basis with respect to which the matrices of the group representation are expressed as direct sums of submatrices that no change of the basis can reduce to matrices of smaller dimensions. In this case it is an **irreducible group representation**.

By introducing linear combinations of group elements into groups, one establishes the **group algebra** $\bar{G}$, which is attached to the group G, with the properties of a vector space. In the group algebra $\bar{G}$ the elements of G form a basis of $\bar{G}$ since they are linearly independent. Analogous to the centre of a group, the **centre of the group algebra** $\bar{G}$ consists of all elements that commute with all elements of G. The class sums, i.e. sums of group elements that belong to classes of conjugate elements, form a basis of the centre of the group algebra. The dimension of the centre equals the number of classes.

The **idempotents** of the centre of a group algebra are linear combinations of class sums, i.e. they belong to the centre of the group algebra. They act as projection operators that generate subspaces of definite symmetry types. There is a relation between projection operators expressed by characters of irreducible group representations, which are classified in character tables, and orthogonal idempotents of the centre of a group algebra. Owing to this relation the idempotents can be written down directly from the character table, since they are definite linear combinations of group elements with characters as their coefficients.

By introducing the elements of the group G as linear operators of V, one obtains the **G-vector space**. Elements of the group algebra $\bar{G}$, as linear combinations of group elements, act as linear operators of V too. In the case of symmetry groups an idempotent π_i of the centre of the group algebra, acting as a linear operator of V, does not change vectors of its symmetry type and nullifies all vectors of other symmetry types. When U_i is a collection of all vectors of the ith symmetry type, U_i is a G-invariant subspace of V and it contains all vectors $\pi_i X$ of all X of V. If $B_1, \ldots, B_n$ is a basis of V, then $\pi_i B_1, \ldots, \pi_i B_n$ span the subspace U_i. Any vector X of a G-vector space V can be expressed as a sum of vectors X_i of G-invariant subspaces U_i in a unique way, i.e. a G-vector space is a direct sum of G-invariant subspaces U_i: $V = U_1 + U_2 + \ldots + U_k$. One definite symmetry type is associated with each irreducible G-vector space or with the corresponding irreducible group representation.

When the symmetry properties of a space object can be described by a group, the object may be represented in the G-vector space. Then the matrix of equations that describe the behaviour of the object will appear in block diagonal form.

1.4.1 Group representations

Definitions

In section 1.3.2 it was shown that the set of all regular linear operators $\alpha, \beta, \ldots, \sigma, \ldots$ in an n-dimensional vector space V forms a group. When a basis in an n-dimensional vector space is established, the linear operators can be described by their matrix representatives. The mapping of the group G on a group of $n \times n$ matrices is called an **n-dimensional group representation of the group G**.

When the basis of V is changed, the matrices $A, B, \ldots$ with respect to the initial basis will be replaced by their conjugates $C^{-1}AC, C^{-1}BC, \ldots$. The new set of matrices also provides a representation of the group G which is equivalent to the original representation.

Two representations of the group by $n \times n$ matrices are equivalent if there is a matrix A, with the inverse A^{-1}, for which

$$D^{(2)}_{(\sigma)} = A D^{(1)}_{(\sigma)} A^{-1}$$

is valid for every element σ of the group, where $D^{(1)}_{(\sigma)}, D^{(2)}_{(\sigma)}$ designate matrices of the first and the second group representations respectively.

In section 1.3.3 it was shown that a basis transformation leaves the trace of the matrix unchanged. In group representations the trace of a matrix representing an operator σ is

Representation theory

called the character of σ and is denoted by $\chi(\sigma)$. It is evident that equivalent matrix representations possess the same set of characters.

Reducible and irreducible representations

Let Γ be an n-dimensional representation of a group of linear operators σ acting on the vectors of an n-dimensional vector space V. It is assumed that it is possible to find a basis $B_1, \ldots, B_n$ of V with the property that the first k of these basis vectors are transformed among themselves by all operators σ, namely for every σ and a fixed k one may write

$$\sigma B_1 = [D_{(\sigma)}]_{11} B_1 + \ldots + [D_{(\sigma)}]_{k1} B_k + 0 \cdot B_{k+1} + \ldots + 0 \cdot B_n$$
$$\sigma B_2 = [D_{(\sigma)}]_{12} B_1 + \ldots + [D_{(\sigma)}]_{k2} B_k + 0 \cdot B_{k+1} + \ldots + 0 \cdot B_n$$
$$\vdots$$
$$\sigma B_k = [D_{(\sigma)}]_{1k} B_1 + \ldots + [D_{(\sigma)}]_{kk} B_k + 0 \cdot B_{k+1} + \ldots + 0 \cdot B_n,$$

where $[D_{(\sigma)}]$ designates an element of the matrix $D_{(\sigma)}$ which is the representative of σ. If such a basis exists, the k basis vectors $B_1, \ldots, B_k$ by themselves form a subspace U_1 of the original space V, with the property that by applying the operator σ any vector of U_1 is transformed into another vector of U_1. In this case U_1 is called an **invariant subspace** of V.

When only orthonormal bases are considered, the remaining basis vectors $B_{k-1}, \ldots, B_n$ form another subspace U_2 of $n - k$ dimensions, where any vector of U_2 is orthogonal to any vector of U_1. If X_1 is any vector of U_1 and X_2 any vector of U_2, $(X_1, X_2) = 0$ must hold.

Since σX_1 is a vector of U_1 (the initial hypothesis), while X_1 and σX_1 may be any vectors of U_1, one finds that σX_2 must be orthogonal to every vector of U_1, because $(\sigma X_1, \sigma X_2) = 0$ if σ is a symmetry operator. A symmetry operator σ applied simultaneously to any two vectors A, B of V will transform (rotate or reflect) them into σA, σB respectively without changing their lengths or the angle between them. Thus σ has no effect on the scalar product: $(\sigma A, \sigma B) = (A, B)$.

It can be concluded that the vector σX_2 must belong to the subspace U_2 and be a linear combination of the basis vectors $B_{k+1}, \ldots, B_n$. The remaining equations $\sigma B_{k+1}, \ldots, \sigma B_n$ are

$$\sigma B_{k+1} = 0 \cdot B_1 + \ldots + 0 \cdot B_k + [D_{(\sigma)}]_{k+1,k+1} B_{k+1} + \ldots + [D_{(\sigma)}]_{n,k+1} B_n,$$
$$\vdots$$
$$\sigma B_n = 0 \cdot B_1 + \ldots + 0 \cdot B_k + [D_{(\sigma)}]_{k+1,n} B_{k+1} + \ldots + [D_{(\sigma)}]_{nn} B_n.$$

If it is possible to find an orthonormal basis which can be divided in this way into two independent sets of basis vectors for k less than n, the original representation Γ is a **reducible representation**. Conversely, an irreducible representation cannot have such a basis.

The above sets of equations $\sigma B_1, \ldots, \sigma B_k$ of the reducible n-dimensional representation Γ can be written in matrix form, so that, with respect to the basis $B_1, \ldots, B_n$, all matrices of Γ have the same reduced form

$$D_{(\sigma)} = \begin{bmatrix} [D_{(\sigma)}]_{11} & \cdots & [D_{(\sigma)}]_{1k} & 0 & \cdots & 0 \\ \vdots & & \vdots & \vdots & & \vdots \\ [D_{(\sigma)}]_{k1} & \cdots & [D_{(\sigma)}]_{kk} & 0 & \cdots & 0 \\ \hline 0 & \cdots & 0 & [D_{(\sigma)}]_{k+1,k+1} & \cdots & [D_{(\sigma)}]_{k+1,n} \\ \vdots & & \vdots & \vdots & & \vdots \\ 0 & \cdots & 0 & [D_{(\sigma)}]_{n,k+1} & \cdots & [D_{(\sigma)}]_{n,n} \end{bmatrix} \begin{matrix} k \text{ rows} \\ \\ \\ n-k \text{ rows} \\ \\ \end{matrix}$$

$$\underbrace{\hspace{4em}}_{k \text{ columns}} \quad \underbrace{\hspace{6em}}_{n-k \text{ columns}}$$

or by denoting the $k \times k$ matrix by $D_{(\sigma)}^{(1)}$ and the $(n-k) \times (n-k)$ matrix by $D_{(\sigma)}^{(2)}$,

$$D_{(\sigma)} = \begin{bmatrix} D_{(\sigma)}^{(1)} & 0 \\ 0 & D_{(\sigma)}^{(2)} \end{bmatrix}.$$

$D_{(\sigma)}$ is the direct sum of $D_{(\sigma)}^{(1)}$ and $D_{(\sigma)}^{(2)}$.

We can summarize as follows.

(1) The original vector space V is divided into two independent subspaces U_1 and U_2 which give rise to two independent representations Γ_1 and Γ_2 of the group of operators σ.

(2) The bases of U_1, U_2 consist of $B_1, \ldots, B_k$ and $B_{k+1}, \ldots, B_n$ respectively.

(3) The subspace U_1 produces the representation Γ_1, with matrices $D_{(\sigma)}^{(1)}$, while U_2 produces Γ_2 with matrices $D_{(\sigma)}^{(2)}$.

(4) The relation between the original representation Γ and representations Γ_1 and Γ_2 is denoted by $\Gamma = \Gamma_1 + \Gamma_2$.

The representations Γ_1 and Γ_2 are reducible if U_1 and U_2 themselves contain invariant subspaces. By suitable change of bases within U_1 and U_2 the representations Γ_1 and Γ_2 may be reduced, so that matrices $D_{(\sigma)}^{(1)}$ and $D_{(\sigma)}^{(2)}$ become direct sums of matrices belonging to representations of smaller dimensions. By continuing this process the original space V will be finally divided into a number of invariant subspaces, so that none of the subspaces can be divided into subspaces of smaller dimensions. Each subspace gives rise to a representation of group operators σ and all matrices of Γ are transformed into block diagonal form

$$D_{(\sigma)} = \begin{bmatrix} D_{(\sigma)}^{(1)} & & & \\ & D_{(\sigma)}^{(2)} & & \\ & & \ddots & \\ & & & D_{(\sigma)}^{(k)} \end{bmatrix}$$

The submatrices $D_{(\sigma)}^{(1)}, D_{(\sigma)}^{(2)}, \ldots, D_{(\sigma)}^{(k)}$ belong to irreducible group representations $\Gamma_1, \Gamma_2, \ldots, \Gamma_k$ respectively. The reducible representation Γ is completely decomposed into its component irreducible representations

$$\Gamma = \Gamma_1 + \Gamma_2 + \ldots + \Gamma_k.$$

Properties of irreducible representations
In the case of an irreducible representation of a group, as has been shown, no change of basis can transform its matrices into direct sums of matrices of smaller dimensions. Matrices of irreducible representations of groups have certain properties that make possible classification of all irreducible representations of a group. Four theorems of representation theory contain some properties of irreducible representations which will be utilized in section 1.4.3.

Since elements of a group G, containing h elements, can be divided into k non-overlapping classes of conjugate elements, the following theorem can be proved.

Theorem 1.1
The group G possesses k different (i.e. non-equivalent) irreducible representations $\Gamma_1, \Gamma_2, \ldots, \Gamma_k$ and their dimensions h_i satisfy the equation

$$h_1^2 + h_2^2 + \ldots + h_k^2 = h.$$

In any representation all group elements which belong to the same class have the same characters, since traces of conjugate elements are equal (see section 1.3.2). Let the k different classes of G, denoted by $K_1, K_2, \ldots, K_k$, contain $g_1, g_2, \ldots, g_k$ group elements respectively and $\chi_i^{(\mu)}$ be the character which belongs to the class K_i in the representation Γ_μ. Then the first orthogonality relation is stated by the next theorem.

Theorem 1.2
The characters $\chi^{(\mu)}$ of different irreducible representations satisfy the relation

$$\sum_{i=1}^{k} g_i \chi_i^{*(\mu)} \chi_i^{(\nu)} = h \delta_{\mu\nu},$$

where $\chi_i^{*(\mu)}$ is the complex conjugate of $\chi_i^{(\mu)}$.

The first orthogonality relation yields the following.

(1) If $\mu \neq \nu, \delta_{\mu\nu} = 0$ and $\sum_{i=1}^{k} g_i \chi_i^{*(\mu)} \chi_i^{(\nu)} = 0$.

(2) If $\mu = \nu, \delta_{\mu\nu} = 1$ and $\sum_{i=1}^{k} g_i \chi_i^{*(\mu)} \chi_i^{(\nu)} = \sum_{i=1}^{k} g_i |\chi_i^{(\mu)}|^2 = h$.

The consequence is that the sets of characters $\chi_i^{(\mu)}$ ($i = 1, \ldots, k$) belonging to different irreducible representations must be different. This property makes it possible to classify irreducible representations of groups by means of characters, which is realized in the form of character tables. The top of the character table has symbols $K_1, K_2, \ldots, K_k$, which denote k different classes of G, while the side of the character table contains symbols of irreducible representations of G:

	K_1	K_2	$\ldots$	K_i	$\ldots$	K_k
Γ_1	$\chi_1^{(1)}$	$\chi_2^{(1)}$	$\ldots$	$\chi_i^{(1)}$	$\ldots$	$\chi_k^{(1)}$
$\vdots$	$\vdots$	$\vdots$		$\vdots$		$\vdots$
Γ_μ	$\chi_1^{(\mu)}$	$\chi_2^{(\mu)}$	$\ldots$	$\chi_i^{(\mu)}$	$\ldots$	$\chi_k^{(\mu)}$
$\vdots$	$\vdots$	$\vdots$		$\vdots$		$\vdots$
Γ_k	$\chi_1^{(k)}$	$\chi_2^{(k)}$	$\ldots$	$\chi_i^{(k)}$	$\ldots$	$\chi_k^{(k)}$

The first orthogonality relation shows that any two rows of the character table, i.e. any two character row vectors, are orthogonal.

The second orthogonality relation is stated by the following theorem.

Theorem 1.3
If Γ_μ is irreducible, then

$$\sum_{\mu=1}^{k} \chi_{(K_i)}^{*(\mu)} \chi_{(K_j)}^{(\mu)} = \frac{h}{h_i} \delta_{ij}.$$

The consequence is that any two columns of the character table, i.e. any two character column vectors, are orthogonal.

When for each irreducible representation of G a definite set of $h_i \times h_i$ matrices, representing the operations σ of G, is determined in an orthonormal basis, then the relation of these representations is stated by the next theorem.

Theorem 1.4
The elements of matrices of any two irreducible representations Γ_μ and Γ_ν satisfy the relation

$$\sum_{\sigma} [D_{(\sigma)}^{(\mu)}]_{ij}^* [D_{(\sigma)}^{(\nu)}]_{pq} = \frac{h}{h_\mu} \delta_{\mu\nu} \delta_{ip} \delta_{jq},$$

where summation runs over all σ of G.

Three cases are considered.

(1) If $\mu \neq \nu$, then $\delta_{\mu\nu} = 0$ and

$$\sum_{\sigma} [D_{(\sigma)}^{(\mu)}]_{ij}^* [D_{(\sigma)}^{(\nu)}]_{pq} = 0.$$

(2) If $\mu = \nu$, then $\delta_{\mu\nu} = 1$, but $i \neq p$ and/or $j \neq q$, so that δ_{ip} and/or δ_{jk} are zero; then

$$\sum_\sigma [D^{(\mu)}_{(\sigma)}]^*_{ij} [D^{(\mu)}_{(\sigma)}]_{pq} = 0.$$

(3) If $\mu = \nu, i = p, j = q$, then $\delta_{\mu\nu} = \delta_{ip} = \delta_{jk} = 1$, and

$$\sum_\sigma [D^{(\mu)}_{(\sigma)}]^*_{ij} [D^{(\mu)}_{(\sigma)}]_{ij} = \sum_\sigma |[D^{(\mu)}_{(\sigma)}]_{ij}|^2 = \frac{h}{h_\mu}.$$

1.4.2 Group algebra

Definitions

When group elements act as linear operators in a vector space a product in the group corresponds to a product of linear operators. Linear operators can be multiplied by scalars and form sums, differences and linear combinations, which is not possible in a group. The group algebra is formed by the introduction of linear combinations of group elements into groups. The group algebra $\bar{G}$ attached to the group G consists of all linear combinations of group elements

$$\xi = x(\epsilon)\epsilon + x(\alpha)\alpha + x(\beta)\beta + \ldots + x(\sigma)\sigma + \ldots = \sum_\sigma x(\sigma)\sigma,$$

where summation runs over all group elements. The coefficients $x(\epsilon), x(\alpha), x(\beta), \ldots$ are real numbers which are called components, while ϵ is the identity element. An element of the group algebra equals a group element σ when $x(\sigma) = 1$ and all other components are zero. Two elements of the group algebra are equal if and only if they coincide in all their components. Therefore, every element of the group algebra is uniquely determined by its h components $x(\sigma)$.

Multiplication of G is preserved in the group algebra

$$\sum_\sigma x(\sigma)\sigma \sum_\tau y(\tau)\tau = \sum_{\sigma,\tau} x(\sigma)y(\tau)\sigma\tau,$$

(where $\sigma\tau$ is a group element). The unit element ϵ of G is also the unit element of $\bar{G}$. The zero vector 0 of $\bar{G}$ is the element of the group algebra, all components of which equal zero. The sum of two elements of $\bar{G}$ is defined by

$$\sum_\sigma x(\sigma)\sigma + \sum_\sigma y(\sigma)\sigma = \sum_\sigma [x(\sigma) + y(\sigma)]\sigma,$$

while multiplication by scalars is defined by

$$a \sum_\sigma x(\sigma)\sigma = \sum_\sigma [ax(\sigma)]\sigma.$$

With these definitions, $\bar{G}$ has the properties of a vector space, its dimension equalling the order of G. Group elements of G form a basis of $\bar{G}$ since they are linearly independent.

The elements and operations of the group algebra are as follows.

elements	$\xi = x(\epsilon)\epsilon + x(\alpha)\alpha + \ldots = \sum_\sigma x(\sigma)\sigma$	
	(Dim $\bar{G} = h$, h is the order of G),	
neutral elements		
addition	0 such that $0 + \xi = \xi$,	
multiplication	ϵ such that $\epsilon\xi = \xi\epsilon = \xi$,	
scalar multiplication	1 such that $1\xi = \xi$,	
inverse elements		
addition	$-\xi$ such that $(-\xi) + \xi = 0$,	
multiplication	ξ^{-1} such that $\xi^{-1}\xi = \xi\xi^{-1} = \epsilon$,	
addition	$\xi_1 + \xi_2 = \xi_3$,	
multiplication		
scalar multiplication	$a\xi$,	
element multiplication	$\xi_1\xi_2 = \xi_3$	
	(apply first ξ_2 then ξ_1),	
associative laws		
addition	$\xi_1 + (\xi_2 + \xi_3) = (\xi_1 + \xi_2) + \xi_3$,	
multiplication	$\xi_1(\xi_2\xi_3) = (\xi_1\xi_2)\xi_3$, $a(b)\xi = (ab)\xi$,	
distributive laws	$\xi_1(\xi_2 + \xi_3) = \xi_1\xi_2 + \xi_1\xi_3$, $(a+b)\xi = a\xi + b\xi$,	
	$a(\xi_1 + \xi_2) = a\xi_1 + a\xi_2$,	
commutative laws		
addition	$\xi_1 + \xi_2 = \xi_2 + \xi_1$,	
multiplication	$\xi_1\xi_2 \neq \xi_2\xi_1$ (in general).	

Centre of the group algebra

An analogy to the centre of a group is the centre of the group algebra, which is defined as a set of all elements that commute with every element of $\bar{G}$. A group element which commutes with all elements of $\bar{G}$ must lie in the centre of $\bar{G}$ (ϵ is always in the centre). The centre of the group algebra $\bar{G}$ attached to an Abelian group is the group itself.

It was shown that every group can be divided into nonoverlapping classes of conjugate elements $\sigma_2 = \rho^{-1}\sigma\rho$. If K is a class containing g_k elements $\sigma_1, \sigma_2, \ldots, \sigma_{g_k}$, the sum of these elements $\kappa = \sigma_1 + \sigma_2 + \ldots + \sigma_{g_k}$ is the corresponding class sum. It can be proved that the centre of a group algebra forms a space.

Theorem 1.5

Class sums form a basis for the centre of a group algebra. The dimension of the centre equals the number of classes of conjugate elements.

The proof of this theorem consists of the following parts.

(1) Class sums lie in the centre of $\bar{G}$.
(2) Every element of the centre of $\bar{G}$ is a linear combination of class sums.
(3) Class sums are linearly independent.

The dimension of the centre of the group algebra $\bar{G}$ always equals the number of class sums, which coincides with the number of classes of conjugate elements of G.

1.4.3 Idempotents of the centre of a group algebra

Definition
An idempotent π of the group algebra is a non-zero element in the algebra which satisfies the equation $\pi^2 = \pi$. It follows that all powers of π are equal: $\pi^n = \pi$ $(n = 1, 2, \ldots)$.

The idempotents π_1, π_2 are orthogonal when $\pi_1 \pi_2 = 0$. A system of orthogonal idempotents satisfies the equation $\pi_i \pi_j = 0$ if $i \neq j$.

Linear independence of idempotents

Theorem 1.6
Orthogonal idempotents are linearly independent.

Theorem 1.7
The sum of orthogonal idempotents is an idempotent.

Consequences:
(1) No system of orthogonal idempotents can have more than h elements (h is the order of the group and the dimension of the group algebra too);
(2) Since the dimension of the centre equals the number of classes of conjugate elements (Theorem 1.5), the number of idempotents cannot surpass the number of classes.

The maximum system of orthogonal idempotents of the centre
It will be shown that the idempotents of the centre of the group algebra act as projection operators which create subspaces of definite symmetry types. Since a vector space usually ought to be decomposed into the greatest possible number of subspaces, it is necessary to determine the maximum number of orthogonal idempotents of the centre.

The following theorem explains the relations of the maximum system of orthogonal idempotents of the centre of the group algebra.

Theorem 1.8
If $S = \pi_1, \pi_2, \ldots, \pi_k$ is a maximum system of orthogonal idempotents of the centre of a group algebra, the following relations are valid.

(1) $\sum_{i=1}^{k} \pi_i = \epsilon$ (ϵ is the unit element of G);

(2) No π_i can be decomposed, i.e. there are no idempotents $\pi_i^{(1)}$ and $\pi_i^{(2)}$ in the centre of $\bar{G}$ which satisfy the equation $\pi_i^{(1)} + \pi_i^{(2)} = \pi_i$.

(3) If π is any idempotent of the centre, then there are idempotents in S with their sum equalling π.
(4) The centre of the group algebra contains one uniquely determined maximum system of orthogonal idempotents.

Relations between characters and idempotents

The relation between characters of irreducible representations of groups, which are classified in character tables, and the orthogonal idempotents of the centre of a group algebra make it possible to obtain the idempotents by utilizing the character tables.

If $D^{(\nu)}_{(\sigma)}$ are matrices of all irreducible representations of the group in orthogonal bases and $B^{(\nu)}_1, \ldots, B^{(\nu)}_{h_\nu}$ the h_ν basis vectors of Γ_ν, then

$$\sigma B^{(\nu)}_q = \sum_{p=1}^{h_\nu} [D^{(\nu)}_{(\sigma)}]_{pq} B^{(\nu)}_p.$$

By multiplying this equation by $[D^{(\mu)}_{(\sigma)}]^*_{ij}$ and by summing over all σ, one finds that

$$\sum_\sigma [D^{(\mu)}_{(\sigma)}]^*_{ij} \sigma B^{(\nu)}_q = \sum_{p=1}^{h_\nu} \sum_\sigma [D^{(\mu)}_{(\sigma)}]^*_{ij} [D^{(\nu)}_{(\sigma)}]_{pq} B^{(\nu)}_p.$$

Theorem 1.4 states that

$$\sum_\sigma [D^{(\mu)}_{(\sigma)}]^*_{ij} [D^{(\nu)}_{(\sigma)}]_{pq} = \frac{h}{h_\mu} \delta_{\mu\nu} \delta_{ip} \delta_{jq},$$

and by substituting it into the right-hand side of the equation, one obtains

$$\sum_\sigma [D^{(\mu)}_{(\sigma)}]^*_{ij} \sigma B^{(\nu)}_q = \sum_{p=1}^{h_\nu} \frac{h}{h_\mu} \delta_{\mu\nu} \delta_{ip} \delta_{kj} B^{(\nu)}_p.$$

The right-hand side of this equation is zero if $\mu \neq \nu$. When $\mu = \nu$ then the right-hand side of the equation is equal to $(h/h_\mu)\delta_{jq} B^{(\mu)}_i$, since in the summation over p the only non-zero term is obtained for $p = i$. The above equation is then

$$\sum_\sigma [D^{(\mu)}_{(\sigma)}]^*_{ij} \sigma B^{(\nu)}_q = \frac{h}{h_\mu} \delta_{\mu\nu} \delta_{jq} B^{(\mu)}_i.$$

For $j = i$ and

$$P^{(\mu)}_{ii} = \frac{h_\mu}{h} \sum_\sigma [D^{(\mu)}_{(\sigma)}]^*_{ij} \sigma$$

the sum of all P_{ii} for $i = 1, \ldots, h_\nu$ is

$$P^{(\mu)} = \sum_{i=1}^{h_\mu} P^{(\mu)}_{ii} = \frac{h_\mu}{h} \sum_\sigma \sum_{i=1}^{h_\mu} [D^{(\mu)}_{(\sigma)}]^*_{ii} \sigma = \frac{h_\mu}{h} \sum_\sigma \chi^{*(\mu)}_{(\sigma)} \sigma.$$

By putting $j = i$ into the equation and by summing over all values of i, one obtains

$$P^{(\mu)}B_q^{(\nu)} = \sum_{i=1}^{h_\mu} \delta_{\mu\nu}\delta_{iq}B_i^{(\mu)}.$$

There are two cases:

(1) $\mu \neq \nu$, $\quad P^{(\mu)}B_q^{(\nu)} = 0$,
(2) $\mu = \nu$, $\quad P^{(\mu)}B_q^{(\mu)} = B_q^{(\mu)}$.

Any vector $X^{(\nu)}$ in the space of Γ_ν is now a linear combination of the basis vectors $B_1^{(\nu)}, \ldots, B_{h_\nu}^{(\nu)}$ and one finds that

(1) $P^{(\mu)}X^{(\nu)} = 0$ if $\mu \neq \nu$ and
(2) $P^{(\mu)}X^{(\mu)} = X^{(\mu)}$ if $\mu = \nu$.

It can be concluded that $P^{(\mu)}$ is an operator which nullifies every vector which does not belong to the space of Γ_μ. Thus, by acting on a set of vectors belonging to various subspaces $U_1, U_2, \ldots, U_\mu, \ldots, U_k$ the operator $P^{(\mu)}$ picks out all vectors $X^{(\mu)}$ belonging to the subspace U_μ and transforms all other vectors into zero vectors. Because of this property $P^{(\mu)}$ is called the **projection operator** of the subspace U_μ.

The equation $P^{(\mu)}X^{(\mu)} = X^{(\mu)}$ shows that $P^{(\mu)}$ is an idempotent, so that it can be written

$$P^{(\mu)} = \frac{h_\mu}{h} \sum_\sigma \chi_{(\sigma)}^{*(\mu)} \sigma = \pi^{(\mu)}.$$

The character of the ith irreducible representation is usually denoted by a subscript, i.e. χ_i, while for real characters $\chi_i(\sigma) = \chi_i(\sigma^{-1})$.

Finally, the relation between characters of irreducible representations of groups and orthogonal idempotents of the centre of group algebra is

$$\pi_i = \frac{h_i}{h} \sum_\sigma \chi_i(\sigma^{-1})\sigma.$$

The idempotent π_i can be written down directly from the character table, since it is a definite linear combination of operators σ with characters χ_i as their coefficients.

Classification of irreducible representations

The dimensions of irreducible group representations of thirty-two point groups are not greater than two, except for T_d, O and O_h which possess three-dimensional representations. The dimension of a character equals $\chi(E)$ if χ is real, while its dimension equals $2\chi(E)$ if it is complex. Evidently, the dimension of the ith character corresponds to the dimension of the space of the ith irreducible representation, which coincides with the trace of the identity matrix in this space.

Mulliken's notation for irreducible representations of symmetry groups contains the following symbols:

(1) one-dimensional characters A, B
     two-dimensional characters E
     three-, four- and five-dimensional characters F, G, H;

(2) when C_n is a rotation about the main rotation axis, then the one-dimensional characters are designated by A if $\chi(C_n) = 1$ and B if $\chi(C_n) = -1$;

(3) the subscript g means that $\chi(i) = 1$, while the subscript u means that $\chi(i) = -1$ (i is the inversion, while g and u denote 'gerade' and 'ungerade' respectively, i.e. even and odd in German);

(4) the superscripts $'$ and $''$ denote the value of the character for the horizontal reflection: A' for $\chi(\sigma_h) = 1$ and A'' for $\chi(\sigma_h) = -1$.

1.4.4 G-vector spaces

Definition
The G-vector spaces are a link between groups and vector spaces. A vector space V is G-vector space if the following axioms are satisfied.

(1) Every group element σ is a linear operator of V, σX is a vector of V, $\sigma(X + Y) = \sigma X + \sigma Y$, $\sigma(aX) = a\sigma X$.
(2) $(\sigma\tau)X = \sigma(\tau X)$.
(3) $\epsilon X = X$.

A subspace U of a G-vector space V is G-invariant if it is a G-vector space, i.e. if σX lie in U for all σ of G and all X in U.

Group representations as G-vector spaces
A group representation of degree n (or an n-dimensional representation) is a mapping which associates an $n \times n$ invertible matrix $A(\sigma)$ with every σ of G, so that

(1) $A(\sigma\tau) = A(\sigma)A(\tau)$ and
(2) $A(\epsilon) = E$.

Since with respect to an established basis every $n \times n$ matrix is a linear operator of the n-dimensional vector space V, σX is defined by $\sigma X = A(\sigma)X$ in a representation of G of degree n, while the axioms for a vector space are satisfied. The association of a matrix representation with a G-vector space is not unique, because changes of bases produce different sets of matrices of the group representation.

The elements and operations of a G-vector space are as follows.

Representation theory

	Symbolic notation	Matrix notation
elements	σX,	AX,
neutral elements		
addition	0 such that $0 + X = X$,	
multiplication	ϵ such that $\epsilon X = X$,	E such that $EX = X$,
scalar multiplication	1 such that $1X = X$,	
inverse elements		
addition	$-\sigma$ such that $-\sigma X + \sigma X = 0$,	$-A$ such that $-AX + AX = 0$,
multiplication	σ^{-1} such that $\sigma^{-1}\sigma X = X$	A^{-1} such that $A^{-1}AX = X$
addition	$\sigma_1 X_1 + \sigma_2 X_2$,	$A_1 X_1 + A_2 X_2$,
multiplication		
scalar multiplication	$a\sigma X$,	aAX,
operator multiplication	$(\sigma_1\sigma_2)X = \sigma_3 X$,	$(A_1 A_2)X = A_3 X$,
associative laws		
addition	$\sigma_1 X_1 + (\sigma_2 X_2 + \sigma_3 X_3) = (\sigma_1 X_1 + \sigma_2 X_2) + \sigma_3 X_3$,	$A_1 X_1 + (A_2 X_2 + A_3 X_3) = (A_1 X_1 + A_2 X_2) + A_3 X_3$,
multiplication	$\sigma_1(\sigma_2\sigma_3)X = (\sigma_1\sigma_2)\sigma_3 X$,	$A_1(A_2 A_3)X = (A_1 A_2)A_3 X$,
distributive laws	$\sigma_1(\sigma_2 + \sigma_3)X = \sigma_1\sigma_2 X + \sigma_1\sigma_3 X$,	$A_1(A_2 + A_3)X = A_1 A_2 X + A_1 A_3 X$,
	$(a + b)\sigma X = a\sigma X + b\sigma X$,	$(a + b)AX = aAX + bAX$,
	$a(\sigma_1 + \sigma_2)X = a\sigma_1 X + a\sigma_2 X$,	$a(A_1 + A_2)X = aA_1 X + aA_2 X$,
commutative laws		
addition	$\sigma_1 X_1 + \sigma_2 X_2 = \sigma_2 X_2 + \sigma_1 X_1$,	$A_1 X_1 + A_2 X_2 = A_2 X_2 + A_1 X_1$,
multiplication	$\sigma_1\sigma_2 X \neq \sigma_2\sigma_1 X$ (in general).	$A_1 A_2 X \neq A_2 A_1 X$ (in general).

Decomposition of a vector space

Since elements of the group algebra $\bar{G}$ are linear combinations of group elements, which act as linear operators in the G-vector space, a linear operator can be associated with every element of $\bar{G}$ and defined by

$$\left(\sum_\sigma x(\sigma)\sigma\right) X = \sum_\sigma x(\sigma)\sigma X.$$

In case of symmetry groups the idempotents π_i of the centre of group algebra, as linear operators acting on V, have the following definiton.

A vector X of V has the ith symmetry type if $\pi_i X = X$.

Theorem 1.9

When U_i is a set of all vectors of the ith symmetry type, its properties are as follows.

(1) U_i is a subspace of V.
(2) U_i is G-invariant.
(3) U_i contains all vectors $\pi_i X$ of all X of V.
(4) If $B_1, \ldots, B_n$ is a basis of V, then the vectors $\pi_i B_1, \ldots, \pi_i B_n$ generate the subspace U_i.

Theorem 1.10

Every vector X of a G-vector space V can be expressed as the sum of the vectors X_i of the G-invariant subspaces U_i in a unique way. Thus the G-vector space V is the direct sum of G-invariant subspaces U_i: $V = U_1 + U_2 + \ldots + U_k$.

Conclusions

(1) A G-vector space is irreducible if it contains no genuine non-zero subspace.
(2) An irreducible G-vector space can contain vectors of only one symmetry type. Therefore, one definite symmetry type is associated with each irreducible G-vector space or with the corresponding irreducible representation of the group. Thus a decomposition of a G-vector space can be considered as a decomposition into irreducible representations of the group.
(3) Since characters are traces of matrices of irreducible representations, the dimension of a character equals the dimension of the corresponding irreducible G-vector space.

Equation for evaluation of dimensions of G-invariant subspaces

The dimension of the subspaces U_i are determined by the expression

$$\text{Dim } U_i = h_i(\text{Tr}, \chi_i),$$

where $h_i = \chi_i(E)$, χ_i is the ith character, while Tr is the trace of the matrix of the symmetry operator (group element).

When a configuration consisting of n points (nodes) is given, the trace of a group element equals the number of points (nodes) which remain in their positions when the group element (the symmetry operation) is applied.

2

G-vector analysis

When a space object possesses symmetry properties which can be described by a group, one can represent it in the G-vector space and obtain a system of equations, with the matrix in block diagonal form, that formulates the behaviour of the object. Thus the system of equations is composed of independent sets of equations, each set belonging to its G-invariant subspace. The procedure, based on these properties and developed by Zloković (1973, 1974, 1976, 1977, 1989) for various fields of the theory of structures (vibration, stability and statics), was called G-vector analysis.

The division of the continuum of a structure into surface or space elements allows various analyses to be made with most efficient applications of group theory and G-vector spaces if elements with complex symmetry are chosen.

G-vector analysis provides a decomposition of the G-vector space of the element into G-invariant subspaces and a solution of the problem with substantial qualitative and quantitative advantages in comparison with conventional analysis. The extent of decomposition depends not only on the form of the element but also, very substantially, on the number of nodal points and on their pattern. These properties can be utilized very fruitfully in the finite element method, where subdivision of the continuum is quite arbitrary. For a given structural continuum it may be convenient to introduce suitable partitions that make the most of the G-vector analysis.

Division of the continuum into elements, with definition of their relations by functions and determination of function values under given conditions, may be regarded as a discrete analysis with a free choice of the form and the size of elements. In the case of space frame structures there is a preference for geometrical configurations with a minimum of different types of nodal joint, which may be the properties of special geometrical patterns that are composed either of regular polygons and polyhedra or of their parts in suitable combinations.

In comparison with applications of symmetry in standard practice, usually in relation to one or two planes of symmetry, the G-vector analysis makes use of 8 symmetry operations for the square, 48 for the cube and 120 for the regular icosahedron. This provides systems of equations with the matrices with much more pronounced block diagonal form than in the cases of application of symmetry in the usual way.

48 G-vector analysis [Ch. 2]

In this chapter the G-vector analysis is applied to some nodal patterns according to the following procedure.

(1) Determination of the symmetry group G of the defined configuration.
(2) Record of elements, classes and the character table, and determination of the idempotents of the centre of the group algebra $\bar{G}$.
(3) Derivation of the basis vectors of the G-invariant subspaces of the space V of the configuration.
(4) Confirmation of dimensions of the subspaces by analytic evaluation.

2.1 NODAL CONFIGURATIONS WITH PATTERNS BASED ON THE SQUARE

The nodal pattern given in Fig. 2.1 has four nodes at the vertices of the square. The origin of coordinates is at the centre of the square and the axes x and y are parallel to its sides, while the axis z is perpendicular to the plane of the square.

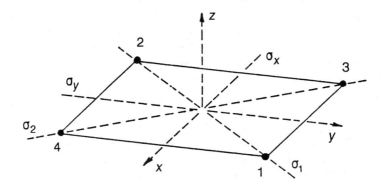

Fig. 2.1. The four-node pattern based on the square.

For description of symmetry properties of the above configuration the group C_{4v} is chosen. This contains the complete set of symmetry operations that transform the square into itself:

E identity operation,
C_4 rotation through 90° about the z axis,
C_4^{-1} rotation through −90° about the z axis,
C_2 rotation through 180° about the z axis,
σ_x reflection in the plane $y = 0$,
σ_y reflection in the plane $x = 0$,
σ_1 reflection in the plane $x = y$,
σ_2 reflection in the plane $x = -y$.

Sec. 2.1] Nodal configurations with patterns based on the square

The classes of group elements (symmetry operations) are

$$K_1 = [E], \quad K_2 = [C_4, C_4^{-1}], \quad K_3 = [C_2], \quad K_4 = [\sigma_x, \sigma_y],$$
$$K_5 = [\sigma_1, \sigma_2].$$

The character table of the group C_{4v} is

C_{4v}	E	$2C_4$	C_2	$2\sigma_v$	$2\sigma_d$
A_1	1	1	1	1	1
A_2	1	1	1	−1	−1
B_1	1	−1	1	1	−1
B_2	1	−1	1	−1	1
E	2	0	−2	0	0

The expression for determination of the idempotents of the centre of the group algebra is

$$\pi_i = \frac{h_i}{h} \sum_\sigma \chi_i(\sigma^{-1})\sigma,$$

where h_i is the dimension of the ith character, given by $h_i = \chi_i(E)$, h is the order, χ_i is the ith character, σ is the element and σ^{-1} its inverse.

The idempotents of the centre of the group algebra of the group C_{4v} are

$$\pi_1 = \tfrac{1}{8}(E + C_4 + C_4^{-1} + C_2 + \sigma_x + \sigma_y + \sigma_1 + \sigma_2)$$
$$\pi_2 = \tfrac{1}{8}(E + C_4 + C_4^{-1} + C_2 - \sigma_x - \sigma_y - \sigma_1 - \sigma_2)$$
$$\pi_3 = \tfrac{1}{8}(E - C_4 - C_4^{-1} + C_2 + \sigma_x + \sigma_y - \sigma_1 - \sigma_2)$$
$$\pi_4 = \tfrac{1}{8}(E - C_4 - C_4^{-1} + C_2 - \sigma_x - \sigma_y + \sigma_1 + \sigma_2)$$
$$\pi_5 = \tfrac{1}{4}(2E - 2C_2).$$

The basis vectors of G-invariant subspaces U_i are as follows:

$U_1(U_1^*)$: $\quad \overline{\varphi}_1^* = \pi_1\varphi_1 = \pi_1\varphi_2 = \pi_1\varphi_3 = \pi_1\varphi_4 = \tfrac{1}{4}(\varphi_1 + \varphi_2 + \varphi_3 + \varphi_4),$

U_2: $\quad \pi_2\varphi_1 = \pi_2\varphi_2 = \pi_2\varphi_3 = \pi_2\varphi_4 = 0,$

U_3: $\quad \pi_3\varphi_1 = \pi_3\varphi_2 = \pi_3\varphi_3 = \pi_3\varphi_4 = 0,$

$U_4(U_2^*)$: $\quad \overline{\varphi}_2^* = \pi_4\varphi_1 = \pi_4\varphi_2 = -\pi_4\varphi_3 = -\pi_4\varphi_4 = \tfrac{1}{4}(\varphi_1 + \varphi_2 - \varphi_3 - \varphi_4),$

$U_5(U_3^*)$: $\quad \overline{\varphi}_3^* = \pi_5\varphi_1 = -\pi_5\varphi_2 = \tfrac{1}{2}(\varphi_1 - \varphi_2),$

$\qquad\qquad \overline{\varphi}_4^* = \pi_5\varphi_3 = -\pi_5\varphi_4 = \tfrac{1}{2}(\varphi_3 - \varphi_4),$

with $\overline{\varphi}_3^*\overline{\varphi}_4^* = 0.$

The above derivation shows that the four-dimensional space of the configuration is decomposed into three G-invariant subspaces: one-dimensional U_1^* and U_2^*, and two-dimensional U_3^*. In Fig. 2.2 the basis vectors of each subspace are designated in such a manner that if the nodal function φ_j takes a positive sign (+), an arrow is placed at the node and directed towards centre of the square (the origina of coordinates), while if it takes a negative sign (−) the arrow has opposite direction. Thus, at the same time, the picture visualizes the symmetry type of the subspace.

The positive directions of displacements or rotations of the nodes, as well as of forces or moments at the nodes, in the G-vector analysis are always defined by basis vectors, which are derived by the first idempotent π_1. Figure 2.3(a) designates positive directions at nodes 1, 2, 3, 4 in the G-vector analysis, where the set of directions corresponds to the symmetry type of the first irreducible representation of the group C_{4v} and the subspace U_1. The standard convention has positive directions at the nodes parallel to and in the same direction as the axes x and y (Fig. 2.3(b)).

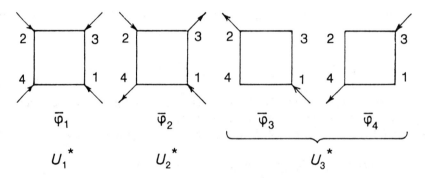

Fig. 2.2. Basis vectors of the subspaces of the four-node pattern based on the square and described by the group C_{4v}.

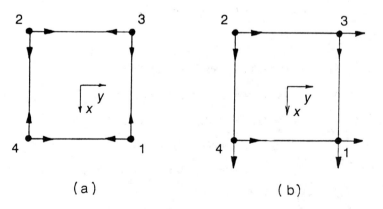

Fig. 2.3. Positive directions of displacements and rotations of the nodes, as well as of forces and moments at the nodes, for the four-node pattern based on the square: (a) in G-vector analysis, (b) in the standard method.

Nodal configurations with patterns based on the square

The system of equations of the configuration is composed of systems of equations of particular independent G-invariant subspaces U_i, which, if given in matrix form, can be represented schematically because of $\overline{\varphi}_3^* \overline{\varphi}_4^* = 0$ as

$$\begin{bmatrix} \boxed{\cdot} & & & \\ & \boxed{\cdot} & & \\ & & \boxed{\cdot} & \\ & & & \boxed{\cdot} \end{bmatrix} \begin{bmatrix} \overline{\varphi}_1^* \\ \overline{\varphi}_2^* \\ \overline{\varphi}_3^* \\ \overline{\varphi}_4^* \end{bmatrix} = \begin{bmatrix} \cdot \\ \cdot \\ \cdot \\ \cdot \end{bmatrix}$$

The nodal functions φ_j ($j = 1, 2, 3, 4$), expressed by the basis vectors $\overline{\varphi}_j$ are

$$\varphi_1 = \overline{\varphi}_1^* + \overline{\varphi}_2^* + \overline{\varphi}_3^*$$
$$\varphi_2 = \overline{\varphi}_1^* + \overline{\varphi}_2^* - \overline{\varphi}_3^*$$
$$\varphi_3 = \overline{\varphi}_1^* - \overline{\varphi}_2^* + \overline{\varphi}_4^*$$
$$\varphi_4 = \overline{\varphi}_1^* - \overline{\varphi}_2^* - \overline{\varphi}_4^*.$$

To conform previous results, the dimensions of the subspaces U_i will be determined using the expression

$$\text{Dim } U_i = h_i (\text{Tr}, \chi_i),$$

given in section 1.4.3, where $h_i = \chi(E)$ is the dimension of the ith character of E, χ_i the ith character and Tr the trace of the matrix of symmetry operation (equals the number of nodal points which are not moved by the action of the symmetry operation).

For the configuration analysed, which is described by the group C_{4v}, one obtains the following traces and dimensions of U_i:

$$\text{Tr}(E) = 4 \qquad \text{Tr}(\sigma_1) = \text{Tr}(\sigma_2) = 2$$
$$\text{Tr}(C_4) = \text{Tr}(C_4^{-1}) = \text{Tr}(C_2) = \text{Tr}(\sigma_x) = \text{Tr}(\sigma_y) = 0$$
$$\text{Dim } U_1 = \tfrac{1}{8}(1 \cdot 4 + 1 \cdot 0 + 1 \cdot 0 + 1 \cdot 0 + 1 \cdot 0 + 1 \cdot 0 + 1 \cdot 2 + 1 \cdot 2) = 1$$
$$\text{Dim } U_2 = \tfrac{1}{8}(1 \cdot 4 + 1 \cdot 0 + 1 \cdot 0 + 1 \cdot 0 - 1 \cdot 0 - 1 \cdot 0 - 1 \cdot 2 - 1 \cdot 2) = 0$$
$$\text{Dim } U_3 = \tfrac{1}{8}(1 \cdot 4 - 1 \cdot 0 - 1 \cdot 0 + 1 \cdot 0 + 1 \cdot 0 + 1 \cdot 0 - 1 \cdot 2 - 1 \cdot 2) = 0$$
$$\text{Dim } U_4 = \tfrac{1}{8}(1 \cdot 4 - 1 \cdot 0 - 1 \cdot 0 + 1 \cdot 0 - 1 \cdot 0 - 1 \cdot 0 + 1 \cdot 2 + 1 \cdot 2) = 1$$
$$\text{Dim } U_5 = \tfrac{1}{4}(2 \cdot 4 + 0 \cdot 0 + 0 \cdot 0 - 2 \cdot 0 + 0 \cdot 0 + 0 \cdot 0 + 0 \cdot 0) = 2.$$

These dimensions correspond to the dimensions of the subspaces U_i, which were derived by means of the idempotents. Since the space V is divided into three subspaces and the pattern analysed has only four nodes, in this case the basis vectors $\overline{\varphi}_3^*$ and $\overline{\varphi}_4^*$ are orthogonal and their two-dimensional subspace U_3^* is decomposed into two one-dimensional subspaces.

The configuration considered, based on the square, has only four nodal points, while the applied group C_{4v} possesses eight elements (symmetry operations), which produces zero subspaces U_2 and U_3 of the second and third symmetry types of the group. Therefore, it is appropriate to apply the group C_{2v}, which is a subgroup of the group C_{4v}. The group C_{2v} describes the symmetry properties of the rectangle and it contains four elements, i.e. the following symmetry operations:

E identity operation,

C_2 rotation through 180° about the z axis

σ_1 reflection in the xz plane

σ_2 reflection in the yz plane.

Each element is in its class alone. The character table of the group C_{2v} is

C_{2v}	E	C_2	σ_1	σ_2
A_1	1	1	1	1
A_2	1	1	−1	−1
B_1	1	−1	1	−1
B_2	1	−1	−1	1

The idempotents π_i ($i = 1, 2, 3, 4$) of the centre of group algebra are determined using the expression given earlier:

$$\pi_1 = \tfrac{1}{4}(E + C_2 + \sigma_1 + \sigma_2)$$
$$\pi_2 = \tfrac{1}{4}(E + C_2 - \sigma_1 - \sigma_2)$$
$$\pi_3 = \tfrac{1}{4}(E - C_2 + \sigma_1 - \sigma_2)$$
$$\pi_4 = \tfrac{1}{4}(E - C_2 - \sigma_1 + \sigma_2).$$

Interchange of positions of x and y axes in Fig. 2.1 gives the nodal numbering in Fig. 3.7 in section 3.5 so that the basis vectors of G-invariant subspaces U_i are as follows:

$$U_1: \quad \bar{\varphi}_1 = \tfrac{1}{4}(\varphi_1 + \varphi_2 + \varphi_3 + \varphi_4)$$
$$U_2: \quad \bar{\varphi}_2 = \tfrac{1}{4}(\varphi_1 + \varphi_2 - \varphi_3 - \varphi_4)$$
$$U_3: \quad \bar{\varphi}_3 = \tfrac{1}{4}(\varphi_1 - \varphi_2 + \varphi_3 - \varphi_4)$$
$$U_4: \quad \bar{\varphi}_4 = \tfrac{1}{4}(\varphi_1 - \varphi_2 - \varphi_3 + \varphi_4).$$

The above derivation shows that the four-dimensional space of the configuration is decomposed into four one-dimensional G-invariant subspaces U_1, U_2, U_3, U_4. As previously in the case of the application of the group C_{4v}, the basis vectors

$\bar{\varphi}_i$ ($i = 1, 2, 3, 4$) are designated by arrows pointing to the centre of the square when the sign is positive, and in the opposite direction when the sign is negative, as shown in Fig. 2.4. Thus this figure gives the picture of the symmetry type of each subspace.

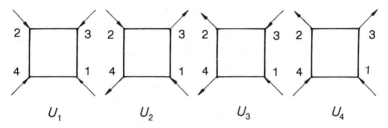

Fig. 2.4. Basis vectors of the subspaces of the four-node pattern based on the square and described by the group C_{2v}.

The system of equations of the configuration is composed of independent systems of equations for each G-invariant subspace separately, which can be presented schematically as

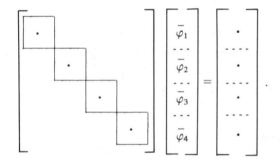

The nodal functions φ_j ($j = 1, 2, 3, 4$), expressed by the basis vectors $\bar{\varphi}_j$, are

$$\varphi_1 = \bar{\varphi}_1 + \bar{\varphi}_2 + \bar{\varphi}_3 + \bar{\varphi}_4$$
$$\varphi_2 = \bar{\varphi}_1 + \bar{\varphi}_2 - \bar{\varphi}_3 - \bar{\varphi}_4$$
$$\varphi_3 = \bar{\varphi}_1 - \bar{\varphi}_2 + \bar{\varphi}_3 - \bar{\varphi}_4$$
$$\varphi_4 = \bar{\varphi}_1 - \bar{\varphi}_2 - \bar{\varphi}_3 + \bar{\varphi}_4.$$

The twelve-node pattern in Fig. 2.5 has four nodes at the vertices and eight nodes at the sides of the square at the same distance from the vertices of the square.

The basis vectors $\bar{\varphi}_1, \bar{\varphi}_2, \ldots, \bar{\varphi}_{12}$ of G-invariant subspaces U_1, U_2, U_3, U_4, U_5 are obtained by applying the idempotents π_i ($i = 1, 2, 3, 4, 5$), determined earlier for the group C_{4v}, to the nodal functions $\varphi_1, \varphi_2, \ldots, \varphi_{12}$

subspace U_1: $\quad \bar{\varphi}_1 = \frac{1}{4}(\varphi_1 + \varphi_2 + \varphi_3 + \varphi_4)$

$\qquad \bar{\varphi}_2 = \frac{1}{8}(\varphi_5 + \varphi_6 + \varphi_7 + \varphi_8 + \varphi_9 + \varphi_{10} + \varphi_{11} + \varphi_{12})$

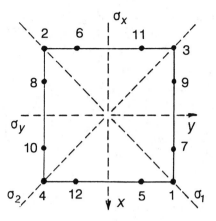

Fig. 2.5. Twelve-node pattern based on the square.

subspace U_2: $\quad \bar{\varphi}_3 = \tfrac{1}{8}(\varphi_5 + \varphi_6 + \varphi_7 + \varphi_8 - \varphi_9 - \varphi_{10} - \varphi_{11} - \varphi_{12})$

subspace U_3: $\quad \bar{\varphi}_4 = \tfrac{1}{8}(\varphi_5 + \varphi_6 - \varphi_7 - \varphi_8 - \varphi_9 - \varphi_{10} + \varphi_{11} + \varphi_{12})$

subspace U_4: $\quad \bar{\varphi}_5 = \tfrac{1}{4}(\varphi_1 + \varphi_2 - \varphi_3 - \varphi_4)$

$\qquad\qquad\quad \bar{\varphi}_6 = \tfrac{1}{8}(\varphi_5 + \varphi_6 - \varphi_7 - \varphi_8 + \varphi_9 + \varphi_{10} - \varphi_{11} - \varphi_{12})$

subspace U_5: $\quad \bar{\varphi}_7 = \tfrac{1}{2}(\varphi_1 - \varphi_2) \qquad \bar{\varphi}_{10} = \tfrac{1}{2}(\varphi_7 - \varphi_8)$

$\qquad\qquad\quad \bar{\varphi}_8 = \tfrac{1}{2}(\varphi_3 - \varphi_4) \qquad \bar{\varphi}_{11} = \tfrac{1}{2}(\varphi_9 - \varphi_{10})$

$\qquad\qquad\quad \bar{\varphi}_9 = \tfrac{1}{2}(\varphi_5 - \varphi_6) \qquad \bar{\varphi}_{12} = \tfrac{1}{2}(\varphi_{11} - \varphi_{12}).$

The twelve-dimensional space of the configuration is decomposed into subspaces U_1, U_2, U_3, U_4, U_5 with 2, 1, 1, 2, 6 dimensions respectively. This is shown in Fig. 2.6 in the same manner as for the previous four-node pattern, so that the picture visualizes the symmetry type of each subspace.

The six-dimensional subspace U_5 can be subdivided into two subspaces if by linear combinations of basis vectors $\bar{\varphi}_7$, $\bar{\varphi}_8$, $\bar{\varphi}_9$, $\bar{\varphi}_{10}$, $\bar{\varphi}_{11}$, $\bar{\varphi}_{12}$ one can form a new basis of the subspace U_5 consisting of two sets of vectors, each set containing three vectors, where each vector of the first set is orthogonal to each vector of the second set, i.e. if they fulfil the following orthogonality conditions:

$$\bar{\varphi}'_7\bar{\varphi}'_{10} = \bar{\varphi}'_7\bar{\varphi}'_{11} = \bar{\varphi}'_7\bar{\varphi}'_{12} = \bar{\varphi}'_8\bar{\varphi}'_{10} = \bar{\varphi}'_8\bar{\varphi}'_{11} = \bar{\varphi}'_8\bar{\varphi}'_{12} = \bar{\varphi}'_9\bar{\varphi}'_{10}$$
$$= \bar{\varphi}'_9\bar{\varphi}'_{11} = \bar{\varphi}'_9\bar{\varphi}'_{12} = 0.$$

This is satisfied by the following two sets of basis vectors $\bar{\varphi}'_7$, $\bar{\varphi}'_8$, $\bar{\varphi}'_9$ and $\bar{\varphi}'_{10}$, $\bar{\varphi}'_{11}$, $\bar{\varphi}'_{12}$, which span the two three-dimensional subspaces of U_5

Sec. 2.1]   Nodal configurations with patterns based on the square   55

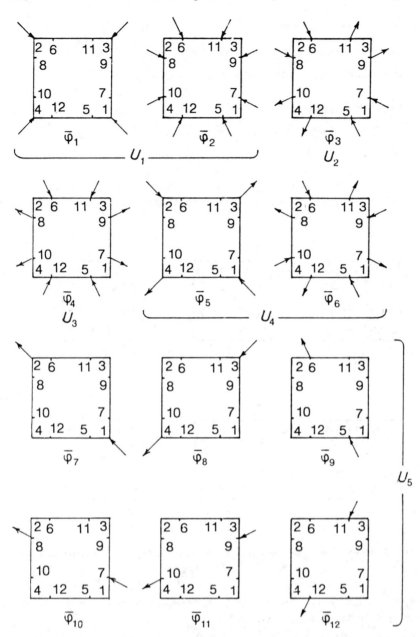

Fig. 2.6. Basis vectors of the subspaces of the twelve-node pattern based on the square and described by the group C_{4v}.

$$\bar{\varphi}_7' = \bar{\varphi}_7 = \tfrac{1}{2}(\varphi_1 - \varphi_2)$$
$$\bar{\varphi}_8' = \bar{\varphi}_9 + \bar{\varphi}_{10} = \tfrac{1}{2}(\varphi_5 - \varphi_6 + \varphi_7 - \varphi_8) \quad \Big\} \; \text{I}$$
$$\bar{\varphi}_9' = \bar{\varphi}_{11} - \bar{\varphi}_{12} = \tfrac{1}{2}(\varphi_9 - \varphi_{10} - \varphi_{11} + \varphi_{12})$$

$$\bar{\varphi}_{10}' = \bar{\varphi}_8 = \tfrac{1}{2}(\varphi_3 - \varphi_4)$$
$$\bar{\varphi}_{11}' = \bar{\varphi}_9 - \bar{\varphi}_{10} = \tfrac{1}{2}(\varphi_5 - \varphi_6 - \varphi_7 + \varphi_8) \quad \Big\} \; \text{II}$$
$$\bar{\varphi}_{12}' = \bar{\varphi}_{11} + \bar{\varphi}_{12} = \tfrac{1}{2}(\varphi_9 - \varphi_{10} + \varphi_{11} - \varphi_{12})$$

These basis vectors are shown in Fig. 2.7, which demonstrates that all these vectors are both symmetric and antisymmetric, and the orthogonality of the vector sets I and II is evident.

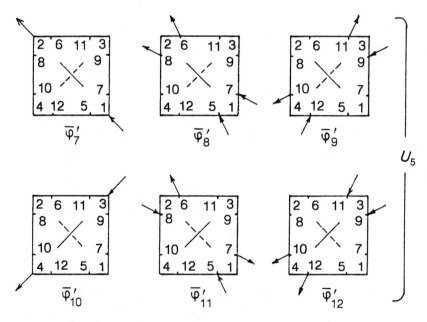

Fig. 2.7. Basis vectors of the subspace U_s of the twelve-node pattern based on the square and described by the group C_{4v}.

Thus, schematically, the following system of equations with the matrix in block diagonal form can be presented as (see p. 57).

Determination of basis vectors $\bar{\varphi}_l$ ($l = 7, 8, 9, 10, 11, 12$) and nodal function φ_j is as follows:

$$\bar{\varphi}_7 = \bar{\varphi}_7', \qquad \varphi_1 = \bar{\varphi}_1 + \bar{\varphi}_3 + \bar{\varphi}_7,$$
$$\bar{\varphi}_8 = \bar{\varphi}_{10}', \qquad \varphi_2 = \bar{\varphi}_1 + \bar{\varphi}_3 - \bar{\varphi}_7,$$
$$\bar{\varphi}_9 = \tfrac{1}{2}(\bar{\varphi}_8' + \bar{\varphi}_{11}'), \qquad \varphi_3 = \bar{\varphi}_1 - \bar{\varphi}_3 + \bar{\varphi}_8,$$

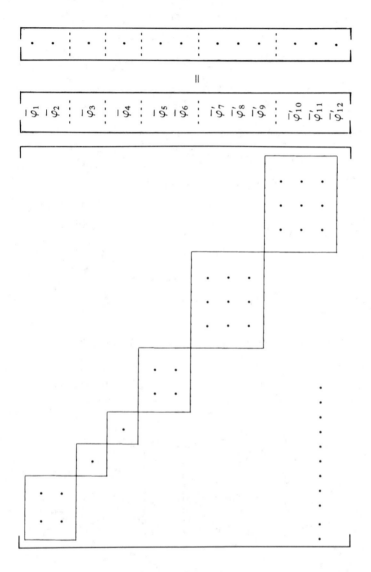

$\bar{\varphi}_{10} = \frac{1}{2}(\bar{\varphi}'_8 - \bar{\varphi}'_{11})$,
$\bar{\varphi}_{11} = \frac{1}{2}(\bar{\varphi}'_9 + \bar{\varphi}'_{12})$,
$\bar{\varphi}_{12} = \frac{1}{2}(-\bar{\varphi}'_9 + \bar{\varphi}'_{12})$,

$\varphi_4 = \bar{\varphi}_1 - \bar{\varphi}_3 - \bar{\varphi}_8$,
$\varphi_5 = \bar{\varphi}_2 + \bar{\varphi}_4 + \bar{\varphi}_5 + \bar{\varphi}_6 + \bar{\varphi}_9$,
$\varphi_6 = \bar{\varphi}_2 + \bar{\varphi}_4 + \bar{\varphi}_5 + \bar{\varphi}_6 - \bar{\varphi}_9$,
$\varphi_7 = \bar{\varphi}_2 + \bar{\varphi}_4 - \bar{\varphi}_5 - \bar{\varphi}_6 + \bar{\varphi}_{10}$,
$\varphi_8 = \bar{\varphi}_2 + \bar{\varphi}_4 - \bar{\varphi}_5 - \bar{\varphi}_6 - \bar{\varphi}_{10}$,
$\varphi_9 = \bar{\varphi}_2 - \bar{\varphi}_4 - \bar{\varphi}_5 + \bar{\varphi}_6 + \bar{\varphi}_{11}$,
$\varphi_{10} = \bar{\varphi}_2 - \bar{\varphi}_4 - \bar{\varphi}_5 + \bar{\varphi}_6 - \bar{\varphi}_{11}$,
$\varphi_{11} = \bar{\varphi}_2 - \bar{\varphi}_4 + \bar{\varphi}_5 - \bar{\varphi}_6 + \bar{\varphi}_{12}$,
$\varphi_{12} = \bar{\varphi}_2 - \bar{\varphi}_4 + \bar{\varphi}_5 - \bar{\varphi}_6 - \bar{\varphi}_{12}$.

Analytic determination of dimensions of subspaces U_i gives:

$\text{Tr}(E) = 12$, $\quad \text{Tr}(\sigma_1) = \text{Tr}(\sigma_2) = 2$,

$\text{Tr}(C_4) = \text{Tr}(C_4^{-1}) = \text{Tr}(C_2) = \text{Tr}(\sigma_x) = \text{Tr}(\sigma_y) = 0$,

$\text{Dim } U_1 = \frac{1}{8}(1 \cdot 12 + 1 \cdot 0 + 1 \cdot 0 + 1 \cdot 0 + 1 \cdot 0 + 1 \cdot 0 + 1 \cdot 2 + 1 \cdot 2) = 2$,

$\text{Dim } U_2 = \frac{1}{8}(1 \cdot 12 + 1 \cdot 0 + 1 \cdot 0 + 1 \cdot 0 - 1 \cdot 0 - 1 \cdot 0 - 1 \cdot 2 - 1 \cdot 2) = 1$,

$\text{Dim } U_3 = \frac{1}{8}(1 \cdot 12 - 1 \cdot 0 - 1 \cdot 0 + 1 \cdot 0 + 1 \cdot 0 + 1 \cdot 0 - 1 \cdot 2 - 1 \cdot 2) = 1$,

$\text{Dim } U_4 = \frac{1}{8}(1 \cdot 12 - 1 \cdot 0 - 1 \cdot 0 + 1 \cdot 0 - 1 \cdot 0 - 1 \cdot 0 + 1 \cdot 2 + 1 \cdot 2) = 2$,

$\text{Dim } U_5 = \frac{1}{4}(2 \cdot 12 + 0 \cdot 0 + 0 \cdot 0 - 2 \cdot 0 + 0 \cdot 0 + 0 \cdot 0 + 0 \cdot 2 + 0 \cdot 2) = 6$.

2.2 NODAL CONFIGURATION WITH A PATTERN BASED ON THE CUBE

The configuration with the pattern in Fig. 2.8 has eight nodes—one node at each vertex of the cube. The origin of coordinates is at the centre of the cube, with coordinate axes x, y, z parallel to the edges of the cube and with a chosen nodal numbering.

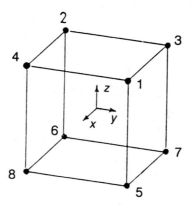

Fig. 2.8. The eight-node pattern based on the cube.

Sec. 2.2] Nodal configurations with patterns based on the cube 59

The group O_h with 48 elements contains all symmetry operations that turn the cube into itself. Since the above pattern has only eight nodes, it is convenient to choose a subgroup of the group O_h with a smaller number of elements (groups O or D_{2h}). In section 2.3 the right parallelepiped is analysed with respect to the group D_{2h}, which is also applicable to the cube. Therefore, the group O is adopted here for the following analysis.

The group O contains the identity operation, 8 threefold rotations about four main diagonals of the cube, 3 twofold rotations about the x, y, z axes, six twofold rotations about the diagonals in coordinate planes and six fourfold rotations about the x, y, z axes. These symmetry elements form the following classes of conjugate elements:

$$K_1 = \{E\},$$
$$K_2 = \{C_3^{xyz}, C_3^{x\bar{y}\bar{z}}, C_3^{\bar{x}y\bar{z}}, C_3^{\bar{x}\bar{y}z}, C_{\bar{3}}^{xyz}, C_{\bar{3}}^{x\bar{y}\bar{z}}, C_{\bar{3}}^{\bar{x}y\bar{z}}, C_{\bar{3}}^{\bar{x}\bar{y}z}\},$$
$$K_3 = \{C_2^x, C_2^y, C_2^z\},$$
$$K_4 = \{C_2^{x=y}, C_2^{x=z}, C_2^{y=z}, C_2^{x=-y}, C_2^{x=-z}, C_2^{y=-z}\},$$
$$K_5 = \{C_{4x}, C_{4y}, C_{4z}, C_{4x}^{-1}, C_{4y}^{-1}, C_{4z}^{-1}\}.$$

The symmetry operations of the group O permute the nodes 1, 2, 3, 4, 5, 6, 7, 8 within a single set of eight elements.

The character table of the group O is as follows:

O	E	$8C_3$	$3C_2$	$6C_2'$	$6C_4$
A_1	1	1	1	1	1
A_2	1	1	1	-1	-1
E_2	2	-1	2	0	0
T_1	3	0	-1	-1	1
T_2	3	0	-1	1	-1

The idempotents of the centre of the group algebra are

$$\pi_1 = \tfrac{1}{24}(\kappa_1 + \kappa_2 + \kappa_3 + \kappa_4 + \kappa_5),$$
$$\pi_2 = \tfrac{1}{24}(\kappa_1 + \kappa_2 + \kappa_3 - \kappa_4 - \kappa_5),$$
$$\pi_3 = \tfrac{2}{24}(2\kappa_1 - \kappa_2 + 2\kappa_3),$$
$$\pi_4 = \tfrac{3}{24}(3\kappa_1 \qquad - \kappa_3 - \kappa_4 + \kappa_5),$$
$$\pi_5 = \tfrac{3}{24}(3\kappa_1 \qquad - \kappa_3 + \kappa_4 - \kappa_5),$$

with

$$\kappa_1 = \{E\},$$
$$\kappa_2 = \{C_3^{xyz} + C_3^{x\bar{y}\bar{z}} + C_3^{\bar{x}y\bar{z}} + C_3^{\bar{x}\bar{y}z} + C_{\bar{3}}^{xyz} + C_{\bar{3}}^{x\bar{y}\bar{z}} + C_{\bar{3}}^{\bar{x}y\bar{z}} + C_{\bar{3}}^{\bar{x}\bar{y}z}\},$$

$$\kappa_3 = \{C_2^x + C_2^y + C_2^z\},$$
$$\kappa_4 = \{C_2^{x=y} + C_2^{x=z} + C_2^{y=z} + C_2^{x=-y} + C_2^{x=-z} + C_2^{y=-z}\},$$
$$\kappa_5 = \{C_{4x} + C_{4y} + C_{4z} + C_{4x}^{-1} + C_{4y}^{-1} + C_{4z}^{-1}\}.$$

Each of the 24 symmetry operations of the group O is represented by a 3 × 3 matrix. The action of a symmetry operation on a nodal configuration is obtained by multiplying the vector of coordinates of the node by the matrix of symmetry operation, which produces a permutation of the nodes. Thus, the application of the idempotents π_i ($i = 1, 2, 3, 4, 5$) produces the following basis vectors of G-invariant subspaces U_i:

$$U_1: \quad \varphi_1 = \tfrac{1}{8}(\varphi_1 + \varphi_2 + \varphi_3 + \varphi_4 + \varphi_5 + \varphi_6 + \varphi_7 + \varphi_8),$$
$$U_2: \quad \varphi_2 = \tfrac{1}{8}(\varphi_1 + \varphi_2 - \varphi_3 - \varphi_4 - \varphi_5 - \varphi_6 + \varphi_7 + \varphi_8),$$

$$U_4: \begin{cases} \varphi_3 = \tfrac{1}{8}(3\varphi_1 - \varphi_2 + \varphi_3 + \varphi_4 + \varphi_5 - 3\varphi_6 - \varphi_7 - \varphi_8), \\ \varphi_4 = \tfrac{1}{8}(-\varphi_1 + 3\varphi_2 + \varphi_3 + \varphi_4 - 3\varphi_5 + \varphi_6 - \varphi_7 - \varphi_8), \\ \varphi_5 = \tfrac{1}{8}(\varphi_1 + \varphi_2 + 3\varphi_3 - \varphi_4 - \varphi_5 - \varphi_6 + \varphi_7 - 3\varphi_8), \end{cases}$$

$$U_5: \begin{cases} \varphi_6 = \tfrac{1}{8}(3\varphi_1 - \varphi_2 - \varphi_3 - \varphi_4 - \varphi_5 + 3\varphi_6 - \varphi_7 - \varphi_8), \\ \varphi_7 = \tfrac{1}{8}(-\varphi_1 + 3\varphi_2 - \varphi_3 - \varphi_4 + 3\varphi_5 - \varphi_6 - \varphi_7 - \varphi_8), \\ \varphi_8 = \tfrac{1}{8}(-\varphi_1 - \varphi_2 + 3\varphi_3 - \varphi_4 - \varphi_5 - \varphi_6 - \varphi_7 + 3\varphi_8). \end{cases}$$

This set of basis vectors shows that the eight-dimensional space of the configuration is decomposed into four G-invariant subspaces U_1, U_2, U_4, U_5 with 1, 1, 3, 3 dimensions respectively (U_3 is a zero subspace). In Fig. 2.9 the basis vectors of particular subspaces are demonstrated, similar to configurations based on the square. In case of the positive sign (+) of the function an arrow is placed at the node and directed towards the centre of the cube (the origin of coordinates), while in case of the negative sign (−) the arrow has the opposite direction. Thus the picture presents the symmetry type of the subspace.

Positive directions of displacements and rotations of the nodes, as well as of forces or moments at the nodes, are always defined in G-vector analysis by basis vectors which are derived by the first idempotent π_1. Figure 2.10(a) shows the positive directions of nodal displacements, parallel to coordinate axes x, y, z in G-vector analysis, where the set of positive directions corresponds to the symmetry type of the first subspace. The standard convention has the positive directions of the nodal displacements parallel to and in the same direction as the axes x, y and z (Fig. 2.10(b)).

The system of equations of the configuration consists of systems of equations of particular independent subspaces U_i, which, if given in matrix form, can be represented schematically as (see p. 61).

The nodal functions φ_j ($j = 1, 2, \ldots, 8$) expressed by the basis vectors $\bar{\varphi}_j$ are

$$\varphi_1 = \bar{\varphi}_1 + \bar{\varphi}_2 + \bar{\varphi}_3 + \bar{\varphi}_6,$$
$$\varphi_2 = \bar{\varphi}_1 + \bar{\varphi}_2 + \bar{\varphi}_3 + \bar{\varphi}_7,$$

Sec. 2.2] Nodal configurations with patterns based on the cube

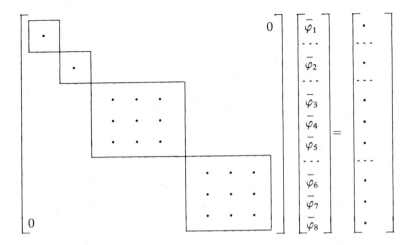

$$\bar{\varphi}_3 = \bar{\varphi}_1 - \bar{\varphi}_2 + \bar{\varphi}_3 + \bar{\varphi}_8,$$

$$\bar{\varphi}_4 = \bar{\varphi}_1 - \bar{\varphi}_2 + \bar{\varphi}_3 + \bar{\varphi}_4 - \bar{\varphi}_5 - \bar{\varphi}_6 - \bar{\varphi}_7 - \bar{\varphi}_8,$$

$$\bar{\varphi}_5 = \bar{\varphi}_1 - \bar{\varphi}_2 - \bar{\varphi}_3 + \bar{\varphi}_7,$$

$$\bar{\varphi}_6 = \bar{\varphi}_1 - \bar{\varphi}_2 - \bar{\varphi}_3 + \bar{\varphi}_6,$$

$$\bar{\varphi}_7 = \bar{\varphi}_1 + \bar{\varphi}_2 - \bar{\varphi}_3 - \bar{\varphi}_4 + \bar{\varphi}_5 - \bar{\varphi}_6 - \bar{\varphi}_7 - \bar{\varphi}_8,$$

$$\bar{\varphi}_8 = \bar{\varphi}_1 + \bar{\varphi}_2 - \bar{\varphi}_3 + \bar{\varphi}_8.$$

Analytic determination of dimensions of subspaces U_i gives

$\text{Tr}(E) = 8,$

$\text{Tr}(C_3^{xyz}) = \text{Tr}(C_3^{x\bar{y}\bar{z}}) = \text{Tr}(C_3^{\bar{x}y\bar{z}}) = \text{Tr}(C_3^{\bar{x}\bar{y}z}) = \text{Tr}(C_3^{xyz})$
$= \text{Tr}(C_3^{x\bar{y}\bar{z}}) = \text{Tr}(C_3^{\bar{x}y\bar{z}}) = \text{Tr}(C_3^{\bar{x}\bar{y}z}) = 2,$

$\text{Tr}(C_2^x) = \text{Tr}(C_2^y) = \text{Tr}(C_2^z) = 0,$

$\text{Tr}(C_2^{x=y}) = \text{Tr}(C_2^{x=z}) = \text{Tr}(C_2^{y=z}) = \text{Tr}(C_2^{x=-y}) = \text{Tr}(C_2^{x=-z})$
$= \text{Tr}(C_2^{y=-z}) = 0,$

$\text{Tr}(C_{4x}) = \text{Tr}(C_{4y}) = \text{Tr}(C_{4z}) = \text{Tr}(C_{4x}^{-1}) = \text{Tr}(C_{4y}^{-1}) = \text{Tr}(C_{4z}^{-1}) = 0.$

Dim $U_1 = \frac{1}{24}(1 \times 8 + 1 \times 2 + 1 \times 2 + 1 \times 2 + 1 \times 2 + 1 \times 2 + 1 \times 2 + 1 \times 2 + 1 \times 2) = 1$

Dim $U_2 = \frac{1}{24}(1 \times 8 + 1 \times 2 + 1 \times 2 + 1 \times 2 + 1 \times 2 + 1 \times 2 + 1 \times 2 + 1 \times 2 + 1 \times 2) = 1$

Dim $U_3 = \frac{2}{24}(2 \times 8 - 1 \times 2 - 1 \times 2 - 1 \times 2 - 1 \times 2 - 1 \times 2 - 1 \times 2 - 1 \times 2 - 1 \times 2) = 0$

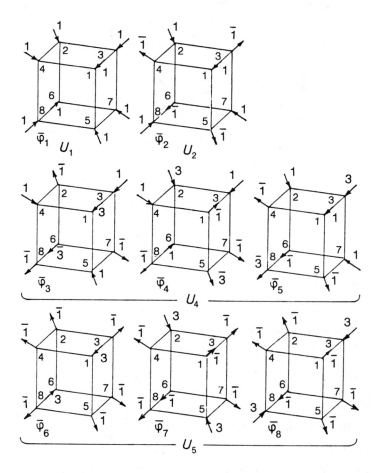

Fig. 2.9. Basis vectors of the subspaces of the eight-node pattern based on the cube and described by the group O.

Dim $U_4 = \frac{3}{24} 3 \times 8 = 3$

Dim $U_5 = \frac{3}{24} 3 \times 8 = 3$.

2.3 NODAL CONFIGURATION WITH A PATTERN BASED ON THE RIGHT PARALLELEPIPED

The pattern in Fig. 2.11 has eight nodes, one at each vertex of the right parallelepiped. The origin of coordinates is at the centre of the parallelepiped, with coordinate axes s, y, z parallel to the edges of the parallelepiped and with the unique nodal numbering derived in section 3.8. The results that will be obtained here are also applicable to the eight-node pattern based on the cube in section 2.2, since the cube is a particular case of the right parallelepiped.

Sec. 2.3]   Nodal configuration with a pattern based on the right parallelepiped   63

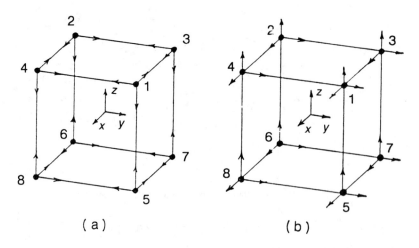

Fig. 2.10. Positive directions of displacements and rotations of the nodes, as well as of forces and moments at the nodes, for the eight-node pattern based on the cube: (a) in G-vector analysis; (b) in the standard method.

To the given pattern based on the right parallelepiped corresponds the group D_{2h} with 8 elements, each element alone in its class:

$$E \quad C_2^z \quad C_2^y \quad C_2^x \quad i \quad \sigma_{xy} \quad \sigma_{xz} \quad \sigma_{yz}.$$

The symmetry operations of the group D_{2h} permute the nodes 1, 2, 3, 4, 5, 6, 7, 8 within a single set.

The character table of the group D_{2h} is as follows:

D_{2h}	E	C_2^z	C_2^y	C_2^x	i	σ_{xy}	σ_{xz}	σ_{yz}
A_g	1	1	1	1	1	1	1	1
B_{1g}	1	1	−1	−1	1	1	−1	−1
B_{2g}	1	−1	1	−1	1	−1	1	−1
B_{3g}	1	−1	−1	1	1	−1	−1	1
A_u	1	1	1	1	−1	−1	−1	−1
B_{1u}	1	1	−1	−1	−1	−1	1	1
B_{2u}	1	−1	1	−1	−1	1	−1	1
B_{3u}	1	−1	−1	1	−1	1	1	−1

By using the expression

$$\pi_i = \frac{h_i}{h} \sum_\sigma \chi_i(\sigma^{-1})\sigma,$$

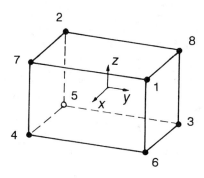

Fig. 2.11. Eight-node pattern based on the right parallelepiped.

given in section 2.1, the idempotents of the centre of group algebra are obtained

$$\pi_1 = \tfrac{1}{8}(E + C_2^z + C_2^y + C_2^x + i + \sigma_{xy} + \sigma_{xz} + \sigma_{yz})$$
$$\pi_2 = \tfrac{1}{8}(E + C_2^z - C_2^y - C_2^x + i + \sigma_{xy} - \sigma_{xz} - \sigma_{yz})$$
$$\pi_3 = \tfrac{1}{8}(E - C_2^z + C_2^y - C_2^x + i - \sigma_{xy} + \sigma_{xz} - \sigma_{yz})$$
$$\pi_4 = \tfrac{1}{8}(E - C_2^z - C_2^y + C_2^x + i - \sigma_{xy} - \sigma_{xz} + \sigma_{yz})$$
$$\pi_5 = \tfrac{1}{8}(E + C_2^z + C_2^y + C_2^x - i - \sigma_{xy} - \sigma_{xz} - \sigma_{yz})$$
$$\pi_6 = \tfrac{1}{8}(E + C_2^z - C_2^y - C_2^x - i - \sigma_{xy} + \sigma_{xz} + \sigma_{yz})$$
$$\pi_7 = \tfrac{1}{8}(E - C_2^z + C_2^y - C_2^x - i + \sigma_{xy} - \sigma_{xz} + \sigma_{yz})$$
$$\pi_8 = \tfrac{1}{8}(E - C_2^z - C_2^y + C_2^x - i + \sigma_{xy} + \sigma_{xz} - \sigma_{yz}).$$

Application of the idempotents π_i ($i = 1, 2, \ldots, 8$) to the nodal functions φ_j ($j = 1, 2, \ldots, 8$) derives the basis vectors $\bar{\varphi}_j$ of G-invariant subspaces U_i:

$$U_1: \bar{\varphi}_1 = \tfrac{1}{8}(\varphi_1 + \varphi_2 + \varphi_3 + \varphi_4 + \varphi_5 + \varphi_6 + \varphi_7 + \varphi_8)$$
$$U_2: \bar{\varphi}_2 = \tfrac{1}{8}(\varphi_1 + \varphi_2 - \varphi_3 - \varphi_4 + \varphi_5 + \varphi_6 - \varphi_7 - \varphi_8)$$
$$U_3: \bar{\varphi}_3 = \tfrac{1}{8}(\varphi_1 - \varphi_2 + \varphi_3 - \varphi_4 + \varphi_5 - \varphi_6 + \varphi_7 - \varphi_8)$$
$$U_4: \bar{\varphi}_4 = \tfrac{1}{8}(\varphi_1 - \varphi_2 - \varphi_3 + \varphi_4 + \varphi_5 - \varphi_6 - \varphi_7 + \varphi_8)$$
$$U_5: \bar{\varphi}_5 = \tfrac{1}{8}(\varphi_1 + \varphi_2 + \varphi_3 + \varphi_4 - \varphi_5 - \varphi_6 - \varphi_7 - \varphi_8)$$
$$U_6: \bar{\varphi}_6 = \tfrac{1}{8}(\varphi_1 + \varphi_2 - \varphi_3 - \varphi_4 - \varphi_5 - \varphi_6 + \varphi_7 + \varphi_8)$$
$$U_7: \bar{\varphi}_7 = \tfrac{1}{8}(\varphi_1 - \varphi_2 + \varphi_3 - \varphi_4 - \varphi_5 + \varphi_6 - \varphi_7 + \varphi_8)$$
$$U_8: \bar{\varphi}_8 = \tfrac{1}{8}(\varphi_1 - \varphi_2 - \varphi_3 + \varphi_4 - \varphi_5 + \varphi_6 + \varphi_7 - \varphi_8).$$

This derivation shows that the eight-dimensional space of the configuration is decomposed into eight one-dimensional subspaces $U_1, U_2, \ldots, U_8$. As previously in the case of the pattern based on the cube in section 2.2, the basis vectors $\bar{\varphi}_j$ are designated by arrows pointing to the centre of the parallelepiped when the sign is positive, while for the

Sec. 2.3]   Nodal configuration with a pattern based on the right parallelepiped   65

negative sign the arrow has the opposite direction, which is given in Fig. 2.12. As in the case of the cube in 2.2, the set of positive directions of nodal displacements suits the symmetry type of the first subspace.

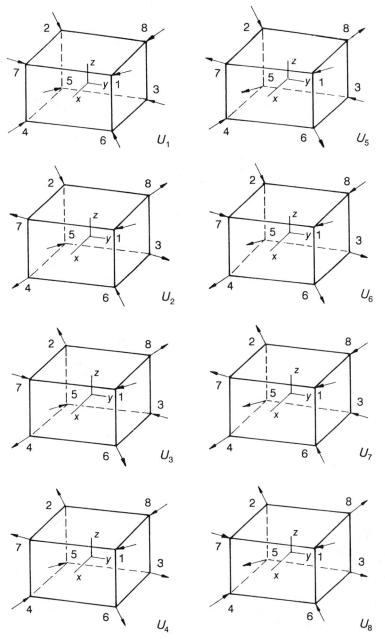

Fig. 2.12. Basis vectors of the subspaces of the eight-node pattern based on the right parallelepiped and described by the group D_{2h}.

The system of equations of the configuration is composed of systems of equations of particular G-invariant subspaces U_i, which, if expressed in matrix form, can be given schematically as

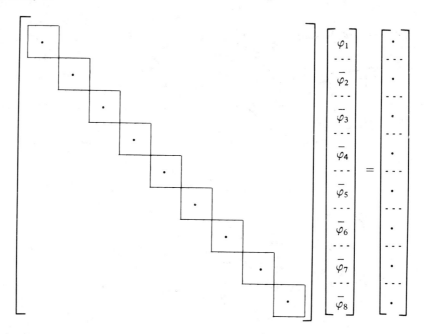

The nodal functions φ_j ($j = 1, 2, \ldots, 8$) expressed by the basis vectors $\bar{\varphi}_j$ are

$$\varphi_1 = \bar{\varphi}_1 + \bar{\varphi}_2 + \bar{\varphi}_3 + \bar{\varphi}_4 + \bar{\varphi}_5 + \bar{\varphi}_6 + \bar{\varphi}_7 + \bar{\varphi}_8,$$
$$\varphi_2 = \bar{\varphi}_1 + \bar{\varphi}_2 - \bar{\varphi}_3 - \bar{\varphi}_4 + \bar{\varphi}_5 + \bar{\varphi}_6 - \bar{\varphi}_7 - \bar{\varphi}_8,$$
$$\varphi_3 = \bar{\varphi}_1 - \bar{\varphi}_2 + \bar{\varphi}_3 - \bar{\varphi}_4 + \bar{\varphi}_5 - \bar{\varphi}_6 + \bar{\varphi}_7 - \bar{\varphi}_8,$$
$$\varphi_4 = \bar{\varphi}_1 - \bar{\varphi}_2 - \bar{\varphi}_3 + \bar{\varphi}_4 + \bar{\varphi}_5 - \bar{\varphi}_6 - \bar{\varphi}_7 + \bar{\varphi}_8,$$
$$\varphi_5 = \bar{\varphi}_1 + \bar{\varphi}_2 + \bar{\varphi}_3 + \bar{\varphi}_4 - \bar{\varphi}_5 - \bar{\varphi}_6 - \bar{\varphi}_7 - \bar{\varphi}_8,$$
$$\varphi_6 = \bar{\varphi}_1 + \bar{\varphi}_2 - \bar{\varphi}_3 - \bar{\varphi}_4 - \bar{\varphi}_5 - \bar{\varphi}_6 + \bar{\varphi}_7 + \bar{\varphi}_8,$$
$$\varphi_7 = \bar{\varphi}_1 - \bar{\varphi}_2 + \bar{\varphi}_3 - \bar{\varphi}_4 - \bar{\varphi}_5 + \bar{\varphi}_6 - \bar{\varphi}_7 + \bar{\varphi}_8,$$
$$\varphi_8 = \bar{\varphi}_1 - \bar{\varphi}_2 - \bar{\varphi}_3 + \bar{\varphi}_4 - \bar{\varphi}_5 + \bar{\varphi}_6 + \bar{\varphi}_7 - \bar{\varphi}_8.$$

Since $\mathrm{Tr}(E) = 8$ and the traces of matrices of all other symmetry operations equal zero, the dimensions of all subspaces equal 1:

$$\mathrm{Dim}\, U_1 = \mathrm{Dim}\, U_2 = \ldots = \mathrm{Dim}\, U_7 = \mathrm{Dim}\, U_8 = \tfrac{1}{8} 8 = 1.$$

2.4 NODAL CONFIGURATIONS WITH PATTERNS BASED ON THE REGULAR ICOSAHEDRON

Geometric models, shown in Fig. 2.13, illustrate the symmetry of a dodecahedral–icosahedral configuration: the dodecahedron (a) superposed on the icosahedron (b) produce together the model (c) with evident fivefold, threefold and twofold rotation axes. Dodecahedral–icosahedral configurations with nodal patterns based on the model (c) possess symmetry properties which are described by the icosahedral group I_h.

The regular icosahedron can be placed into the orthogonal coordinate system by putting three rectangles into the coordinate planes, with their centres at the origin and with sides $a/b = 2/2\varphi$, where $\varphi = \frac{1}{2}(\sqrt{5} + 1)$ — see Fig. 2.14. The corners of these rectangles coincide with the vertices of the icosahedron and their coordinates are as shown in Table 2.1.

The positions of symmetry elements with regard to the icosahedron can be visualized on its orthogonal projections (Fig. 2.15): (a) in the direction of the fivefold axis, (b) in the direction of the twofold axis, (c) in the direction of the threefold axis. The icosahedral group I_h contains 10 classes of conjugate symmetry elements

$$E \quad 12C_5 \quad 12C_5^2 \quad 20C_3 \quad 15C_2$$
$$i \quad 12S_{10} \quad 12S_{10}^2 \quad 20S_6 \quad 15\sigma ,$$

with

E	identity operation
$12C_5$	rotations C_5, C_5^{-1} through $72°$, $-72°$ (2 rotations for each of 6 axes)
$12C_5^2$	rotations C_5^2, C_5^{-2} through $144°$, $-144°$ (2 rotations for each of 6 axes)
$20C_3$	rotations C_3, C_3^{-4} through $120°$, $-120°$ (2 rotations for each of 10 axes)
$15C_2$	rotations C_2 through $180°$ (15 rotations for 15 axes)
i	inverse symmetry operation (inversion)
$12S_{10}$	rotary reflections S_{10}, S_{10}^{-1} through $36°$, $-36°$ (2 rotary reflections for each of 6 axes)
$12S_{10}^2$	rotary reflections S_{10}^2, S_{10}^{-2} through $72°$, $-72°$ (2 rotary reflections for each of 6 axes)
$20S_6$	rotary reflections S_6, S_6^{-1} through $60°$, $-60°$ (2 rotary reflections for each of 10 axes)
15σ	reflections σ (15 reflections in 15 symmetry planes)
120	symmetry operations in total.

68   **G-vector analysis**   [Ch. 2

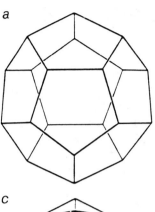
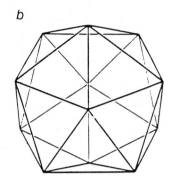
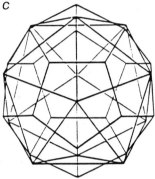

Fig. 2.13. Geometric models that illustrate the symmetry of a dodecahedral–icosahedral configuration: (a) the dodecahedron, (b) the icosahedron, (c) superposition of the dodecahedron and the icosahedron.

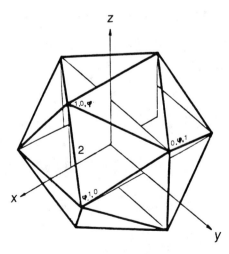

Fig. 2.14. The regular icosahedron with its vertices coinciding with the corners of three orthogonal rectangles with sides $a/b = 2/2\varphi$.

Nodal configurations with patterns based on the regular icosahedron

Table 2.1. Coordinates of the corners of rectangles

x	y	z
0	$\pm\varphi$	± 1
± 1	0	$\pm\varphi$
$\pm\varphi$	± 1	0

The character table of the group I_h [with $\varphi = \frac{1}{2}(\sqrt{5}+1)$] is as follows:

I_h	E	$12C_5$	$12C_5^2$	$20C_3$	$15C_2$	i	$12S_{10}$	$12S_{10}^2$	$20S_6$	15σ
A_g	1	1	1	1	1	1	1	1	1	1
A_u	1	1	1	1	1	-1	-1	-1	-1	-1
T_{1g}	3	φ	$-\frac{1}{\varphi}$	0	-1	3	$-\frac{1}{\varphi}$	φ	0	-1
T_{1u}	3	φ	$-\frac{1}{\varphi}$	0	-1	-3	$\frac{1}{\varphi}$	$-\varphi$	0	1
T_{2g}	3	$-\frac{1}{\varphi}$	φ	0	-1	3	φ	$-\frac{1}{\varphi}$	0	-1
T_{2u}	3	$-\frac{1}{\varphi}$	φ	0	-1	-3	$-\varphi$	$\frac{1}{\varphi}$	0	1
G_g	4	-1	-1	1	0	4	-1	-1	1	0
G_u	4	-1	-1	1	0	-4	1	1	-1	0
H_g	5	0	0	-1	1	5	0	0	-1	1
H_u	5	0	0	-1	1	-5	0	0	1	-1

The idempotents of the centre of the group algebra are determined by

$$\pi_i = \frac{h_i}{h} \sum_\sigma \chi_i(\sigma^{-1})\sigma,$$

with h_i the dimension of the ith character given by $h_i = \chi_i(E)$, h the order of the group, χ_i the ith character of the group, σ group element and σ^{-1} the inversion of the element σ.
When the class sums are designated by

$\kappa_1 = E \quad \kappa_2 = \sum 12C_5 \quad \kappa_3 = \sum 12C_5^2 \quad \kappa_4 = \sum 20C_3 \quad \kappa_5 = \sum 15C_2$

$\kappa_6 = i \quad \kappa_7 = \sum 12S_{10} \quad \kappa_8 = \sum 12S_{10}^2 \quad \kappa_9 = \sum 20S_6 \quad \kappa_{10} = \sum 15\sigma,$

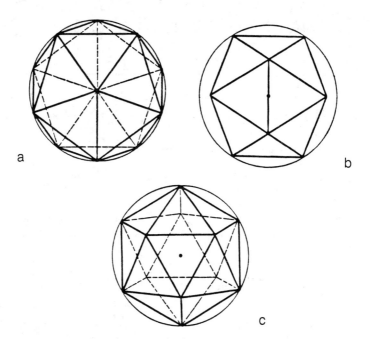

Fig. 2.15. Orthogonal projections of the regular icosahedron: (a) in the direction of the fivefold axis, (b) in the direction of the twofold axis, (c) in the direction of the threefold axis.

the idempotents of the centre of group algebra are

$$\pi_1 = \tfrac{1}{120}(\kappa_1 + \kappa_2 + \kappa_3 + \kappa_4 + \kappa_5 + \kappa_6 + \kappa_7 + \kappa_8 + \kappa_9 + \kappa_{10})$$

$$\pi_2 = \tfrac{1}{120}(\kappa_1 + \kappa_2 + \kappa_3 + \kappa_4 + \kappa_5 - \kappa_6 - \kappa_7 - \kappa_8 - \kappa_9 - \kappa_{10})$$

$$\pi_3 = \tfrac{3}{120}\left(3\kappa_1 + \varphi\kappa_2 - \tfrac{1}{\varphi}\kappa_3 - \kappa_5 + 3\kappa_6 - \tfrac{1}{\varphi}\kappa_7 + \varphi\kappa_8 - \kappa_{10}\right)$$

$$\pi_4 = \tfrac{3}{120}\left(3\kappa_1 + \varphi\kappa_2 - \tfrac{1}{\varphi}\kappa_3 - \kappa_5 - 3\kappa_6 + \tfrac{1}{\varphi}\kappa_7 - \varphi\kappa_8 + \kappa_{10}\right)$$

$$\pi_5 = \tfrac{3}{120}\left(3\kappa_1 - \tfrac{1}{\varphi}\kappa_2 + \varphi\kappa_3 - \kappa_5 + 3\kappa_6 + \varphi\kappa_7 - \tfrac{1}{\varphi}\kappa_8 - \kappa_{10}\right)$$

$$\pi_6 = \tfrac{3}{120}\left(3\kappa_1 - \tfrac{1}{\varphi}\kappa_2 + \varphi\kappa_3 - \kappa_5 - 3\kappa_6 - \varphi\kappa_7 + \tfrac{1}{\varphi}\kappa_8 + \kappa_{10}\right)$$

$$\pi_7 = \tfrac{4}{120}(4\kappa_1 - \kappa_2 - \kappa_3 + \kappa_4 + 4\kappa_6 - \kappa_7 - \kappa_8 + \kappa_9)$$

$$\pi_8 = \tfrac{4}{120}(4\kappa_1 - \kappa_2 - \kappa_3 + \kappa_4 - 4\kappa_6 + \kappa_7 + \kappa_8 - \kappa_9)$$

Sec. 2.4] Nodal configurations with patterns based on the regular icosahedron

$$\pi_9 = \tfrac{5}{120}(5\kappa_1 \quad -\kappa_4 + \kappa_5 + 5\kappa_6 - \kappa_9 + \kappa_{10})$$
$$\pi_{10} = \tfrac{5}{120}(5\kappa_1 \quad -\kappa_4 + \kappa_5 - 5\kappa_6 + \kappa_9 - \kappa_{10}).$$

Application of the idempotents π_i to functions at the nodes derives the basis vectors of G-invariant subspaces of the system. The number of basis vectors that span a subspace denotes the dimension of the subspace. In this analysis there will be no derivation of basis vectors, while the dimensions of subspaces U_i will be determined by

$$\text{Dim } U_i = \frac{h_i}{h} \sum_j \chi_{ij} n_{fj} n_j \, \text{Tr } \sigma_j,$$

derived from $\text{Dim } U_i = h_i \, (\text{Tr}, \chi_i)$, where

$i = 1, 2, \ldots, 10$ and indices of subspaces U_i and representations Γ_i,

χ_{ij} is the character pertaining to Γ_i and K_j

Tr σ_j is the trace of the matrix of an element σ from the class K_j,

$j = 1, 2, \ldots, 10$ are indices of classes of conjugate elements of the group I_h,

h_i is the dimension of the ith character given by $h_i = \chi_i(E)$,

h is the number of group elements in the group,

n_j is the number of elements in the class K_j,

n_{fj} is the number of nodes which are not moved by action of an element from the jth class.

The matrices representing elements from a single class have the same trace, since the trace is invariant with respect to conjugation. Traces of matrices of symmetry elements are determined by means of general forms of matrices for rotations and rotary reflections:

$$C(\alpha) = \begin{bmatrix} \cos \alpha & -\sin \alpha & 0 \\ \sin \alpha & \cos \alpha & 0 \\ 0 & 0 & 1 \end{bmatrix}, \quad S(\alpha) = \begin{bmatrix} \cos \alpha & -\sin \alpha & 0 \\ \sin \alpha & \cos \alpha & 0 \\ 0 & 0 & -1 \end{bmatrix}$$

K_1: $\text{Tr}(E) = C(0°) = 2 \cos 0° + 1 = 2 \cdot 1 + 1 = 3$ (for three degrees of freedom)

K_2: $\text{Tr}(C_5) = 2 \cos 72° + 1 = 2 \dfrac{1}{2\varphi} + 1 = \varphi$

K_3: $\text{Tr}(C_5^2) = 2 \cos 144° + 1 = 2 \left(-\dfrac{\varphi}{2}\right) + 1 = -\dfrac{1}{\varphi}$

K_4: $\text{Tr}(C_3) = 2 \cos 120° + 1 = 2 \left(-\tfrac{1}{2}\right) + 1 = 0$

K_5: $\text{Tr}(C_2) = 2 \cos 180° + 1 = 2(-1) + 1 = -1$

K_6: $\text{Tr}(i) = S(180°) = 2\cos 180° - 1 = 2(-1) - 1 = -3$

K_7: $\text{Tr}(S_{10}) = 2\cos 36° - 1 = 2\left(\dfrac{\varphi}{2}\right) - 1 = \varphi - 1 = \dfrac{1}{\varphi}$

K_8: $\text{Tr}(S_{10}^2) = 2\cos 72° - 1 = 2\left(\dfrac{1}{2\varphi}\right) - 1 = \dfrac{1}{\varphi} - 1$

K_9: $\text{Tr}(S_6) = 2\cos 60° - 1 = 2(\tfrac{1}{2}) - 1 = 0$

K_{10}: $\text{Tr}(\sigma) = S(0°) = 2\cos 0° - 1 = 2 \cdot 1 - 1 = 1$

$$\left[\text{where } \varphi = \frac{1}{2}(\sqrt{5}+1), \quad \frac{1}{\varphi} = \frac{1}{2}(\sqrt{5}-1), \quad \varphi = 1 + \frac{1}{\varphi}\right].$$

The chosen dodecahedral–icosahedral configuration contains 182 nodes, with three degrees of freedom in each node. Arrangement of nodes and their linking by line elements are given in Fig. 2.16: (a) projection in the direction of the threefold axis, (b) projection in the direction of the fivefold axis. According to their position the nodes can be classified into five types, shown in Table 2.2.

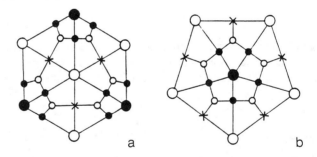

Fig. 2.16. Details of the scheme of the nodal pattern of the dodecahedral–icosahedral configuration: (a) projection in the direction of the threefold axis, (b) projection in the direction of the fivefold axis.

The illustration of the dodecahedral–icosahedral configuration is given in Fig. 2.17.

Since it is presumed that each node has three degrees of freedom, the total number of degrees of freedom amounts to

$$n = 182 \times 3 = 546.$$

Number of nodes which are not moved by action of particular symmetry operations can be visualized by analysis of the node pattern in Fig. 2.16(a) and Fig. 2.16(b). For every symmetry operation from the same class this number is the same and it amounts to

Sec. 2.4]   Nodal configurations with patterns based on the regular icosahedron   73

Table 2.2 Node classification

Type	Designation	Total number	Position of nodes with respect to the inscribed icosahedron and dodecahedron
1	●	12	in vertices of the icosahedron
2	○	20	in vertices of the dodecahedron
3	X	30	in midpoints of edges of the icosahedron
4	•	60	on altitudes of sides of the icosahedron
5	○	60	on edges of the icosahedron
	Total	182	

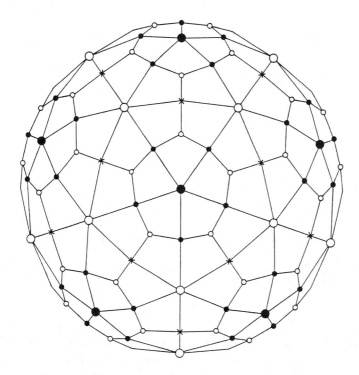

Fig. 2.17. Illustration of the configuration of the dodecahedral–icosahedral configuration with 182 nodes.

74 G-vector analysis   [Ch. 2

K_1 of E : $n_{f1} = 182$ K_6 of i : $n_{f6} = 0$
K_2 of $12C_5$: $n_{f2} = 2$ K_7 of $12S_{10}$: $n_{f7} = 0$
K_3 of $12C_5^2$: $n_{f3} = 2$ K_8 of $12S_{10}^2$: $n_{f8} = 0$
K_4 of $20C_3$: $n_{f4} = 2$ K_9 of $20S_6$: $n_{f9} = 0$
K_5 of $15C_2$: $n_{f5} = 2$ K_{10} of 15σ : $n_{f10} = 20$.

When all necessary previously found values are introduced into the expression for determination of dimensions of subspaces

$$\text{Dim } U_i = \frac{h_i}{h} \sum_j \chi_{ij} n_{fj} n_j \text{ Tr } \sigma_j,$$

for subspaces U_1 to U_{10} one obtains the dimensions 7, 2, 36, 51, 33, 48, 72, 72, 125 and 100 respectively, which is given explicitly in Table 2.3.

The system of equations with its matrix in block diagonal form is illustrated graphically by sizes of submatrices given proportionally with regard to dimensions of corresponding subspaces (see Fig. 2.18).

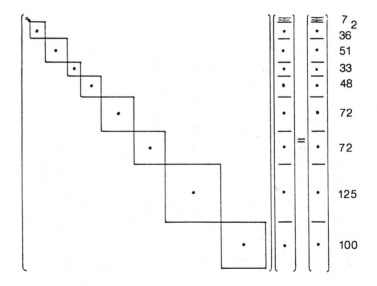

Fig. 2.18. Illustration of the system of equations with the matrix in block diagonal form for the dodecahedral–icosahedral configuration with 182 nodes and 546 degrees of freedom.

Table 2.4 contains data on dimensions of irreducible representations of the group I_h and dimensions of matrices of G-invariant subspaces of the G-vector space of the system with 546 dimensions. In addition, there are results showing that, with regard to the standard method of solution of the system of equations having a general matrix, the number of multiplications in Gauss' algorithm in G-vector analysis is reduced to 2.6% of the original, and in the case of a symmetric matrix to 2.7%. The number of different matrix elements is reduced to 14.5% in the case of a general matrix and to 14.7% by symmetric matrix.

2.4.1 General configuration of nodal patterns compatible with the symmetry of the icosahedral group I_h

When a sphere is circumscribed around the regular icosahedron so that its vertices lie in a spherical surface, while through the edges of the icosahedron and the centre of the sphere planes are inserted, a network of 20 congruent equilateral spherical triangles is formed. By drawing altitudes of each triangle by arcs of the principal circle of the sphere, the triangle is divided into six rectangular spherical triangles. When this division is accomplished on the whole sphere, one obtains 120 triangles. Schematically, the spherical triangle T_{20}, the 20th part of the sphere (Fig. 2.19a), is divided into six spherical tirangles T_{120}, 120th parts of the sphere (Fig. 2.19b). Application of symmetry elements of the icosahedral group as generators on the triangle T_{120} generates the spherical network on the whole sphere.

In regard to its position with respect to the triangle T_{120} nodes can be classified into five types as in Table 2.5.

On the basis of these nodal types it is possible to formulate the general expression for the total number of nodes in an arbitrary configuration compatible with the icosahedral group I_h:

$$n = 12\delta_I + 20\delta_D + 30\delta_{I'} + 60n_s + 120n_u,$$

where

$\delta_I, \delta_D, \delta_{I'} = 0$ or 1 (data on existence of the node);

n_s number of nodes on sides of the triangle T_{120}, excluding its vertices,

n_u number of nodes in interior of the triangle T_{120}.

For illustration the chosen pattern of nodes with respect to the triangle T_{120} (Fig. 2.20) with parameters $\delta_I = 1$, $\delta_D = 1$, $\delta_{I'} = 0$, $n_S = 6$ and $n_U = 3$ defines a configuration, compatible with the icosahedral group I_h, containing in total

$$n = 12 \times 1 + 20 \times 1 + 30 \times 0 + 60 \times 6 + 120 \times 3 = 752 \text{ nodes}.$$

The dodecahedral–icosahedral structure with 182 nodes (Fig. 2.16), analysed previously, is defined by the following parameters:

$$\delta_I = 1, \quad \delta_D = 1, \quad \delta_{I'} = 1, \quad n_s = 2, \quad n_u = 0,$$

and by using the expression for the total number of nodes one obtains

$$n = 12 \times 1 + 20 \times 1 + 30 \times 1 + 60 \times 2 + 120 \times 0 = 182.$$

Classification into nodal types and the general expression for the total number of nodes in a configuration are also valid when nodes do not lie in the spherical surface (Fig. 2.21), since moving of a node along a ray from the centre of the sphere changes neither the characteristics of the nodal type nor the relations in the expression for the total

Table 2.3 Determining dimensions of subspaces

	E	$12C_5$	$12C_5^2$	$15C_2$	15σ	
$\text{Dim } U_1 = \frac{1}{120}$	$[1 \times 182 \times 1 \times 3 +$	$1 \times 2 \times 12\varphi +$	$1 \times 2 \times 12\left(-\frac{1}{\varphi}\right) +$	$1 \times 2 \times 15(-1) +$	$1 \times 20 \times 15 \times 1]$	$= 7$
$\text{Dim } U_2 = \frac{1}{120}$	$1 \times 182 \times 1 \times 3 +$	$1 \times 2 \times 12\varphi +$	$1 \times 2 \times 12\left(-\frac{1}{\varphi}\right) +$	$1 \times 2 \times 15(-1) + (-1)20 \times 15 \times 1$		$= 2$
$\text{Dim } U_3 = \frac{3}{120}$	$[3 \times 182 \times 1 \times 3 +$	$\varphi \times 2 \times 12\varphi +$	$\left(-\frac{1}{\varphi}\right)2 \times 12\left(-\frac{1}{\varphi}\right) +$	$(-1)2 \times 15(-1) + (-1)20 \times 15 \times 1]$		$= 36$
$\text{Dim } U_4 = \frac{3}{120}$	$[3 \times 182 \times 1 \times 3 +$	$\varphi \times 2 \times 12\varphi +$	$\left(-\frac{1}{\varphi}\right)2 \times 12\left(-\frac{1}{\varphi}\right) +$	$(-1)2 \times 15(-1) +$	$1 \times 20 \times 15 \times 1]$	$= 51$
$\text{Dim } U_5 = \frac{3}{120}$	$[3 \times 182 \times 1 \times 3 +$	$\left(-\frac{1}{\varphi}\right)2 \times 12\varphi +$	$\varphi \times 2 \times 12\left(-\frac{1}{\varphi}\right) +$	$(-1)2 \times 15(-1) + (-1)20 \times 15 \times 1]$		$= 33$
$\text{Dim } U_6 = \frac{3}{120}$	$[3 \times 182 \times 1 \times 3 +$	$\left(-\frac{1}{\varphi}\right)2 \times 12\varphi +$	$\varphi \times 2 \times 12\left(-\frac{1}{\varphi}\right) +$	$(-1)2 \times 15(-1) +$	$1 \times 20 \times 15 \times 1]$	$= 48$
$\text{Dim } U_7 = \frac{4}{120}$	$[4 \times 182 \times 1 \times 3 + (-1)2 \times 12\varphi + (-1)2 \times 12\left(-\frac{1}{\varphi}\right) + 0 \times 2 \times 15(-1) +$				$0 \times 20 \times 15 \times 1]$	$= 72$

Table 2.3 (continued)

	E	$12C_5$	$12C_5^2$	$15C_2$	15σ	
$\mathrm{Dim}\, U_8 = \frac{4}{120}$	$\left[4 \times 182 \times 1 \times 3 + \right.$	$(-1)2 \times 12\varphi +$	$(-1)2 \times 12 \left(-\dfrac{1}{\varphi}\right) +$	$0 \times 2 \times 15(-1) +$	$\left. 0 \times 20 \times 15 \times 1 \right] =$	72
$\mathrm{Dim}\, U_9 = \frac{5}{120}$	$\left[5 \times 182 \times 1 \times 3 + \right.$	$0 \times 2 \times 12\varphi +$	$0 \times 2 \times 12 \left(-\dfrac{1}{\varphi}\right) +$	$1 \times 2 \times 15(-1) +$	$\left. 1 \times 20 \times 15 \times 1 \right] =$	125
$\mathrm{Dim}\, U_{10} = \frac{5}{120}$	$\left[5 \times 182 \times 1 \times 3 + \right.$	$0 \times 2 \times 12\varphi +$	$0 \times 2 \times 12 \left(-\dfrac{1}{\varphi}\right) +$	$1 \times 2 \times 15(-1) +$	$\left. (-1)20 \times 15 \times 1 \right] =$	100

Total number of dimension: 546

Applied transformations: $\varphi = \tfrac{1}{2}(\sqrt{5}+1)$, $\quad \dfrac{1}{\varphi} = \tfrac{1}{2}(\sqrt{5}-1)$, $\quad \varphi = 1 + \dfrac{1}{\varphi}$, $\quad \varphi^2 + \dfrac{1}{\varphi^2} = 3$, $\quad -\varphi + \dfrac{1}{\varphi} = -1$.

Table 2.4. Results of G-vector analysis of the dodecahedral–icosahedral configuration with 182 nodes and 546 degrees of freedom

Number of group elements: 120				group I_h
Number of nodes: 182		degrees of freedom		546

(1)	dimensions of irreducible representations	U_1 U_2			1
		U_3 U_4	U_5	U_6	3
		U_7 U_8			4
		U_9 U_{10}			5
(2)	dimensions of matrices of G-invariant subspaces		U_1		7
			U_2		2
			U_3		36
			U_4		51
			U_5		33
			U_6		48
			U_7		72
			U_8		72
			U_9		125
			U_{10}		100
(3)	number of multiplications	standard procedure	M		54 854 797
			M'		27 725 787
		G-vector analysis	M_G		1 428 034
			M'_G		758 621
	ratios		M_G/M		0,026
			M'_G/M'		0,027
(4)	number of different matrix elements	standard procedure	N		298 116
			N'		149 331
		G-vector analysis	N_G		43 336
			N'_G		21 943
	ratios		N_G/N		0,145
			N'_G/N'		0,147

number of nodes. Thus, nodes in or on the trihedron determined by the spherical triangle T_{120}, are projected by central rays onto the triangle, and classification of nodes is accomplished according to the positions of their central projections on the spherical triangle T_{120}.

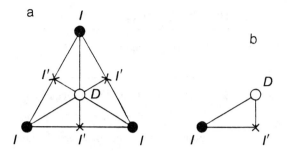

Fig. 2.19. Schematical illustration of the general configuration of nodal patterns with the symmetry of the icosahedral group I_h: (a) the spherical triangle T_{20} divided into six spherical triangles T_{120}; (b) the spherical triangle T_{120}.

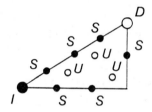

Fig. 2.20. Example of a nodal pattern with positions of nodes defined in regard to the spherical triangle T_{120}.

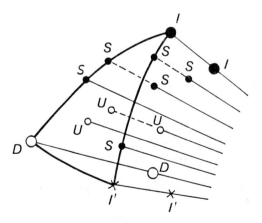

Fig. 2.21. Nodes in or on the trihedron determined by the spherical triangle T_{120}.

Table 2.5. T_{120} node classification

Nodal type	Designation	Position of the node
I	●	in vertex of circumscribed icosahedron
D	○	in vertex of circumscribed dodecahedron
I'	×	in midpoint of side of spherical triangle T_{120}
S	●	on contour of spherical triangle T_{120}
U	○	in spherical triangle T_{120}

2.4.2 Derivation of the general expression for determination of dimensions of G-invariant subspaces for configurations compatible with the icosahedral group

Classification of nodes into types makes possible derivation of the general expression for determination of the number of nodes which are not moved by the action of symmetry operations. So far, the number of 'fixed' nodes is determined by visual analysis of the structure, which, in the case of a great number of nodes, is not easy, since it is necessary to observe the complete configuration in space.

The number of 'fixed' nodes for each class of elements of the icosahedral group I_h will be expressed by parameters $\delta_I, \delta_D, \delta_{I'}, n_s$ and n_u:

$K_1: n_f(E) = \delta_I + \delta_D + \delta_{I'} + n_s + n_u = n$

$K_2: n_f(C_5) = 2\delta_I$

$K_3: n_f(C_5^2) = 2\delta_I$

$K_4: n_f(C_3) = 2\delta_D$

$K_5: n_f(C_2) = 2\delta_{I'}$

$K_6: n_f(i) = 0$

$K_7: n_f(S_{10}) = 0$

$K_8: n_f(S_{10}^2) = 0$

$K_9: n_f(S_6) = 0$

$K_{10}: n_f(\sigma) = 4(\delta_I + \delta_D + \delta_{I'} + n_s)$.

Values of traces of matrices of symmetry elements (found previously):

$$\text{Tr}(E) = 3, \quad \text{Tr}(C_5) = \varphi, \quad \text{Tr}(C_5^2) = -\frac{1}{\varphi}, \quad \text{Tr}(C_3) = 0, \quad \text{Tr}(C_2) = -1,$$

$$\text{Tr}(i) = -3, \quad \text{Tr}(S_{10}) = \frac{1}{\varphi}, \quad \text{Tr}(S_{10}^2) = \frac{1}{\varphi} - 1, \quad \text{Tr}(S_6) = 0 \quad \text{and} \quad \text{Tr}(\sigma) = 1.$$

Sec. 2.4] Nodal configurations with patterns based on the regular icosahedron

When the values found are introduced into the expression for determination of dimensions of subspaces,

$$\text{Dim } U_i = \frac{h_i}{h} \sum_j \chi_{ij} n_{fj} n_j \text{ Tr } \sigma_j,$$

one obtains

$$\text{Dim } U_i = \frac{\chi_{i1}}{120} [\chi_{i1} n_f(E) n(E) \text{ Tr } (E) + \chi_{i2} n_f(C_5) n(C_5) \text{ Tr } (C_5)$$

$$+ \chi_{i3} n_f(C_5^2) n(C_5^2) \text{ Tr } (C_5^2) + \chi_{i5} n_f(C_2) n(C_2) \text{ Tr } (C_2)$$

$$+ \chi_{i10} n_f(\sigma) n(\sigma) \text{ Tr } (\sigma)],$$

then

$$\text{Dim } U_i = \frac{\chi_{i1}}{120} \left[\chi_{i1} n(1)(3) + \chi_{i2} 2\delta_I 12\varphi + \chi_{i3} 2\delta_I 12 \left(-\frac{1}{\varphi}\right) \right.$$

$$\left. + \chi_{i5} 2\delta_{I'} 15(-1) + \chi_{i10} 4(\delta_I + \delta_D + \delta_{I'} + n_s) 15(1) \right]$$

and finally

$$\text{Dim } U_i = \frac{\chi_{i1}}{120} \left[3\chi_{i1} n + 24\delta_I \left(\chi_{i2} \varphi - \frac{h_{i3}}{\varphi} \right) - 30 \chi_{i5} \delta_{I'} \right.$$

$$\left. + 60 \chi_{i10} (\delta_I + \delta_D + \delta_{I'} + n_s) \right].$$

The expression derived for the determination of dimensions of subspaces does not require a visual analysis of the configuration to be made in order to determine the number of nodes not moved by action of particular symmetry operations, since the number of 'fixed' nodes n_f is contained in the expression for the dimensions of subspaces. It is necessary to determine only the values of the parameters δ_I, δ_D, $\delta_{I'}$, n_s and n_u depending on the positions of central projections of nodes onto the spherical triangle T_{120} pertaining to the trihedron encompassing these nodes.

This analysis will be illustrated by two configurations having 120 nodes with three degrees of freedom in each node, but with different positions of nodes with regard to the spherical triangle T_{120}. Configuration A has one node in the interior of the triangle T_{120}, while configuration B has two nodes on sides of this triangle. On the basis of this data all parameters are determined and the expressions for the dimensions of subspaces are obtained (Table 2.6).

Results of analyses of configurations A and B are given in Table 2.7: dimensions of irreducible representations, dimensions of matrices of G-invariant subspaces, number of multiplications in Gauss' algorithm for solving of equations and number of different

Table 2.6 Dimensions of subspaces

Configuration A	Configuration B
$\delta_I = \delta_D = \delta_{I'} = 0$, $n_s = 0$, $n_u = 1$	$\delta_I = \delta_D = \delta_{I'} = 0$, $n_s = 2$, $n_u = 0$
$n = 12 \times 0 + 20 \times 0 + 30 \times 0 + 60 \times 0$ $+ 120 \times 1 = 120$	$n = 12 \times 0 + 20 \times 0 + 30 \times 0 + 60 \times 2$ $+ 120 \times 0 = 120$
$\text{Dim } U_i = \dfrac{\chi_{i1}}{120} 3\chi_{i1} n$	$\text{Dim } U_i = \dfrac{\chi_{i1}}{120}(3\chi_{i1} n + 60\chi_{i10} n_s)$
$= \dfrac{\chi_{i1}}{120} 3\chi_{i1} 120 = 3\chi_{i1}^2$	$= \dfrac{\chi_{i1}}{120}(3\chi_{i1} 120 + 60 \chi_{i10} 2)$
	$= 3\chi_{i1}^2 + \chi_{i1}\chi_{i10}.$

matrix elements (in standard procedure and in G-vector analysis). In comparison with the standard procedure in G-vector analysis the number of multiplications is reduced to 2.7% of the original number and the number of different matrix elements to 14.5%.

Comparison of dimensions of G-invariant subspaces of configurations A and B is given by the diagram in Fig. 2.22, which shows that dimensions of subspaces of configurations A and B differ a little. The reason can be found by analysis of the expression for determination of dimensions of subspaces, where the first term with the identity element has a dominant influence on the size of dimensions, while influence of other classes of elements is relatively small.

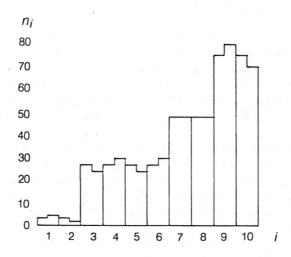

Fig. 2.22. Graphical comparison of dimensions of G-invariant subspaces of configurations A and B with 120 nodes and 360 degrees of freedom.

Table 2.7 Results of G-vector analysis of dodecahedral–icosahedral configurations A and B with 120 nodes and 360 degrees of freedom

					group I_h	
Number of group elements: 120					A	B
number of nodes: 120		configuration			$n_s = 0$	$n_s = 2$
degrees of freedom: 360,		$\delta_I = \delta_D = \delta_{I'} = 0$			$n_u = 1$	$n_u = 0$
(1)	dimensions of irreducible representations	U_1 U_2 U_3 U_4 U_5 U_6 U_7 U_8 U_9 U_{10}			1 3 4 5	1 3 4 5
(2)	dimensions of matrices of G-invariant subspaces		U_1		3	4
			U_2		3	2
			U_3		27	24
			U_4		27	30
			U_5		27	24
			U_6		27	30
			U_7		48	48
			U_8		48	48
			U_9		75	80
			U_{10}		75	70
(3)	number of multiplications	standard procedure	M M'		15 812 157 8 035 858	15 812 157 8 035 858
		G-vector analysis	M_G M'_G		419 754 228 844	424 658 231 384
	ratios		M_G/M		0,028	0,027
			M'_G/M'		0,028	0,029
(4)	number of different matrix elements	standard procedure	N N'		129 600 64 980	129 600 64 980
		G-vector analysis	N_G N'_G		18 792 9 576	18 880 9 620
	ratios		N_G/N		0,146	0,146
			N'_G/N'		0,147	0,148

2.5 ALGORITHMIC SCHEME OF G-VECTOR ANALYSIS

G-vector analysis can be systematized in the form of an algorithmic scheme, which may serve as a basis for computer programs. The data base should contain the set of isomorphic groups given in Table 2.8 and the group elements and their total number, the number of classes of elements and the number of irreducible group representations given in Tables 2.9 and 2.10. The matrix operators of symmetry operations (elements) of the groups should also be included.

The algorithmic scheme for G-vector analysis, given in Table 2.11, contains the following steps.

(1) Start of the program for G-vector analysis (survey of options).
(2) Input of data that describe the configuration: nodal coordinates x, y, z in the right coordinate system, with the z axis as the main rotation axis and the reflection planes positioned in relation to the coordinate axes, as well as the connections of the nodes (existence of bars).
(3) Three-dimensional graphic presentation of the configuration with the nodes and the bars.
(4) Identification of groups which correspond to the configuration (transformation of the configuration into itself examined by application of matrices of symmetry operations of the group on coordinate vectors).
(5) Date base for the symmetry groups (group elements, classes of group elements, irreducible group representations and character tables).
(6) List of the groups which correspond to the configuration.
(7) Selection of the group for G-vector analysis with respect to the following criteria:
   – the group with the greatest number of irreducible representations (decomposition of the matrix of the system into the greatest number of independent submatrices);
   – the group with the smallest dimensions of irreducible group representations (decomposition of the matrix of the system into submatrices with the smallest dimensions);
   – the group with its order (number of group elements) amounting to n in relation to the n-fold number of nodes of the configuration ($n = 1, 2, \ldots$).
(8) Derivation of the idempotents of the centre of the group algebra of the chosen group (utilization of the character table).
(9) Derivation of the basis vectors of G-invariant subspaces by application of idempotents on functions at the nodes.
(10) Verification of the orthogonality of basis vectors of different subspaces.
(11) Selection from three options (steps (12), (13) or (14)).
(12) List of the basis vectors of G-invariant subspaces of the configuration accepted as the final result of the analysis and the end of the program (step (17)).
(13) Derivation of a new set of basis vectors of G-invariant subspaces by application of another group from the list of groups which correspond to the configuration (return to step (7) and carry out a new analysis).

(14) Determination of elements of matrices of G-invariant subspaces and the free members (in applications of the stiffness method, the force method, the finite-difference method and other options).
(15) Solution of the system of equations (for each G-invariant subspace separately).
(16) Final results expressed by the group variables or by the initial variables.
(17) End of the program.

Table 2.8. Isomorphic groups (k is the number of irreducible group representations)

Groups			k	Groups				k
C_2	C_i	C_s	2	C_{4h}				6
C_3			2	C_{5h}				6
C_{3v}	D_3		3	C_{6v}	D_6	D_{3h}	D_{3d}	6
C_4	S_4		3	D_{4d}				7
C_5			3	C_{6h}				8
T			3	D_{2h}				8
C_{2v}	C_{2h}	D_2	4	D_{5h}	D_{5d}			8
C_{5v}	D_5		4	D_{6d}				9
C_{3h}	C_6	S_6	4	D_{4h}				10
C_{4v}	D_4	D_{2d}	5	O_h				10
O	T_d		5	I_h				10
I			5	D_{6h}				12

Table 2.9. Data on groups C_n, D_n, C_{nv} and C_{nh}

Group	Classes of elements							Number of group elements	Number of classes of elements	Sum of dimensions of irreducible representations
C_s	E	σ_h						2	2	2
C_i	E	i						2	2	2
Groups C_n										
C_2	E	C_2						2	2	2
C_3	E	$2C_3$						3	2	3
C_4	E	$2C_4$	C_2					4	3	4
C_5	E	$2C_5$	$2C_5^2$					5	3	5
C_6	E	$2C_6$	$2C_3$	C_2				6	4	6
Groups D_n										
D_2	E	C_2^x	C_2^y	C_2^z				4	4	4
D_3	E	$2C_3$	$3C_2$					6	3	4
D_4	E	$2C_4$	C_2	$2C_2'$	$2C_2''$			8	5	6
D_5	E	$2C_5$	$2C_5^2$	$5C_2$				10	4	6
D_6	E	$2C_6$	$2C_3$	C_2	$3C_2'$	$3C_2''$		12	6	8
Groups C_{nv}										
C_{2v}	E	C_2	σ_{xz}	σ_{yz}				4	4	4
C_{3v}	E	$2C_3$	$3\sigma_v$					6	3	4
C_{4v}	E	$2C_4$	C_2	$2\sigma_v$	$2\sigma_d$			8	5	6
C_{5v}	E	$2C_5$	$2C_5^2$	$5\sigma_v$				10	4	6
C_{6v}	E	$2C_6$	$2C_3$	C_2	$3\sigma_v$	$3\sigma_d$		12	6	8
Groups C_{nh}										
C_{2h}	E	C_2	i	σ_h				4	4	4
C_{3h}	E	$2C_3$	σ_h	$2S_3$				6	4	6
C_{4h}	E	$2C_4$	C_2	σ_h	$2S_4$	i		8	6	8
C_{5h}	E	$2C_5$	$2C_5^2$	σ_h	$2S_5^3$	$2S_5^2$		10	6	10
C_{6h}	E	$2C_6$	$2C_3$	C_2	σ_h	$2S_6$	$2S_3$ i	12	8	12

Table 2.10. Data on groups D_{nh}, D_{nd}, S_n, and the cubic and icosahedral groups

Group	Classes of elements									Number of group elements	Number of classes of elements	Sum of dimensions of irreducible representations
Groups D_{nh}												
D_{2h}	E	C_2^x	C_2^y	C_2^z	i	σ_{xy}	σ_{xz}	σ_{yz}		8	8	8
D_{3h}	E	$2C_3$	$3C_2$	σ_h	$2S_3$		$3\sigma_v$			12	6	8
D_{4h}	E	$2C_4$	C_2	$2C_2'$	$2C_2''$							
	i	$2S_4$	σ_h	$2\sigma_v$	$2\sigma_d$					16	10	12
D_{5h}	E	$2C_5$	$2C_5^2$	$5C_2$	σ_h	$2S_5$	$2S_5^3$	$5\sigma_v$		20	8	12
D_{6h}	E	$2C_6$	$2C_3$	C_2	$3C_2'$	$3C_2''$						
	i	$2S_3$	$2S_6$	σ_h	$3\sigma_d$	$3\sigma_v$				24	12	16
Groups D_{nd}												
D_{2d}	E	$2S_4$	C_2	$2C_2'$	$2\sigma_d$					8	5	6
D_{3d}	E	$2C_3$	$3C_2$	i	$2S_6$	$3\sigma_d$				12	6	8
D_{4d}	E	$2S_8$	$2C_4$	$2S_8^3$	C_2	$4C_2'$	$4\sigma_d$			16	7	10
D_{5d}	E	$2C_5$	$2C_5^2$	$5C_2$	i	$2S_{10}^3$	$2S_{10}$	$5\sigma_d$		20	8	12
D_{6d}	E	$2S_{12}$	$2C_6$	$2S_4$	$2C_3$	$2S_{12}^5$	C_2	$6C_2'$	$6\sigma_d$	24	9	14

Table 2.10 (continued)

Group	Classes of elements							Number of group elements	Number of classes of elements	Sum of dimensions of irreducible representations
Groups S_n										
S_4	E	$2S_4$	C_2					4	3	5
S_6	E	$2C_3$	i	$2S_6$				6	4	6
Cubic groups										
T	E	$4C_3$	$4C_3^2$	$3C_2$				12	4	6
T_d	E	$8C_3$	$3C_2$	$6S_4$	$6\sigma_d$			24	5	10
O	E	$8C_3$	$3C_2$	$6C_2'$	$6C_4$			24	5	10
O_h	E	$8C_3$	$6C_2$	$6C_4$	$3C_4^2$			48	10	20
	i	$6S_4$	$8S_6$	$3\sigma_h$	$6\sigma_d$					
Icosahedral groups										
I	E	$12C_5$	$12C_5^2$	$20C_3$	$15C_2$			60	5	16
I_h	E	$12C_5$	$12C_5^2$	$20C_3$	$15C_2$			120	10	32
	i	$12S_{10}$	$12S_{10}^3$	$20S_6$	15σ					

Table 2.11. Algorithmic scheme of G-vector analysis

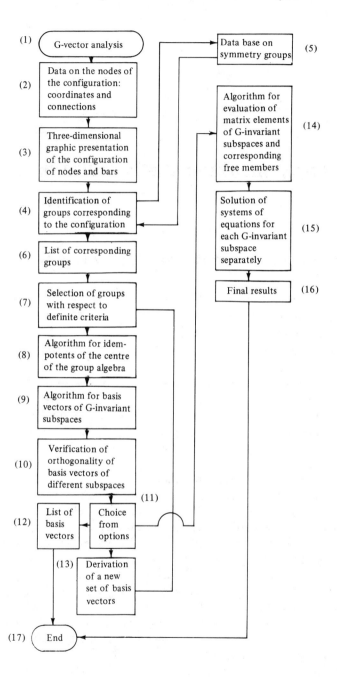

3

Group supermatrix transformations

3.1 DEFINITION OF GROUP SUPERMATRICES AND FORMULATION OF THEIR RELATIONS

Let the symmetry group G, having h elements, describe the symmetry properties of an element with a pattern of h nodes, which are permuted within a single nodal set by actions of group elements.

The nodal function sets $\Phi^{(1)}, \Phi^{(2)}, \ldots, \Phi^{(h)}$ pertaining to the nodes $1, 2, \ldots, h$ contain l ($l = 1, 2, \ldots$) nodal functions at each node.

The choice of the optimum nodal numbering is based here on the concept of preserving the structure of the character matrix of the character table in derivations of basis vectors of G-invariant subspaces of the G-vector space of the function field of the nodal configuration. This is achieved by assuming the optimum position for the first node of the nodal pattern and by applying to this node group elements in the order determined by the sequence of classes of conjugate elements in the character table.

It is assumed that the position of the first node is on the positive branch of the x axis for line elements, in the first quadrant of the xy coordinate system for plane elements and in the first octant of the xyz coordinate system for space elements.

The sequence of classes of conjugate elements $K_1, K_2, \ldots, K_k$ existing in the character table of G, and the evident sequence of elements in each class, provide a unique sequence of h elements of the group.

Application of h group elements in this unique sequence to the first node, numbered by 1, moves this node to nodal positions which will be numbered $1, 2, \ldots, h$ according to the order of actions of the elements.

Actions of group elements (symmetry operations) on the initial nodal sequence $[1\ 2 \ldots h]$ permute the nodes of this nodal set, producing h permutations, which appear as rows in the permutation table.

When in the $h \times h$ matrix of this permutation table the nodal numbers $1, 2, \ldots, h$ are replaced by matrices $A, B, \ldots, L$, one obtains the supermatrix G, called the *group supermatrix in normal form* pertaining to the group G.

Sec. 3.1]   Definition of group supermatrices and formulation of their relations   91

The character table of the group G is presented in general form as follows:

G	K_1	K_2	...	K_j	...	K_k
Γ_1	$\chi_1^{(1)}$	$\chi_2^{(1)}$	...	$\chi_j^{(1)}$	...	$\chi_k^{(1)}$
Γ_2	$\chi_1^{(2)}$	$\chi_2^{(2)}$	...	$\chi_j^{(2)}$	...	$\chi_k^{(2)}$
...	...	...	...	...	...	...
Γ_i	$\chi_1^{(i)}$	$\chi_2^{(i)}$	...	$\chi_j^{(i)}$	...	$\chi_k^{(i)}$
...	...	...	...	...	...	...
Γ_k	$\chi_1^{(k)}$	$\chi_2^{(k)}$	...	$\chi_j^{(k)}$	...	$\chi_k^{(k)}$

where K_j is the jth class of conjugate elements, Γ_i the ith irreducible representation of the group G and $\chi_j^{(i)}$ the character pertaining to the jth class and the ith group representation, while k is the number of group representations and the number of classes.

As stated in Theorems 1.2 and 1.3, i.e. by the first and second orthogonality relations, the matrix of characters has orthogonal rows and orthogonal columns.

When at each nodal point of the configuration there exist l functions φ_j, each character in the character table of G will be replaced by the product of the character and the $l \times l$ unit matrix E. Thus the character table of G in supermatrix form is formulated as follows:

G	K_1	K_2	...	K_j	...	K_k
Γ_1	$\chi_1^{(1)}E$	$\chi_2^{(1)}E$	...	$\chi_j^{(1)}E$	...	$\chi_k^{(1)}E$
Γ_2	$\chi_1^{(2)}E$	$\chi_2^{(2)}E$	...	$\chi_j^{(2)}E$	...	$\chi_k^{(2)}E$
...	...	...	...	...	...	...
Γ_i	$\chi_1^{(i)}E$	$\chi_2^{(i)}E$	...	$\chi_j^{(i)}E$	...	$\chi_k^{(i)}E$
...	...	...	...	...	...	...
Γ_k	$\chi_1^{(k)}E$	$\chi_2^{(k)}E$	...	$\chi_j^{(k)}E$	...	$\chi_k^{(k)}E$

The idempotents of the centre of the group algebra $\bar{G}$ will be determined by utilizing the character table in supermatrix form, so that the characters in matrix form will be inserted into the expression

$$\pi_i = \frac{h_i}{h} \sum_\sigma \chi_i(\sigma^{-1})\sigma,$$

where h_i is the dimension of the ith character $\chi_i(E)$, h is the order, χ_i is the ith character, σ the element and σ^{-1} the inverse of the element of the group.

Thus, the idempotents are expressed by coefficients in the form of products of characters and unit matrices:

$$\pi_1 = \frac{h_1}{h}(\chi_1^{(1)}E\kappa_1 + \chi_2^{(1)}E\kappa_2 + \ldots + \chi_k^{(1)}E\kappa_k)$$

$$\pi_2 = \frac{h_2}{h}(\chi_1^{(2)}E\kappa_1 + \chi_2^{(2)}E\kappa_2 + \ldots + \chi_k^{(2)}E\kappa_k)$$

$$\vdots$$

$$\pi_k = \frac{h_k}{h}(\chi_1^{(k)}E\kappa_1 + \chi_2^{(k)}E\kappa_2 + \ldots + \chi_k^{(k)}E\kappa_k),$$

where $\kappa_1, \kappa_2, \ldots, \kappa_k$ are the class sums, i.e. the sums of elements of the classes $K_1, K_2, \ldots, K_k$ respectively.

In supermatrix form the idempotents can be expressed as

$$\begin{bmatrix} \pi_1 \\ \pi_2 \\ \vdots \\ \pi_k \end{bmatrix} = \begin{bmatrix} \frac{h_1}{h}\chi_1^{(1)}E & \frac{h_1}{h}\chi_2^{(1)}E & \ldots & \frac{h_1}{h}\chi_k^{(1)}E \\ \frac{h_2}{h}\chi_1^{(2)}E & \frac{h_2}{h}\chi_2^{(2)}E & \ldots & \frac{h_2}{h}\chi_k^{(2)}E \\ \vdots & \vdots & & \vdots \\ \frac{h_k}{h}\chi_1^{(k)}E & \frac{h_k}{h}\chi_2^{(k)}E & \ldots & \frac{h_k}{h}\chi_k^{(k)}E \end{bmatrix} \begin{bmatrix} \kappa_1 \\ \kappa_2 \\ \vdots \\ \kappa_k \end{bmatrix}$$

or $\Pi = T_c\kappa$, where T_c is the transformation supermatrix for formulation of the idempotents.

Because of the orthogonality of the rows of the character matrix, the inverse of the transformation supermatrix T_c is $T_c^{-1} = T_c^T$, i.e.

$$T_c^{-1} = \begin{bmatrix} \frac{h}{h_1}\chi_1^{(1)}E & \frac{h}{h_2}\chi_1^{(2)}E & \ldots & \frac{h}{h_k}\chi_1^{(k)}E \\ \frac{h}{h_1}\chi_2^{(1)}E & \frac{h}{h_2}\chi_2^{(2)}E & \ldots & \frac{h}{h_k}\chi_2^{(k)}E \\ \vdots & \vdots & & \vdots \\ \frac{h}{h_1}\chi_k^{(1)}E & \frac{h}{h_2}\chi_k^{(2)}E & \ldots & \frac{h}{h_k}\chi_k^{(k)}E \end{bmatrix}.$$

Applying the idempotents $\pi_1, \pi_2, \ldots, \pi_k$ to the sets of nodal functions $\Phi^{(1)}, \Phi^{(2)}, \ldots, \Phi^{(h)}$ derives the basis vectors $\bar{\Phi}^{(1)}, \bar{\Phi}^{(2)}, \ldots, \bar{\Phi}^{(k)}$ of G-invariant subspaces $U_1, U_2, \ldots, U_k$ respectively, which is expressed by

$$\bar{\Phi} = T\Phi,$$

where T is the *basis transformation supermatrix* in the relation of the basis vector sets $\bar{\Phi}$ to the nodal function sets Φ.

Conversely,

$$\Phi = T^{-1}\bar{\Phi},$$

where $T^{-1} = T^T$.

The system of equations

$$G\Phi = P,$$

where G is the group supermatrix in normal form (derived earlier), Φ the nodal function sets $\Phi^{(1)}, \Phi^{(2)}, \ldots, \Phi^{(k)}$ and P the nodal free member sets $P^{(1)}, P^{(2)}, \ldots, P^{(k)}$, is transformed into the system of equations

$$\bar{G}\bar{\Phi} = \bar{P},$$

where $\bar{G}$ is the *group supermatrix in diagonal form*, $\bar{\Phi}$ the basis vectors of G-invariant subspaces U_i and $\bar{P}$ the symmetry-adapted free member vectors pertaining to the subspaces U_i, by the *group supermatrix transformation*

$$TGT^{-1}T\Phi = TP,$$

where T is the basis transformation supermatrix, resulting in

$$\begin{bmatrix} \bar{G}_1 & & & \\ & \bar{G}_2 & & \\ & & \ddots & \\ & & & \bar{G}_k \end{bmatrix} \begin{bmatrix} \bar{\Phi}^{(1)} \\ \bar{\Phi}^{(2)} \\ \vdots \\ \bar{\Phi}^{(k)} \end{bmatrix} = \begin{bmatrix} \bar{P}^{(1)} \\ \bar{P}^{(2)} \\ \vdots \\ \bar{P}^{(k)} \end{bmatrix} \quad \text{or} \quad \bar{G}\bar{\Phi} = \bar{P},$$

where supermatrices $\bar{G}_1, \bar{G}_2, \ldots, \bar{G}_k$, basis vector sets $\bar{\Phi}^{(1)}, \bar{\Phi}^{(2)}, \ldots, \bar{\Phi}^{(k)}$, free member sets $\bar{P}^{(1)}, \bar{P}^{(2)}, \ldots, \bar{P}^{(k)}$ pertain to the subspaces $U_1, U_2, \ldots, U_k$ respectively.

Conversely, the system of equations

$$\bar{G}\bar{\Phi} = \bar{P}$$

is transformed into the system of equations

$$G\Phi = P$$

by the inverse group supermatrix transformation

$$T^{-1}\bar{G}TT^{-1}\bar{\Phi} = T^{-1}\bar{P}.$$

3.2 GROUP SUPERMATRICES OF THE GROUP C_2

The character table of the group C_2

C_2	E	C_2
A	1	1
B	1	-1

has the sequence of the symmetry operations E (identity), C_2 (rotation through 180° about the z axis) that is used to derive the nodal numbering of the two-node line element. The position of the first node, numbered by 1, is on the positive branch of the x axis, as given in Fig. 3.1. According to the rule given in 3.1, the unique nodal numbering is obtained by applying E, C_2 (in this order) to the initial node 1, which provides the nodes 1, 2 (in this sequence).

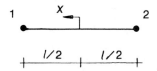

Fig. 3.1. The two-node line element described by the group C_2.

This definition determines the unique nodal numbering in Fig. 3.1, which is stated by the following table:

node	x axis
1	$+$
2	$-$

Expressed as permutations of the nodes, the results of actions of symmetry operations on the initial sequence [1 2] are

$$E = \begin{bmatrix} 1 & 2 \\ 1 & 2 \end{bmatrix}, \quad C_2 = \begin{bmatrix} 1 & 2 \\ 2 & 1 \end{bmatrix},$$

or presented in the form of a permutation table

Node	E	C_2
1	1	2
2	2	1

By replacing the numbers 1, 2 by matrices A, B in the above permutation table one obtains the supermatrix

$$G = \begin{bmatrix} A & B \\ B & A \end{bmatrix},$$

which is the 2×2 group supermatrix in normal form pertaining to the group C_2.

The idempotents π_i ($i = 1, 2$) of the centre of the group algebra are determined by

$$\pi_i = \frac{h_i}{h} \sum_\sigma \chi_i(\sigma^{-1}) \sigma,$$

where $h_i = \chi_i(E)$, h is the order, χ_i is the ith character, σ is the element and σ^{-1} its inverse, so that

$$\begin{bmatrix} \pi_1 \\ \pi_2 \end{bmatrix} = \frac{1}{2} \begin{bmatrix} 1 & 1 \\ 1 & -1 \end{bmatrix} \begin{bmatrix} E \\ C_2 \end{bmatrix} \quad \text{or} \quad \Pi = T_c \Sigma,$$

with the transformation matrix

$$T_c = \frac{1}{2} \begin{bmatrix} 1 & 1 \\ 1 & -1 \end{bmatrix} \quad \text{and} \quad T_c^{-1} = \begin{bmatrix} 1 & 1 \\ 1 & -1 \end{bmatrix} \quad \text{or} \quad T_c^{-1} = 2T_c.$$

When in the matrix T_c the unit values are replaced by $l \times l$ unit matrix E, the transformation supermatrix is obtained:

$$T_c = \frac{1}{2} \begin{bmatrix} E & E \\ E & -E \end{bmatrix} \quad \text{and} \quad T_c^{-1} = \begin{bmatrix} E & E \\ E & -E \end{bmatrix} \quad \text{or} \quad T_c^{-1} = 2T_c.$$

The sets of nodal functions $\Phi^{(1)}$, $\Phi^{(2)}$ pertaining to nodes 1, 2 may contain one or more functions φ_j at a node. When l is the number of nodal functions at each node, the following numbering of nodal functions φ_j ($j = 1, 2, \ldots, n$); $n = 2l$) is introduced:

for $l = 1$:
$$\begin{bmatrix} \Phi^{(1)} \\ \Phi^{(2)} \end{bmatrix} = \begin{bmatrix} \varphi_1 \\ \varphi_2 \end{bmatrix}$$

for $l = 2$:
$$\begin{bmatrix} \Phi^{(1)} \\ \Phi^{(2)} \end{bmatrix} = \begin{bmatrix} \varphi_1 & \varphi_3 \\ \varphi_2 & \varphi_4 \end{bmatrix}$$

for $l = 3$:
$$\begin{bmatrix} \Phi^{(1)} \\ \Phi^{(2)} \end{bmatrix} = \begin{bmatrix} \varphi_1 & \varphi_3 & \varphi_5 \\ \varphi_2 & \varphi_4 & \varphi_6 \end{bmatrix}$$

for $l = 4$:
$$\begin{bmatrix} \Phi^{(1)} \\ \Phi^{(2)} \end{bmatrix} = \begin{bmatrix} \varphi_1 & \varphi_3 & \varphi_5 & \varphi_7 \\ \varphi_2 & \varphi_4 & \varphi_6 & \varphi_8 \end{bmatrix}$$

and similarly for bigger value of l.

Applying the idempotents π_1, π_2 to the sets of nodal functions $\Phi^{(1)}, \Phi^{(2)}$ derives the sets of basis vectors $\bar{\Phi}^{(1)}, \bar{\Phi}^{(2)}$ of G-invariant subspaces U_1, U_2:

$$\begin{bmatrix} \bar{\Phi}^{(1)} \\ \bar{\Phi}^{(2)} \end{bmatrix} = \frac{1}{2}\begin{bmatrix} E & E \\ E & -E \end{bmatrix}\begin{bmatrix} \Phi^{(1)} \\ \Phi^{(2)} \end{bmatrix} = \frac{1}{2}\begin{bmatrix} \Phi^{(1)}+\Phi^{(2)} \\ \Phi^{(2)}-\Phi^{(2)} \end{bmatrix} \quad \text{or} \quad \bar{\Phi} = T\Phi.$$

As previously for $\Phi^{(1)}, \Phi^{(2)}$, where l is the number of nodal functions at each node, the numbering of the basis vectors $\bar{\varphi}_j$ ($j = 1, 2, \ldots, n$; $n = 2l$) corresponds to the numbering of the nodal functions φ_j:

for $l = 1$: $\begin{bmatrix} \bar{\Phi}^{(1)} \\ \bar{\Phi}^{(2)} \end{bmatrix} = \begin{bmatrix} \bar{\varphi}_1 \\ \bar{\varphi}_2 \end{bmatrix}$

for $l = 2$: $\begin{bmatrix} \bar{\Phi}^{(1)} \\ \bar{\Phi}^{(2)} \end{bmatrix} = \begin{bmatrix} \bar{\varphi}_1 & \bar{\varphi}_3 \\ \bar{\varphi}_2 & \bar{\varphi}_4 \end{bmatrix}$

for $l = 3$: $\begin{bmatrix} \bar{\Phi}^{(1)} \\ \bar{\Phi}^{(2)} \end{bmatrix} = \begin{bmatrix} \bar{\varphi}_1 & \bar{\varphi}_3 & \bar{\varphi}_5 \\ \bar{\varphi}_2 & \bar{\varphi}_4 & \bar{\varphi}_6 \end{bmatrix}$

for $l = 4$: $\begin{bmatrix} \bar{\Phi}^{(1)} \\ \bar{\Phi}^{(2)} \end{bmatrix} = \begin{bmatrix} \bar{\varphi}_1 & \bar{\varphi}_3 & \bar{\varphi}_5 & \bar{\varphi}_7 \\ \bar{\varphi}_2 & \bar{\varphi}_4 & \bar{\varphi}_6 & \bar{\varphi}_8 \end{bmatrix}$

and similarly for bigger values of l.

Conversely, the relation of the sets of nodal functions $\Phi^{(1)}, \Phi^{(2)}$ to the sets of basis vectors $\bar{\Phi}^{(1)}, \bar{\Phi}^{(2)}$ is

$$\begin{bmatrix} \Phi^{(1)} \\ \Phi^{(2)} \end{bmatrix} = \begin{bmatrix} E & E \\ E & -E \end{bmatrix}\begin{bmatrix} \bar{\Phi}^{(1)} \\ \bar{\Phi}^{(2)} \end{bmatrix} = \begin{bmatrix} \bar{\Phi}^{(1)}+\bar{\Phi}^{(2)} \\ \bar{\Phi}^{(1)}-\bar{\Phi}^{(2)} \end{bmatrix} \quad \text{or} \quad \Phi = T^{-1}\bar{\Phi} = 2T\bar{\Phi}.$$

The sets of basis vectors $\bar{\Phi}_1, \bar{\Phi}_2, \ldots, \bar{\Phi}_l$, written as

$$[\bar{\Phi}_1 \ \bar{\Phi}_2 \ \ldots \ \bar{\Phi}_l] = \begin{bmatrix} \bar{\varphi}_1 & \bar{\varphi}_3 & \cdots & \bar{\varphi}_{n-1} \\ \bar{\varphi}_2 & \bar{\varphi}_4 & \cdots & \bar{\varphi}_n \end{bmatrix},$$

are related to the sets of nodal functions $\Phi_1, \Phi_2, \ldots, \Phi_l$ as follows:

$$\begin{bmatrix} \bar{\Phi}_1 \\ \bar{\Phi}_2 \\ \vdots \\ \bar{\Phi}_l \end{bmatrix} = \begin{bmatrix} T & & & \\ & T & & \\ & & \ddots & \\ & & & T \end{bmatrix}\begin{bmatrix} \Phi_1 \\ \Phi_2 \\ \vdots \\ \Phi_l \end{bmatrix} \quad \text{or} \quad \bar{\Phi} = \bar{T}\Phi,$$

where

$$T = \frac{1}{2}\begin{bmatrix} E & E \\ E & -E \end{bmatrix} \quad \text{and} \quad [\Phi_1 \ \Phi_2 \ \ldots \ \Phi_l] = \begin{bmatrix} \varphi_1 & \varphi_3 & \ldots & \varphi_{n-1} \\ \varphi_2 & \varphi_4 & \ldots & \varphi_n \end{bmatrix}.$$

Conversely, the relation of the sets of nodal functions $\Phi_1, \Phi_2, \ldots, \Phi_l$ to the sets of basis vectors $\bar{\Phi}_1, \bar{\Phi}_2, \ldots, \bar{\Phi}_l$ is

$$\begin{bmatrix} \Phi_1 \\ \Phi_2 \\ \vdots \\ \Phi_l \end{bmatrix} = \begin{bmatrix} T^{-1} & & & \\ & T^{-1} & & \\ & & \ddots & \\ & & & T^{-1} \end{bmatrix} \begin{bmatrix} \bar{\Phi}_1 \\ \bar{\Phi}_2 \\ \vdots \\ \bar{\Phi}_l \end{bmatrix} = 2 \begin{bmatrix} T & & & \\ & T & & \\ & & \ddots & \\ & & & T \end{bmatrix} \begin{bmatrix} \bar{\Phi}_1 \\ \bar{\Phi}_2 \\ \vdots \\ \bar{\Phi}_l \end{bmatrix}$$

or $\Phi = 2\bar{T}\bar{\Phi}$, since $T^{-1} = 2T$.

The system of equations $G\Phi = P$, with the group supermatrix in normal form derived earlier, expressed as

$$\begin{bmatrix} A & B \\ B & A \end{bmatrix} \begin{bmatrix} \Phi^{(1)} \\ \Phi^{(2)} \end{bmatrix} = \begin{bmatrix} P^{(1)} \\ P^{(2)} \end{bmatrix},$$

can be transformed into the system of equations $\bar{G}\bar{\Phi} = \bar{P}$, with the group supermatrix $\bar{G}$ in diagonal form, by the group supermatrix transformation

$$TGT^{-1}T\Phi = TP.$$

Thus

$$\frac{1}{2}\begin{bmatrix} E & E \\ E & -E \end{bmatrix}\begin{bmatrix} A & B \\ B & A \end{bmatrix}\begin{bmatrix} E & E \\ E & -E \end{bmatrix}\frac{1}{2}\begin{bmatrix} E & E \\ E & -E \end{bmatrix}\begin{bmatrix} \Phi^{(1)} \\ \Phi^{(2)} \end{bmatrix} = \frac{1}{2}\begin{bmatrix} E & E \\ E & -E \end{bmatrix}\begin{bmatrix} P^{(1)} \\ P^{(2)} \end{bmatrix}$$

results in

$$\begin{bmatrix} A+B & \\ & A-B \end{bmatrix}\begin{bmatrix} \bar{\Phi}^{(1)} \\ \bar{\Phi}^{(2)} \end{bmatrix} = \begin{bmatrix} \bar{P}^{(1)} \\ \bar{P}^{(2)} \end{bmatrix} \quad \text{or} \quad \bar{G}\bar{\Phi} = \bar{P},$$

with

$$\begin{bmatrix} \bar{\Phi}^{(1)} \\ \bar{\Phi}^{(2)} \end{bmatrix} = \frac{1}{2}\begin{bmatrix} \Phi^{(1)} + \Phi^{(2)} \\ \Phi^{(1)} - \Phi^{(2)} \end{bmatrix} \quad \text{and} \quad \begin{bmatrix} \bar{P}^{(1)} \\ \bar{P}^{(2)} \end{bmatrix} = \frac{1}{2}\begin{bmatrix} P^{(1)} + P^{(2)} \\ P^{(2)} - P^{(2)} \end{bmatrix}.$$

Conversely, the system of equations $\bar{G}\bar{\Phi} = \bar{P}$, with the group supermatrix in diagonal form, is transformed into the system of equations $G\Phi = P$ with the group supermatrix in normal form by the inverse group supermatrix transformation

$$T^{-1}\bar{G}TT^{-1}\bar{\Phi} = T^{-1}\bar{P}$$

or

$$\begin{bmatrix} E & E \\ E & -E \end{bmatrix} \begin{bmatrix} A+B & \\ & A-B \end{bmatrix} \frac{1}{2} \begin{bmatrix} E & E \\ E & -E \end{bmatrix} \begin{bmatrix} E & E \\ E & -E \end{bmatrix} \begin{bmatrix} \bar{\Phi}^{(1)} \\ \bar{\Phi}^{(2)} \end{bmatrix} = \begin{bmatrix} E & E \\ E & -E \end{bmatrix} \begin{bmatrix} \bar{P}^{(1)} \\ \bar{P}^{(2)} \end{bmatrix}$$

resulting in

$$\begin{bmatrix} A & B \\ B & A \end{bmatrix} \begin{bmatrix} \Phi^{(1)} \\ \Phi^{(2)} \end{bmatrix} = \begin{bmatrix} P^{(1)} \\ P^{(2)} \end{bmatrix}.$$

The sequence of symmetry operations E, C_2, as existing at the top of the character table of the group C_2, is used to derive the nodal numbering of the four-node line element. The nodes are grouped into two nodal sets $S_1(1, 2), S_2(3, 4)$, where the nodes in each set are permuted by symmetry operations of the group. The positions of the first nodes in these nodal sets, i.e. 1 and 3, are on the positive branch of the x axis, as given in Fig. 3.2, so that application of E, C_2 (in this order) to the nodal sequences [1 2], [3 4] provides the order 1, 2 and 3, 4 (in this sequence).

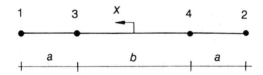

Fig. 3.2. The four-node line element with the nodal sequence 1, 3, 4, 2 described by the group C_2.

Expressed as permutations of the nodes, the results of actions of symmetry operations to the initial sequences [1 2] and [3 4] are

$$E = \begin{bmatrix} 1 & 2 & 3 & 4 \\ 1 & 2 & 3 & 4 \end{bmatrix}, \quad C_2 = \begin{bmatrix} 1 & 2 & 3 & 4 \\ 2 & 1 & 4 & 3 \end{bmatrix},$$

or presented in the form of a permutation table

node	E	C_2
1	1	2
2	2	1
3	3	4
4	4	3

Applying the idempotents π_1, π_2 to the sets of nodal functions $[\Phi^{(1)}, \Phi^{(2)}]$, $[\Phi^{(3)}, \Phi^{(4)}]$, pertaining to the nodal sets $S_1(1, 2)$, $S_2(3, 4)$, derives the sets of basis vectors of G-invariant subspaces U_i

$$U_1: \quad \bar{\Phi}^{(1)} = \pi_1 \Phi^{(1)} = \pi_1 \Phi^{(2)} = \tfrac{1}{2}(\Phi^{(1)} + \Phi^{(2)})$$
$$\bar{\Phi}^{(3)} = \pi_1 \Phi^{(3)} = \pi_1 \Phi^{(4)} = \tfrac{1}{2}(\Phi^{(3)} + \Phi^{(4)})$$
$$U_2: \quad \bar{\Phi}^{(2)} = \pi_2 \Phi^{(1)} = -\pi_2 \Phi^{(2)} = \tfrac{1}{2}(\Phi^{(1)} - \Phi^{(2)})$$
$$\bar{\Phi}^{(4)} = \pi_2 \Phi^{(3)} = -\pi_2 \Phi^{(4)} = \tfrac{1}{2}(\Phi^{(3)} - \Phi^{(4)}).$$

The relation of the sets of basis vectors $\bar{\Phi}$ to the sets of nodal functions Φ is

$$\begin{bmatrix} \bar{\Phi}^{(1)} \\ \bar{\Phi}^{(2)} \\ \bar{\Phi}^{(3)} \\ \bar{\Phi}^{(4)} \end{bmatrix} = \frac{1}{2} \begin{bmatrix} E & E & & \\ E & -E & & \\ & & E & E \\ & & E & -E \end{bmatrix} \begin{bmatrix} \Phi^{(1)} \\ \Phi^{(2)} \\ \Phi^{(3)} \\ \Phi^{(4)} \end{bmatrix}.$$

Conversely, the relation of the sets of nodal functions Φ to the sets of basis vectors $\bar{\Phi}$ is

$$\begin{bmatrix} \Phi^{(1)} \\ \Phi^{(2)} \\ \Phi^{(3)} \\ \Phi^{(4)} \end{bmatrix} = \begin{bmatrix} E & E & & \\ E & -E & & \\ & & E & E \\ & & E & -E \end{bmatrix} \begin{bmatrix} \bar{\Phi}^{(1)} \\ \bar{\Phi}^{(2)} \\ \bar{\Phi}^{(3)} \\ \bar{\Phi}^{(4)} \end{bmatrix}.$$

The transformation supermatrix

$$T = \frac{1}{2} \begin{bmatrix} E & E \\ E & -E \end{bmatrix}, \quad \text{with} \quad T^{-1} = 2T,$$

is applied twice, for each nodal set separately.

When in the case of the two-node line element, shown previously in Fig. 3.1, the node 1 is replaced by the nodal pair 1, 2 and the node 2 by the nodal pair 3, 4, the four-node line element in Fig. 3.3 is obtained. In this case the results of actions of symmetry operations on the initial sequence [1 2 3 4] are

$$E = \begin{bmatrix} 1 & 2 & 3 & 4 \\ 1 & 2 & 3 & 4 \end{bmatrix}, \quad C_2 = \begin{bmatrix} 1 & 2 & 3 & 4 \\ 4 & 3 & 2 & 1 \end{bmatrix}.$$

These permutations formulate the group supermatrix in normal form

$$G = \begin{bmatrix} A_1 & A_2 & A_3 & A_4 \\ B_1 & B_2 & B_3 & B_4 \\ B_4 & B_3 & B_2 & B_1 \\ A_4 & A_3 & A_2 & A_1 \end{bmatrix}.$$

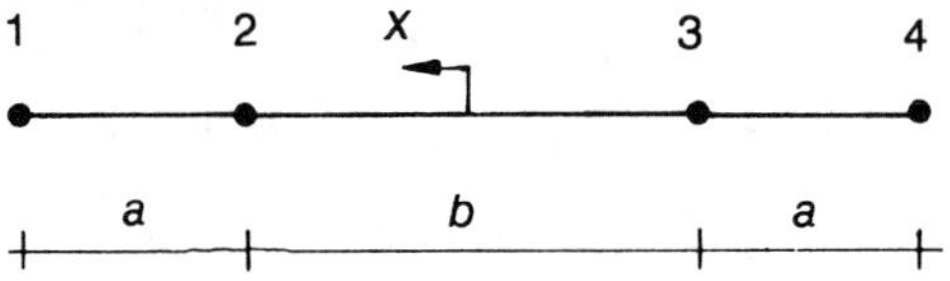

```
 1       2      X      3       4
 •———————•—————┐ ┌—————•———————•
              └─┘
    a              b              a
 ├───────┼───────────────┼───────┤
```

Fig. 3.3. The four-node line element with the nodal sequence 1, 2, 3, 4 described by the group C_2.

Applying the idempotents π_1, π_2 to the sets of nodal functions $\Phi^{(1)}, \Phi^{(2)}, \Phi^{(3)}, \Phi^{(4)}$ derives the sets of basis vectors $\bar{\Phi}^{(1)}, \bar{\Phi}^{(2)}, \bar{\Phi}^{(3)}, \bar{\Phi}^{(4)}$ of G-invariant subspaces U_1, U_2:

$$U_1: \quad \bar{\Phi}^{(1)} = \pi_1 \Phi^{(1)} = \pi_1 \Phi^{(4)} = \tfrac{1}{2}(\Phi^{(1)} + \Phi^{(4)})$$

$$\bar{\Phi}^{(2)} = \pi_1 \Phi^{(2)} = \pi_1 \Phi^{(3)} = \tfrac{1}{2}(\Phi^{(2)} + \Phi^{(3)})$$

$$U_2: \quad \bar{\Phi}^{(3)} = \pi_2 \Phi^{(2)} = -\pi_2 \Phi^{(3)} = \tfrac{1}{2}(\Phi^{(2)} - \Phi^{(3)})$$

$$\bar{\Phi}^{(4)} = \pi_2 \Phi^{(1)} = -\pi_2 \Phi^{(4)} = \tfrac{1}{2}(\Phi^{(1)} - \Phi^{(4)}),$$

or in supermatrix form

$$\begin{bmatrix} \bar{\Phi}^{(1)} \\ \bar{\Phi}^{(2)} \\ \bar{\Phi}^{(3)} \\ \bar{\Phi}^{(4)} \end{bmatrix} = \frac{1}{2} \begin{bmatrix} E & & & E \\ & E & E & \\ & E & -E & \\ E & & & -E \end{bmatrix} \begin{bmatrix} \Phi^{(1)} \\ \Phi^{(2)} \\ \Phi^{(3)} \\ \Phi^{(4)} \end{bmatrix} = \frac{1}{2} \begin{bmatrix} \Phi^{(1)} + \Phi^{(4)} \\ \Phi^{(2)} + \Phi^{(3)} \\ \Phi^{(2)} - \Phi^{(3)} \\ \Phi^{(1)} - \Phi^{(4)} \end{bmatrix},$$

or $T\Phi = \bar{\Phi}$, with $T^{-1} = 2T$.

The system of equations $G\Phi = P$, or

$$\begin{bmatrix} A_1 & A_2 & A_3 & A_4 \\ B_1 & B_2 & B_3 & B_4 \\ B_4 & B_3 & B_2 & B_1 \\ A_4 & A_3 & A_2 & A_1 \end{bmatrix} \begin{bmatrix} \Phi^{(1)} \\ \Phi^{(2)} \\ \Phi^{(3)} \\ \Phi^{(4)} \end{bmatrix} = \begin{bmatrix} P^{(1)} \\ P^{(2)} \\ P^{(3)} \\ P^{(4)} \end{bmatrix},$$

with the group supermatrix G in normal form given earlier, is transformed into the system of equations $\bar{G}\bar{\Phi} = \bar{P}$, with the group supermatrix in diagonal form, by the group supermatrix transformation $TGT^{-1}T\Phi = TP$:

$$\frac{1}{2} \begin{bmatrix} E & & & E \\ & E & E & \\ & E & -E & \\ E & & & -E \end{bmatrix} \begin{bmatrix} A_1 & A_2 & A_3 & A_4 \\ B_1 & B_2 & B_3 & B_4 \\ B_4 & B_3 & B_2 & B_1 \\ A_4 & A_3 & A_2 & A_1 \end{bmatrix} \begin{bmatrix} E & & & E \\ & E & E & \\ & E & -E & \\ E & & & -E \end{bmatrix}$$

Sec. 3.2] **Group supermatrices of the group C_2** 101

$$\frac{1}{2}\begin{bmatrix} E & & & E \\ & E & E & \\ & E & -E & \\ E & & & -E \end{bmatrix}\begin{bmatrix} \Phi^{(1)} \\ \Phi^{(2)} \\ \Phi^{(3)} \\ \Phi^{(4)} \end{bmatrix} = \frac{1}{2}\begin{bmatrix} E & & & E \\ & E & E & \\ & E & -E & \\ E & & & -E \end{bmatrix}\begin{bmatrix} P^{(1)} \\ P^{(2)} \\ P^{(3)} \\ P^{(4)} \end{bmatrix},$$

resulting in

$$\begin{bmatrix} A_1 + A_4 & A_2 + A_3 & & \\ B_1 + B_4 & B_2 + B_3 & & \\ & & B_2 - B_3 & B_1 - B_4 \\ & & A_2 - A_3 & A_1 - A_4 \end{bmatrix}\begin{bmatrix} \bar{\Phi}^{(1)} \\ \bar{\Phi}^{(2)} \\ \bar{\Phi}^{(3)} \\ \bar{\Phi}^{(4)} \end{bmatrix} = \begin{bmatrix} \bar{P}^{(1)} \\ \bar{P}^{(2)} \\ \bar{P}^{(3)} \\ \bar{P}^{(4)} \end{bmatrix}$$

or $\bar{G}\bar{\Phi} = \bar{P}$, with the relation

$$\begin{bmatrix} \Phi^{(1)} \\ \Phi^{(2)} \\ \Phi^{(3)} \\ \Phi^{(4)} \end{bmatrix} = \begin{bmatrix} E & & & E \\ & E & E & \\ & E & -E & \\ E & & & -E \end{bmatrix}\begin{bmatrix} \bar{\Phi}^{(1)} \\ \bar{\Phi}^{(2)} \\ \bar{\Phi}^{(3)} \\ \bar{\Phi}^{(4)} \end{bmatrix}.$$

The system of equations $\bar{G}\bar{\Phi} = \bar{P}$ is transformed into the system of equations $G\Phi = P$, with the group supermatrix G in normal form, by the inverse group supermatrix transformation $T^{-1}\bar{G}TT^{-1}\bar{\Phi} = T^{-1}\bar{P}$

$$\begin{bmatrix} E & & & E \\ & E & E & \\ & E & -E & \\ E & & & -E \end{bmatrix}\begin{bmatrix} A_1 + A_4 & A_2 + A_3 & & \\ B_1 + B_4 & B_2 + B_3 & & \\ & & B_2 - B_3 & B_1 - B_4 \\ & & A_2 - A_3 & A_1 - A_4 \end{bmatrix}$$

$$\frac{1}{2}\begin{bmatrix} E & & & E \\ & E & E & \\ & E & -E & \\ E & & & -E \end{bmatrix}\begin{bmatrix} E & & & E \\ & E & E & \\ & E & -E & \\ E & & & -E \end{bmatrix}\frac{1}{2}\begin{bmatrix} \Phi^{(1)} + \Phi^{(4)} \\ \Phi^{(2)} + \Phi^{(3)} \\ \Phi^{(2)} - \Phi^{(3)} \\ \Phi^{(1)} - \Phi^{(4)} \end{bmatrix}$$

$$= \begin{bmatrix} E & & & E \\ & E & E & \\ & E & -E & \\ E & & & -E \end{bmatrix}\frac{1}{2}\begin{bmatrix} P^{(1)} + P^{(4)} \\ P^{(2)} + P^{(3)} \\ P^{(2)} - P^{(3)} \\ P^{(1)} - P^{(4)} \end{bmatrix}$$

resulting in

$$\begin{bmatrix} A_1 & A_2 & A_3 & A_4 \\ B_1 & B_2 & B_3 & B_4 \\ B_4 & B_3 & B_2 & B_1 \\ A_4 & A_3 & A_2 & A_1 \end{bmatrix} \begin{bmatrix} \Phi^{(1)} \\ \Phi^{(2)} \\ \Phi^{(3)} \\ \Phi^{(4)} \end{bmatrix} = \begin{bmatrix} P^{(1)} \\ P^{(2)} \\ P^{(3)} \\ P^{(4)} \end{bmatrix}.$$

When in the case of the four-node element, shown previously in Fig. 3.3, the distance $b = 0$ and the nodes 2 and 3 become one node designated by 2, while the node 4 is numbered by 3, the three-node line element in Fig. 3.4 is obtained. In this way the results of actions of symmetry operations to the initial sequence [1 2 3] are

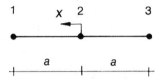

Fig. 3.4. The three-node line element described by the group C_2.

$$E = \begin{bmatrix} 1 & 2 & 3 \\ 1 & 2 & 3 \end{bmatrix}, \quad C_2 = \begin{bmatrix} 1 & 2 & 3 \\ 3 & 2 & 1 \end{bmatrix}.$$

These permutations formulate the group supermatrix in normal form:

$$G = \begin{bmatrix} A & B & C \\ D & F & D \\ C & B & A \end{bmatrix}.$$

Applying the idempotents π_1, π_2 to the sets of nodal functions $\Phi^{(1)}$, $\Phi^{(2)}$, $\Phi^{(3)}$ derives the sets of basis vectors $\bar{\Phi}^{(1)}$, $\bar{\Phi}^{(2)}$, $\bar{\Phi}^{(3)}$ of G-invariant subspaces U_1, U_2:

$$U_1: \quad \bar{\Phi}^{(1)} = \pi_1 \Phi^{(1)} = \pi_1 \Phi^{(3)} = \tfrac{1}{2}(\Phi^{(1)} + \Phi^{(3)})$$

$$\bar{\Phi}^{(2)} = \pi_1 \Phi^{(2)} = \Phi^{(2)}$$

$$U_2: \quad \bar{\Phi}^{(3)} = \pi_2 \Phi^{(1)} = -\pi_2 \Phi^{(3)} = \tfrac{1}{2}(\Phi^{(1)} - \Phi^{(3)}), \quad \pi_2 \Phi^{(2)} = 0,$$

or in supermatrix form

$$\begin{bmatrix} \bar{\Phi}^{(1)} \\ \bar{\Phi}^{(2)} \\ \bar{\Phi}^{(3)} \end{bmatrix} = \frac{1}{2} \begin{bmatrix} E & & E \\ & 2E & \\ E & & -E \end{bmatrix} \begin{bmatrix} \Phi^{(1)} \\ \Phi^{(2)} \\ \Phi^{(3)} \end{bmatrix} = \frac{1}{2} \begin{bmatrix} \Phi^{(1)} + \Phi^{(3)} \\ 2\Phi^{(2)} \\ \Phi^{(1)} - \Phi^{(3)} \end{bmatrix} \quad \text{or} \quad \bar{\Phi} = T\Phi,$$

with

$$T = \frac{1}{2}\begin{bmatrix} E & & E \\ & 2E & \\ E & & -E \end{bmatrix}, \quad T^{-1} = \begin{bmatrix} E & & E \\ & E & \\ E & & -E \end{bmatrix}.$$

The system of equations $G\Phi = P$, i.e.

$$\begin{bmatrix} A & B & C \\ D & F & D \\ C & B & A \end{bmatrix} \begin{bmatrix} \Phi^{(1)} \\ \Phi^{(2)} \\ \Phi^{(3)} \end{bmatrix} = \begin{bmatrix} P^{(1)} \\ P^{(2)} \\ P^{(3)} \end{bmatrix},$$

with the group supermatrix G in normal form given earlier, is transformed into the system of equations $\bar{G}\bar{\Phi} = \bar{P}$, with the group supermatrix $\bar{G}$ in diagonal form, by the group supermatrix transformation $TGT^{-1}T\Phi = TP$:

$$\frac{1}{2}\begin{bmatrix} E & & E \\ & 2E & \\ E & & -E \end{bmatrix} \begin{bmatrix} A & B & C \\ D & F & D \\ C & B & A \end{bmatrix} \begin{bmatrix} E & & E \\ & E & \\ E & & -E \end{bmatrix} \frac{1}{2}\begin{bmatrix} E & & E \\ & 2E & \\ E & & -E \end{bmatrix} \begin{bmatrix} \Phi^{(1)} \\ \Phi^{(2)} \\ \Phi^{(3)} \end{bmatrix}$$

$$= \frac{1}{2}\begin{bmatrix} E & & E \\ & 2E & \\ E & & -E \end{bmatrix} \begin{bmatrix} P^{(1)} \\ P^{(2)} \\ P^{(3)} \end{bmatrix},$$

resulting in

$$\begin{bmatrix} A+C & B & \\ 2D & F & \\ & & A-C \end{bmatrix} \begin{bmatrix} \bar{\Phi}^{(1)} \\ \bar{\Phi}^{(2)} \\ \bar{\Phi}^{(3)} \end{bmatrix} = \begin{bmatrix} \bar{P}^{(1)} \\ \bar{P}^{(2)} \\ \bar{P}^{(3)} \end{bmatrix}$$

or $\bar{G}\bar{\Phi} = \bar{P}$, with the relation

$$\begin{bmatrix} \Phi^{(1)} \\ \Phi^{(2)} \\ \Phi^{(3)} \end{bmatrix} = \begin{bmatrix} E & & E \\ & E & \\ E & & -E \end{bmatrix} \begin{bmatrix} \bar{\Phi}^{(1)} \\ \bar{\Phi}^{(2)} \\ \bar{\Phi}^{(3)} \end{bmatrix}.$$

The system of equations $\bar{G}\bar{\Phi} = \bar{P}$ is transformed back into the system of equations $G\Phi = P$, with the group supermatrix in normal form, by the inverse group supermatrix transformation $T^{-1}\bar{G}TT^{-1}\bar{\Phi} = T^{-1}\bar{P}$

$$\begin{bmatrix} E & & E \\ & E & \\ E & & -E \end{bmatrix} \begin{bmatrix} A+C & B & \\ 2D & F & \\ & & A-C \end{bmatrix} \frac{1}{2}\begin{bmatrix} E & & E \\ & 2E & \\ E & & -E \end{bmatrix} \begin{bmatrix} E & & E \\ & E & \\ E & & -E \end{bmatrix}$$

$$\frac{1}{2}\begin{bmatrix}\Phi^{(1)}+\Phi^{(3)}\\ 2\Phi^{(2)}\\ \Phi^{(1)}-\Phi^{(3)}\end{bmatrix}=\begin{bmatrix}E & & E\\ & E & \\ E & & -E\end{bmatrix}\frac{1}{2}\begin{bmatrix}P^{(1)}+P^{(3)}\\ 2P^{(2)}\\ P^{(1)}-P^{(3)}\end{bmatrix}$$

resulting in

$$\begin{bmatrix}A & B & C\\ D & F & D\\ C & B & A\end{bmatrix}\begin{bmatrix}\Phi^{(1)}\\ \Phi^{(2)}\\ \Phi^{(3)}\end{bmatrix}=\begin{bmatrix}P^{(1)}\\ P^{(2)}\\ P^{(3)}\end{bmatrix}.$$

When in an analysis both the group supermatrix G in normal form and the group supermatrix $\bar{G}$ in diagonal form are used, i.e.

$$G=\begin{bmatrix}A & B\\ B & A\end{bmatrix}\quad\text{and}\quad\bar{G}=\begin{bmatrix}A+B & \\ & A-B\end{bmatrix},$$

it is essential that columns of A and B correspond to respective nodes and variables which are permuted by the symmetry operations E, C_2 of the group C_2.

In the case of a linear element with n nodes, n being an even number, the above requirement is fulfilled by the nodal numbering derived by

$$C_2(j)=\frac{n}{2}+j,$$

with

$$j=1,2,\ldots,\frac{n}{2}.$$

This nodal numbering is developed in section 6.2 for the stiffness group supermatrix in normal form, with the objective that the columns $1,2,\ldots,n-1,n$ correspond to the nodes $1,2,\ldots,n-1,n$ respectively.

3.3 GROUP SUPERMATRICES OF THE GROUP C_3

The character table of the group C_3

C_3	E	$2C_3$
A	1	1
B	2	-1

has the sequence of symmetry operations E, C_3, C_3^{-1} that is used to define the nodal numbering of the three-node equilateral triangular element. The position of the first node, numbered by 1, is at the left corner of the equilateral triangle shown in Fig. 3.5. As previously, it is assumed that application of E, C_3, C_3^{-1} (in this order) on the initial node 1 provides the nodes 1, 2, 3 (in this sequence).

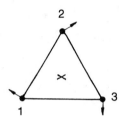

Fig. 3.5. The three-node equilateral triangular element described by the group C_3.

Expressed as permutations of the nodes, the results of actions of symmetry operations on the nodal sequence [1 2 3] are

$$E = \begin{bmatrix} 1 & 2 & 3 \\ 1 & 2 & 3 \end{bmatrix}, \quad C_3 = \begin{bmatrix} 1 & 2 & 3 \\ 2 & 3 & 1 \end{bmatrix}, \quad C_3^{-1} = \begin{bmatrix} 1 & 2 & 3 \\ 3 & 1 & 2 \end{bmatrix},$$

or, presented in the form of a permutation table,

node	E	C_3	C_3^{-1}
1	1	2	3
2	2	3	1
3	3	1	2

By replacing the numbers 1, 2, 3 by matrices A, B, C in the above permutation table, one obtains the supermatrix

$$G = \begin{bmatrix} A & B & C \\ B & C & A \\ C & A & B \end{bmatrix}.$$

which is the group supermatrix in normal form pertaining to the group C_3.

The idempotents π_i ($i = 1, 2$) of the centre of the group algebra are determined by

$$\pi_i = \frac{h_i}{h} \sum_\sigma \chi_i(\sigma^{-1})\sigma,$$

with explanations concerning this expression given in section 3.2, so that

$$\pi_1 = \tfrac{1}{3}(E + C_3 + C_3^{-1}), \qquad \pi_2 = \tfrac{2}{3}(2E - C_3 - C_3^{-1}).$$

Applying the idempotents π_1, π_2 to the sets of nodal functions $\Phi^{(1)}, \Phi^{(2)}, \Phi^{(3)}$ derives the sets of basis vectors $\bar{\Phi}^{(1)}, \bar{\Phi}^{(2)}, \bar{\Phi}^{(3)}$ of G-invariant subspaces U_1, U_2:

$$U_1: \quad \bar{\Phi}^{(1)} = \pi_1 \Phi^{(1)} = \pi_1 \Phi^{(2)} = \pi_1 \Phi^{(3)} = \tfrac{1}{3}(\Phi^{(1)} + \Phi^{(2)} + \Phi^{(3)})$$

$$U_2: \quad \bar{\Phi}^{(2)} = \pi_2 \Phi^{(1)} = \tfrac{2}{3}(2\Phi^{(1)} - \Phi^{(2)} - \Phi^{(3)})$$

$$\bar{\Phi}^{(3)} = \pi_2 \Phi^{(2)} = \tfrac{2}{3}(-\Phi^{(1)} + 2\Phi^{(2)} - \Phi^{(3)})$$

$$[\bar{\Phi}^{(4)} = \pi_2 \Phi^{(3)} = \tfrac{2}{3}(-\Phi^{(1)} - \Phi^{(2)} + 2\Phi^{(3)}) \text{ is linearly dependent since}$$
$$\bar{\Phi}^{(2)}, \bar{\Phi}^{(3)} \text{ span the subspace } U_2].$$

In supermatrix form

$$\begin{bmatrix} \bar{\Phi}^{(1)} \\ \bar{\Phi}^{(2)} \\ \bar{\Phi}^{(3)} \end{bmatrix} = \frac{1}{3} \begin{bmatrix} E & E & E \\ 4E & -2E & -2E \\ -2E & 4E & -2E \end{bmatrix} \begin{bmatrix} \Phi^{(1)} \\ \Phi^{(2)} \\ \Phi^{(3)} \end{bmatrix} \quad \text{or} \quad \bar{\Phi} = T\Phi,$$

with

$$T^{-1} = \frac{1}{2} \begin{bmatrix} 2E & E & \\ 2E & & E \\ 2E & -E & -E \end{bmatrix}.$$

The system of equations $G\Phi = P$, i.e.

$$\begin{bmatrix} A & B & C \\ B & C & A \\ C & A & B \end{bmatrix} \begin{bmatrix} \Phi^{(1)} \\ \Phi^{(2)} \\ \Phi^{(3)} \end{bmatrix} = \begin{bmatrix} P^{(1)} \\ P^{(2)} \\ P^{(3)} \end{bmatrix},$$

with the group supermatrix G in normal form given earlier, is transformed into the system of equations $\bar{G}\bar{\Phi} = \bar{P}$, with the group supermatrix $\bar{G}$ in diagonal form, by the group supermatrix transformation $TGT^{-1}T\Phi = TP$:

$$\frac{1}{3} \begin{bmatrix} E & E & E \\ 4E & -2E & -2E \\ -2E & 4E & -2E \end{bmatrix} \begin{bmatrix} A & B & C \\ B & C & A \\ C & A & B \end{bmatrix} \frac{1}{2} \begin{bmatrix} 2E & E & \\ 2E & & E \\ 2E & -E & -E \end{bmatrix}$$

$$\frac{1}{3} \begin{bmatrix} E & E & E \\ 4E & -2E & -2E \\ -2E & 4E & -2E \end{bmatrix} \begin{bmatrix} \Phi^{(1)} \\ \Phi^{(2)} \\ \Phi^{(3)} \end{bmatrix} = \frac{1}{3} \begin{bmatrix} E & E & E \\ 4E & -2E & -2E \\ -2E & 4E & -2E \end{bmatrix} \begin{bmatrix} P^{(1)} \\ P^{(2)} \\ P^{(3)} \end{bmatrix},$$

resulting in

$$\begin{bmatrix} A+B+C & & \\ & A-C & B-C \\ & -A+B & -A+C \end{bmatrix} \begin{bmatrix} \bar{\Phi}^{(1)} \\ \bar{\Phi}^{(2)} \\ \bar{\Phi}^{(3)} \end{bmatrix} = \begin{bmatrix} \bar{P}^{(1)} \\ \bar{P}^{(2)} \\ \bar{P}^{(3)} \end{bmatrix},$$

with the relation

$$\begin{bmatrix} \Phi^{(1)} \\ \Phi^{(2)} \\ \Phi^{(3)} \end{bmatrix} = \frac{1}{2} \begin{bmatrix} 2E & E & \\ 2E & & E \\ 2E & -E & -E \end{bmatrix} \begin{bmatrix} \bar{\Phi}^{(1)} \\ \bar{\Phi}^{(2)} \\ \bar{\Phi}^{(3)} \end{bmatrix} \quad \text{or} \quad \Phi = T^{-1}\bar{\Phi}.$$

The system of equations $\bar{G}\bar{\Phi} = \bar{P}$ is transformed into the system of equations $G\Phi = P$, with the group supermatrix G in normal form, by the inverse group supermatrix transformation $T^{-1}\bar{G}TT^{-1}\bar{\Phi} = T^{-1}\bar{P}$

$$\frac{1}{2}\begin{bmatrix} 2E & E & \\ 2E & & E \\ 2E & -E & -E \end{bmatrix} \begin{bmatrix} A+B+C & & \\ & A-C & B-C \\ & -A+B & -A+C \end{bmatrix} \frac{1}{3}\begin{bmatrix} E & E & E \\ 4E & -2E & -2E \\ -2E & 4E & -2E \end{bmatrix}$$

$$\frac{1}{2}\begin{bmatrix} 2E & E & \\ 2E & & E \\ 2E & -E & -E \end{bmatrix} \frac{1}{3}\begin{bmatrix} \Phi^{(1)} + \Phi^{(2)} + \Phi^{(3)} \\ 4\Phi^{(1)} - 2\Phi^{(2)} - 2\Phi^{(3)} \\ -2\Phi^{(1)} + 4\Phi^{(2)} - 2\Phi^{(3)} \end{bmatrix} = \frac{1}{2}\begin{bmatrix} 2E & E & \\ 2E & & E \\ 2E & -E & -E \end{bmatrix}$$

$$\frac{1}{3}\begin{bmatrix} P^{(1)} + P^{(2)} + P^{(3)} \\ 4P^{(1)} - 2P^{(2)} - 2P^{(3)} \\ -2P^{(1)} + 4P^{(2)} - 2P^{(3)} \end{bmatrix}$$

resulting in

$$\begin{bmatrix} A & B & C \\ B & C & A \\ C & A & B \end{bmatrix} \begin{bmatrix} \Phi^{(1)} \\ \Phi^{(2)} \\ \Phi^{(3)} \end{bmatrix} = \begin{bmatrix} P^{(1)} \\ P^{(2)} \\ P^{(3)} \end{bmatrix}.$$

3.4 GROUP SUPERMATRICES OF THE GROUP C_4

The character table of the group C_4

C_4	E	$2C_4$	C_2
A	1	1	1
B	1	−1	1
E	2	0	−2

has the sequence of symmetry operations E, C_4, C_4^{-1}, C_2 which is applied for definition of the nodal numbering of the four-node square element. The position of the first node, numbered by 1, is at the lower left corner of the square shown in Fig. 3.6. As in preceding cases, it is assumed that application of E, C_4, C_4^{-1}, C_2 (in this order) on the initial node 1 provides the nodes 1, 2, 3, 4 (in this order).

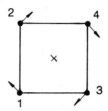

Fig. 3.6. The four-node square element described by the group C_4.

Expressed as permutations of the nodes, the results of actions of symmetry operations on the nodal sequence [1 2 3 4] are

$$E = \begin{bmatrix} 1 & 2 & 3 & 4 \\ 1 & 2 & 3 & 4 \end{bmatrix}, \quad C_4 = \begin{bmatrix} 1 & 2 & 3 & 4 \\ 2 & 4 & 1 & 3 \end{bmatrix}, \quad C_4^{-1} = \begin{bmatrix} 1 & 2 & 3 & 4 \\ 3 & 1 & 4 & 2 \end{bmatrix},$$

$$C_2 = \begin{bmatrix} 1 & 2 & 3 & 4 \\ 4 & 3 & 2 & 1 \end{bmatrix},$$

or, presented in the form of a permutation table,

Node	E	C_4	C_4^{-1}	C_2
1	1	2	3	4
2	2	4	1	3
3	3	1	4	2
4	4	3	2	1

By replacing the numbers 1, 2, 3, 4 by matrices A, B, C, D in the above permutation table one obtains the supermatrix

$$G = \begin{bmatrix} A & B & C & D \\ B & D & A & C \\ C & A & D & B \\ D & C & B & A \end{bmatrix},$$

which is the group supermatrix in normal form pertaining to the group C_4.

The idempotents π_i ($i = 1, 2, 3$) of the centre of the group algebra are determined by

$$\pi_i = \frac{h_i}{h} \sum_\sigma \chi_i(\sigma^{-1})\sigma,$$

with explanations concerning this expression given in section 3.2, so that

$$\pi_1 = \tfrac{1}{4}(E + C_4 + C_4^{-1} + C_2)$$
$$\pi_2 = \tfrac{1}{4}(E - C_4 - C_4^{-1} + C_2)$$
$$\pi_3 = \tfrac{2}{4}(2E - 2C_2) = E - C_2.$$

Applying the idempotents π_1, π_2, π_3 to the sets of nodal functions $\Phi^{(1)}, \Phi^{(2)}, \Phi^{(3)}, \Phi^{(4)}$ derives the sets of basis vectors $\bar{\Phi}^{(1)}, \bar{\Phi}^{(2)}, \bar{\Phi}^{(3)}, \bar{\Phi}^{(4)}$ of G-invariant subspaces U_1, U_2, U_3:

$U_1:$ $\quad \bar{\Phi}^{(1)} = \pi_1 \Phi^{(1)} = \pi_1 \Phi^{(2)} = \pi_1 \Phi^{(3)} = \pi_1 \Phi^{(4)}$
$\qquad\qquad = \tfrac{1}{4}(\Phi^{(1)} + \Phi^{(2)} + \Phi^{(3)} + \Phi^{(4)})$

$U_2:$ $\quad \bar{\Phi}^{(2)} = \pi_2 \Phi^{(1)} = -\pi_2 \Phi^{(2)} = -\pi_2 \Phi^{(3)} = \pi_2 \Phi^{(4)}$
$\qquad\qquad = \tfrac{1}{4}(\Phi^{(1)} - \Phi^{(2)} - \Phi^{(3)} + \Phi^{(4)})$

$U_3:$ $\quad \bar{\Phi}^{(3)} = \pi_3 \Phi^{(1)} = -\pi_3 \Phi^{(4)} = \Phi^{(1)} - \Phi^{(4)}$
$\qquad\quad \bar{\Phi}^{(4)} = \pi_3 \Phi^{(2)} = -\pi_3 \Phi^{(3)} = \Phi^{(2)} - \Phi^{(3)}.$

In supermatrix form

$$\begin{bmatrix} \bar{\Phi}^{(1)} \\ \bar{\Phi}^{(2)} \\ \bar{\Phi}^{(3)} \\ \bar{\Phi}^{(4)} \end{bmatrix} = \frac{1}{4} \begin{bmatrix} E & E & E & E \\ E & -E & -E & E \\ 4E & & & -4E \\ & 4E & -4E & \end{bmatrix} \begin{bmatrix} \Phi^{(1)} \\ \Phi^{(2)} \\ \Phi^{(3)} \\ \Phi^{(4)} \end{bmatrix} \quad \text{or} \quad \bar{\Phi} = T\Phi,$$

with

$$T^{-1} = \frac{1}{2} \begin{bmatrix} 2E & 2E & E & \\ 2E & -2E & & E \\ 2E & -2E & & -E \\ 2E & 2E & -E & \end{bmatrix}.$$

The system of equations $G\Phi = P$

$$\begin{bmatrix} A & B & C & D \\ B & D & A & C \\ C & A & D & B \\ D & C & B & A \end{bmatrix} \begin{bmatrix} \Phi^{(1)} \\ \Phi^{(2)} \\ \Phi^{(3)} \\ \Phi^{(4)} \end{bmatrix} = \begin{bmatrix} P^{(1)} \\ P^{(2)} \\ P^{(3)} \\ P^{(4)} \end{bmatrix},$$

with the group supermatrix G in normal form given earlier, is transformed into the system of equations $\bar{G}\Phi = \bar{P}$, with the group supermatrix $\bar{G}$ in diagonal form, by the group supermatrix transformation $TGT^{-1}T\Phi = TP$:

$$\frac{1}{4}\begin{bmatrix} E & E & E & E \\ E & -E & -E & E \\ 4E & & & -4E \\ & 4E & -4E & \end{bmatrix} \begin{bmatrix} A & B & C & D \\ B & D & A & C \\ C & A & D & B \\ D & C & B & A \end{bmatrix} \frac{1}{2}\begin{bmatrix} 2E & 2E & & E \\ 2E & -2E & & -E \\ 2E & -2E & & -E \\ 2E & 2E & -E & \end{bmatrix}.$$

$$\cdot \frac{1}{4}\begin{bmatrix} E & E & E & E \\ E & -E & -E & E \\ 4E & & & -4E \\ & 4E & -4E & \end{bmatrix} \begin{bmatrix} \Phi^{(1)} \\ \Phi^{(2)} \\ \Phi^{(3)} \\ \Phi^{(4)} \end{bmatrix}$$

$$= \frac{1}{4}\begin{bmatrix} E & E & E & E \\ E & -E & -E & E \\ 4E & & & -4E \\ & 4E & -4E & \end{bmatrix} \begin{bmatrix} P^{(1)} \\ P^{(2)} \\ P^{(3)} \\ P^{(4)} \end{bmatrix}$$

resulting in

$$\begin{bmatrix} A+B+C+D & & & \\ & A-B-C+D & & \\ & & A-D & B-C \\ & & B-C & -A+D \end{bmatrix} \begin{bmatrix} \bar{\Phi}^{(1)} \\ \bar{\Phi}^{(2)} \\ \bar{\Phi}^{(3)} \\ \bar{\Phi}^{(4)} \end{bmatrix} = \begin{bmatrix} \bar{P}^{(1)} \\ \bar{P}^{(2)} \\ \bar{P}^{(3)} \\ \bar{P}^{(4)} \end{bmatrix},$$

with the relation

$$\begin{bmatrix} \Phi^{(1)} \\ \Phi^{(2)} \\ \Phi^{(3)} \\ \Phi^{(4)} \end{bmatrix} = \frac{1}{2}\begin{bmatrix} 2E & 2E & & E \\ 2E & -2E & & E \\ 2E & -2E & & -E \\ 2E & 2E & -E & \end{bmatrix} \begin{bmatrix} \bar{\Phi}^{(1)} \\ \bar{\Phi}^{(2)} \\ \bar{\Phi}^{(3)} \\ \bar{\Phi}^{(4)} \end{bmatrix}.$$

The system of equations $\bar{G}\bar{\Phi} = \bar{P}$ is transformed back into the system of equations $G\Phi = P$, with the group supermatrix G in normal form, by the inverse group supermatrix transformation $T^{-1}\bar{G}TT^{-1}\bar{\Phi} = T^{-1}\bar{P}$:

$$\frac{1}{2}\begin{bmatrix} 2E & 2E & E & \\ 2E & -2E & & E \\ 2E & -2E & & -E \\ 2E & 2E & -E & \end{bmatrix} \begin{bmatrix} A+B+C+D & & & \\ & A-B-C+D & & \\ & & A-D & B-C \\ & & B-C & -A+D \end{bmatrix}$$

$$\cdot \frac{1}{4}\begin{bmatrix} E & E & E & E \\ E & -E & -E & E \\ 4E & & -4E & \\ & 4E & -4E & \end{bmatrix} \frac{1}{2}\begin{bmatrix} 2E & 2E & E & \\ 2E & -2E & & E \\ 2E & -2E & & -E \\ 2E & 2E & -E & \end{bmatrix}$$

$$\cdot \frac{1}{4}\begin{bmatrix} \Phi^{(1)} + \Phi^{(2)} + \Phi^{(3)} + \Phi^{(4)} \\ \Phi^{(1)} - \Phi^{(2)} - \Phi^{(3)} + \Phi^{(4)} \\ 4\Phi^{(1)} \qquad\qquad -4\Phi^{(4)} \\ 4\Phi^{(2)} - 4\Phi^{(3)} \end{bmatrix}$$

$$= \frac{1}{2}\begin{bmatrix} 2E & 2E & E & \\ 2E & -2E & & E \\ 2E & -2E & & -E \\ 2E & 2E & -E & \end{bmatrix} \frac{1}{4}\begin{bmatrix} P^{(1)} + P^{(2)} + P^{(3)} + P^{(4)} \\ P^{(1)} - P^{(2)} - P^{(3)} + P^{(4)} \\ 4P^{(1)} \qquad\qquad -4P^{(4)} \\ 4P^{(2)} - 4P^{(3)} \end{bmatrix}$$

resulting in

$$\begin{bmatrix} A & B & C & D \\ B & D & A & C \\ C & A & D & B \\ D & C & B & A \end{bmatrix} \begin{bmatrix} \Phi^{(1)} \\ \Phi^{(2)} \\ \Phi^{(3)} \\ \Phi^{(4)} \end{bmatrix} = \begin{bmatrix} P^{(1)} \\ P^{(2)} \\ P^{(3)} \\ P^{(4)} \end{bmatrix}.$$

3.5 GROUP SUPERMATRICES OF THE GROUP C_{2v}

The character table of the group C_{2v},

C_{2v}	E	C_2	σ_1 (xz)	σ_2 (yz)
A_1	1	1	1	1
A_2	1	1	−1	−1
B_1	1	−1	1	−1
B_2	1	−1	−1	1

has the sequence of symmetry operations $E, C_2, \sigma_1, \sigma_2$ that will be used for derivation of the nodal numbering of the four-node rectangular element.

The position of the first node, numbered by 1, is in the first quadrant of the xy coordinate system, where branches of both x and y axis are positive, as given in Fig. 3.7.

According to the rule, explained in section 3.1, it is assumed that application of E (identity), C_2 (rotation through 180° about the z axis), σ_1 and σ_2 (reflections in xz and yz planes), in this order, to the initial node 1 provides the nodes 1, 2, 3, 4 (in this sequence). This definition determines the unique nodal numbering in Fig. 3.7, which is stated by the following table:

Node	x axis	y axis
1	+	+
2	−	−
3	+	−
4	−	+

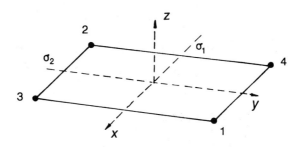

Fig. 3.7. The four-node rectangular element described by the group C_{2v}.

Expressed as permutations of the nodes, the results of actions of symmetry operations on the initial sequence [1 2 3 4] are

$$E = \begin{bmatrix} 1 & 2 & 3 & 4 \\ 1 & 2 & 3 & 4 \end{bmatrix}, \quad C_2 = \begin{bmatrix} 1 & 2 & 3 & 4 \\ 2 & 1 & 4 & 3 \end{bmatrix}, \quad \sigma_1 = \begin{bmatrix} 1 & 2 & 3 & 4 \\ 3 & 4 & 1 & 2 \end{bmatrix},$$

$$\sigma_2 = \begin{bmatrix} 1 & 2 & 3 & 4 \\ 4 & 3 & 2 & 1 \end{bmatrix},$$

or presented in form of a permutation table

Group supermatrices of the group C_{2v}

Node	E	C_2	σ_1 (xz)	σ_2 (yz)
1	1	2	3	4
2	2	1	4	3
3	3	4	1	2
4	4	3	2	1

By replacing the numbers 1, 2, 3, 4 by matrices A, B, C, D in the above permutation table, one obtains the supermatrix

$$G = \begin{bmatrix} A & B & C & D \\ B & A & D & C \\ C & D & A & B \\ D & C & B & A \end{bmatrix},$$

which is the group supermatrix in normal form pertaining to the group C_{2v}

The supermatrix G is symmetrical with respect to both diagonals and it can be also expressed as

$$G = \begin{bmatrix} F & H \\ H & F \end{bmatrix}, \quad \text{with} \quad F = \begin{bmatrix} A & B \\ B & A \end{bmatrix} \quad H = \begin{bmatrix} C & D \\ D & C \end{bmatrix}.$$

The idempotents π_i ($i = 1, 2, 3, 4$) of the centre of the group algebra are determined by

$$\pi_i = \frac{h_i}{h} \sum_\sigma \chi_i(\sigma^{-1})\sigma,$$

with $h_i = \chi_i(E)$, h the order, χ_i the ith character, σ the element and σ^{-1} the inverse of the element of the group, so that

$$\begin{bmatrix} \pi_1 \\ \pi_2 \\ \pi_3 \\ \pi_4 \end{bmatrix} = \frac{1}{4} \begin{bmatrix} 1 & 1 & 1 & 1 \\ 1 & 1 & -1 & -1 \\ 1 & -1 & 1 & -1 \\ 1 & -1 & -1 & 1 \end{bmatrix} \begin{bmatrix} E \\ C_2 \\ \sigma_1 \\ \sigma_2 \end{bmatrix} \quad \text{or} \quad \Pi = T_c \Sigma$$

with the transformation matrix

$$T_c = \frac{1}{4}\begin{bmatrix} 1 & 1 & 1 & 1 \\ 1 & 1 & -1 & -1 \\ 1 & -1 & 1 & -1 \\ 1 & -1 & -1 & 1 \end{bmatrix} \quad \text{and} \quad T_c^{-1} = \begin{bmatrix} 1 & 1 & 1 & 1 \\ 1 & 1 & -1 & -1 \\ 1 & -1 & 1 & -1 \\ 1 & -1 & -1 & 1 \end{bmatrix},$$

where $T_c^{-1} = 4T_c$.

When in the matrix T_c the unit values are replaced by the $l \times l$ unit matrix E, the transformation supermatrix T_c is obtained:

$$T_c = \frac{1}{4}\begin{bmatrix} E & E & E & E \\ E & E & -E & -E \\ E & -E & E & -E \\ E & -E & -E & E \end{bmatrix} \quad \text{and} \quad T_c^{-1} = \begin{bmatrix} E & E & E & E \\ E & E & -E & -E \\ E & -E & E & -E \\ E & -E & -E & E \end{bmatrix},$$

with $T_c^{-1} = 4T_c$.

The sets of nodal functions $\Phi^{(1)}$, $\Phi^{(2)}$, $\Phi^{(3)}$, $\Phi^{(4)}$, pertaining to the nodes 1, 2, 3, 4, may contain one or more functions at a node. Thus, if l is the number of nodal functions at each node, the following numbering of nodal functions φ_j ($j = 1, 2, \ldots, n$, $n = 4l$) is introduced:

for $l = 1$: $\begin{bmatrix} \Phi^{(1)} \\ \Phi^{(2)} \\ \Phi^{(3)} \\ \Phi^{(4)} \end{bmatrix} = \begin{bmatrix} \varphi_1 \\ \varphi_2 \\ \varphi_3 \\ \varphi_4 \end{bmatrix}$

for $l = 2$: $\begin{bmatrix} \Phi^{(1)} \\ \Phi^{(2)} \\ \Phi^{(3)} \\ \Phi^{(4)} \end{bmatrix} = \begin{bmatrix} \varphi_1 & \varphi_5 \\ \varphi_2 & \varphi_6 \\ \varphi_3 & \varphi_7 \\ \varphi_4 & \varphi_8 \end{bmatrix}$

for $l = 3$: $\begin{bmatrix} \Phi^{(1)} \\ \Phi^{(2)} \\ \Phi^{(3)} \\ \Phi^{(4)} \end{bmatrix} = \begin{bmatrix} \varphi_1 & \varphi_5 & \varphi_9 \\ \varphi_2 & \varphi_6 & \varphi_{10} \\ \varphi_3 & \varphi_7 & \varphi_{11} \\ \varphi_4 & \varphi_8 & \varphi_{12} \end{bmatrix}$

for $l = 4$: $\begin{bmatrix} \Phi^{(1)} \\ \Phi^{(2)} \\ \Phi^{(3)} \\ \Phi^{(4)} \end{bmatrix} = \begin{bmatrix} \varphi_1 & \varphi_5 & \varphi_9 & \varphi_{13} \\ \varphi_2 & \varphi_6 & \varphi_{10} & \varphi_{14} \\ \varphi_3 & \varphi_7 & \varphi_{11} & \varphi_{15} \\ \varphi_4 & \varphi_8 & \varphi_{12} & \varphi_{16} \end{bmatrix}$,

and similarly for larger values of l.

Applying the idempotents $\pi_1, \pi_2, \pi_3, \pi_4$ to the sets of nodal functions $\Phi^{(1)}, \Phi^{(2)}, \Phi^{(3)}, \Phi^{(4)}$ derives the sets of basis vectors $\bar{\Phi}^{(1)}, \bar{\Phi}^{(2)}, \bar{\Phi}^{(3)}, \bar{\Phi}^{(4)}$ of G-invariant subspaces U_1, U_2, U_3, U_4 respectively:

$$\begin{bmatrix} \bar{\Phi}^{(1)} \\ \bar{\Phi}^{(2)} \\ \bar{\Phi}^{(3)} \\ \bar{\Phi}^{(4)} \end{bmatrix} = \frac{1}{4} \begin{bmatrix} E & E & E & E \\ E & E & -E & -E \\ E & -E & E & -E \\ E & -E & -E & E \end{bmatrix} \begin{bmatrix} \Phi^{(1)} \\ \Phi^{(2)} \\ \Phi^{(3)} \\ \Phi^{(4)} \end{bmatrix} = \frac{1}{4} \begin{bmatrix} \Phi^{(1)} + \Phi^{(2)} + \Phi^{(3)} + \Phi^{(4)} \\ \Phi^{(1)} + \Phi^{(2)} - \Phi^{(3)} - \Phi^{(4)} \\ \Phi^{(1)} - \Phi^{(2)} + \Phi^{(3)} - \Phi^{(4)} \\ \Phi^{(1)} - \Phi^{(2)} - \Phi^{(3)} + \Phi^{(4)} \end{bmatrix},$$

or $\bar{\Phi} = T\Phi$.

As previously for $\Phi^{(1)}, \Phi^{(2)}, \Phi^{(3)}, \Phi^{(4)}$, where l is the number of nodal functions at each node, the numbering of the basis vectors $\bar{\varphi}_j$ ($j = 1, 2, \ldots, n; n = 4l$) corresponds to the numbering of the nodal functions:

for $l = 1$: $\begin{bmatrix} \bar{\Phi}^{(1)} \\ \bar{\Phi}^{(2)} \\ \bar{\Phi}^{(3)} \\ \bar{\Phi}^{(4)} \end{bmatrix} = \begin{bmatrix} \bar{\varphi}_1 \\ \bar{\varphi}_2 \\ \bar{\varphi}_3 \\ \bar{\varphi}_4 \end{bmatrix}$

for $l = 2$: $\begin{bmatrix} \bar{\Phi}^{(1)} \\ \bar{\Phi}^{(2)} \\ \bar{\Phi}^{(3)} \\ \bar{\Phi}^{(4)} \end{bmatrix} = \begin{bmatrix} \bar{\varphi}_1 & \bar{\varphi}_5 \\ \bar{\varphi}_2 & \bar{\varphi}_6 \\ \bar{\varphi}_3 & \bar{\varphi}_7 \\ \bar{\varphi}_4 & \bar{\varphi}_8 \end{bmatrix}$

for $l = 3$: $\begin{bmatrix} \bar{\Phi}^{(1)} \\ \bar{\Phi}^{(2)} \\ \bar{\Phi}^{(3)} \\ \bar{\Phi}^{(4)} \end{bmatrix} = \begin{bmatrix} \bar{\varphi}_1 & \bar{\varphi}_5 & \bar{\varphi}_9 \\ \bar{\varphi}_2 & \bar{\varphi}_6 & \bar{\varphi}_{10} \\ \bar{\varphi}_3 & \bar{\varphi}_7 & \bar{\varphi}_{11} \\ \bar{\varphi}_4 & \bar{\varphi}_8 & \bar{\varphi}_{12} \end{bmatrix}$

for $l = 4$: $\begin{bmatrix} \bar{\Phi}^{(1)} \\ \bar{\Phi}^{(2)} \\ \bar{\Phi}^{(3)} \\ \bar{\Phi}^{(4)} \end{bmatrix} = \begin{bmatrix} \bar{\varphi}_1 & \bar{\varphi}_5 & \bar{\varphi}_9 & \bar{\varphi}_{13} \\ \bar{\varphi}_2 & \bar{\varphi}_6 & \bar{\varphi}_{10} & \bar{\varphi}_{14} \\ \bar{\varphi}_3 & \bar{\varphi}_7 & \bar{\varphi}_{11} & \bar{\varphi}_{15} \\ \bar{\varphi}_4 & \bar{\varphi}_8 & \bar{\varphi}_{12} & \bar{\varphi}_{16} \end{bmatrix}$,

and similarly for larger values of l.

Conversely, the relation of the set of nodal functions $\Phi^{(1)}, \Phi^{(2)}, \Phi^{(3)}, \Phi^{(4)}$ to the sets of basis vectors $\bar{\Phi}^{(1)}, \bar{\Phi}^{(2)}, \bar{\Phi}^{(3)}, \bar{\Phi}^{(4)}$ is

$$\begin{bmatrix} \Phi^{(1)} \\ \Phi^{(2)} \\ \Phi^{(3)} \\ \Phi^{(4)} \end{bmatrix} = \begin{bmatrix} E & E & E & E \\ E & E & -E & -E \\ E & -E & E & -E \\ E & -E & -E & E \end{bmatrix} \begin{bmatrix} \bar{\Phi}^{(1)} \\ \bar{\Phi}^{(2)} \\ \bar{\Phi}^{(3)} \\ \bar{\Phi}^{(4)} \end{bmatrix} = \begin{bmatrix} \bar{\Phi}^{(1)} + \bar{\Phi}^{(2)} + \bar{\Phi}^{(3)} + \bar{\Phi}^{(4)} \\ \bar{\Phi}^{(1)} + \bar{\Phi}^{(2)} - \bar{\Phi}^{(3)} - \bar{\Phi}^{(4)} \\ \bar{\Phi}^{(1)} - \bar{\Phi}^{(2)} + \bar{\Phi}^{(3)} - \bar{\Phi}^{(4)} \\ \bar{\Phi}^{(1)} - \bar{\Phi}^{(2)} - \bar{\Phi}^{(3)} + \bar{\Phi}^{(4)} \end{bmatrix},$$

or $\Phi = 4T\bar{\Phi}$.

The sets of basis vectors $\bar{\Phi}_1, \bar{\Phi}_2, \ldots, \bar{\Phi}_l$, written as

$$[\bar{\Phi}_1 \ \bar{\Phi}_2 \ \ldots \ \bar{\Phi}_l] = \begin{bmatrix} \bar{\varphi}_1 & \bar{\varphi}_5 & \cdots & \bar{\varphi}_{n-3} \\ \bar{\varphi}_2 & \bar{\varphi}_6 & \cdots & \bar{\varphi}_{n-2} \\ \bar{\varphi}_3 & \bar{\varphi}_7 & \cdots & \bar{\varphi}_{n-1} \\ \bar{\varphi}_4 & \bar{\varphi}_8 & \cdots & \bar{\varphi}_n \end{bmatrix},$$

are related to the sets of nodal functions $\Phi_1, \Phi_2, \ldots, \Phi_l$ as follows:

$$\begin{bmatrix} \bar{\Phi}_1 \\ \bar{\Phi}_2 \\ \cdots \\ \bar{\Phi}_l \end{bmatrix} = \begin{bmatrix} T & & & \\ & T & & \\ & & \cdots & \\ & & & T \end{bmatrix} \begin{bmatrix} \Phi_1 \\ \Phi_2 \\ \cdots \\ \Phi_l \end{bmatrix} \quad \text{or} \quad \bar{\Phi} = \bar{T}\Phi,$$

with

$$T = \frac{1}{4} \begin{bmatrix} E & E & E & E \\ E & E & -E & -E \\ E & -E & E & -E \\ E & -E & -E & E \end{bmatrix}$$

and

$$[\Phi_1 \ \Phi_2 \ \ldots \ \Phi_l] = \begin{bmatrix} \varphi_1 & \varphi_5 & \cdots & \varphi_{n-3} \\ \varphi_2 & \varphi_6 & \cdots & \varphi_{n-2} \\ \varphi_3 & \varphi_7 & \cdots & \varphi_{n-1} \\ \varphi_4 & \varphi_8 & \cdots & \varphi_n \end{bmatrix}.$$

Conversely, the relation of the sets of nodal functions $\Phi_1, \Phi_2, \ldots, \Phi_l$ to the sets of basis vectors $\bar{\Phi}_1, \bar{\Phi}_2, \ldots, \bar{\Phi}_l$ is

Sec. 3.5]   **Group supermatrices of the group C_{2v}**   117

$$\begin{bmatrix} \Phi_1 \\ \Phi_2 \\ \vdots \\ \Phi_l \end{bmatrix} = \begin{bmatrix} T^{-1} & & & \\ & T^{-1} & & \\ & & \ddots & \\ & & & T^{-1} \end{bmatrix} \begin{bmatrix} \bar{\Phi}_1 \\ \bar{\Phi}_2 \\ \vdots \\ \bar{\Phi}_l \end{bmatrix} = 4 \begin{bmatrix} T & & & \\ & T & & \\ & & \ddots & \\ & & & T \end{bmatrix} \begin{bmatrix} \bar{\Phi}_1 \\ \bar{\Phi}_2 \\ \vdots \\ \bar{\Phi}_l \end{bmatrix},$$

or $\Phi = 4\bar{T}\bar{\Phi}$, since $T^{-1} = 4T$.

The system of equations $G\Phi = P$, with the group supermatrix G in normal form derived earlier, expressed as

$$\begin{bmatrix} A & B & C & D \\ B & A & D & C \\ C & D & A & B \\ D & C & B & A \end{bmatrix} \begin{bmatrix} \Phi^{(1)} \\ \Phi^{(2)} \\ \Phi^{(3)} \\ \Phi^{(4)} \end{bmatrix} = \begin{bmatrix} P^{(1)} \\ P^{(2)} \\ P^{(3)} \\ P^{(4)} \end{bmatrix},$$

can be transformed into the system of equations $\bar{G}\bar{\Phi} = \bar{P}$, with the group supermatrix $\bar{G}$ in diagonal form, by the group supermatrix transformation

$$TGT^{-1}T\Phi = TP.$$

Thus

$$\frac{1}{4}\begin{bmatrix} E & E & E & E \\ E & E & -E & -E \\ E & -E & E & -E \\ E & -E & -E & E \end{bmatrix} \begin{bmatrix} A & B & C & D \\ B & A & D & C \\ C & D & A & B \\ D & C & B & A \end{bmatrix} \begin{bmatrix} E & E & E & E \\ E & E & -E & -E \\ E & -E & E & -E \\ E & -E & -E & E \end{bmatrix}$$

$$\cdot \frac{1}{4}\begin{bmatrix} E & E & E & E \\ E & E & -E & -E \\ E & -E & E & -E \\ E & -E & -E & E \end{bmatrix} \begin{bmatrix} \Phi^{(1)} \\ \Phi^{(2)} \\ \Phi^{(3)} \\ \Phi^{(4)} \end{bmatrix} = \frac{1}{4}\begin{bmatrix} E & E & E & E \\ E & E & -E & -E \\ E & -E & E & -E \\ E & -E & -E & E \end{bmatrix} \begin{bmatrix} P^{(1)} \\ P^{(2)} \\ P^{(3)} \\ P^{(4)} \end{bmatrix}$$

results in

$$\begin{bmatrix} A+B+C+D & & & \\ & A+B-C-D & & \\ & & A-B+C-D & \\ & & & A-B-C+D \end{bmatrix} \begin{bmatrix} \bar{\Phi}^{(1)} \\ \bar{\Phi}^{(2)} \\ \bar{\Phi}^{(3)} \\ \bar{\Phi}^{(4)} \end{bmatrix} = \begin{bmatrix} \bar{P}^{(1)} \\ \bar{P}^{(2)} \\ \bar{P}^{(3)} \\ \bar{P}^{(4)} \end{bmatrix}$$

or $\bar{G}\bar{\Phi} = \bar{P}$, with

$$\begin{bmatrix} \bar{\Phi}^{(1)} \\ \bar{\Phi}^{(2)} \\ \bar{\Phi}^{(3)} \\ \bar{\Phi}^{(4)} \end{bmatrix} = \frac{1}{4} \begin{bmatrix} \Phi^{(1)} + \Phi^{(2)} + \Phi^{(3)} + \Phi^{(4)} \\ \Phi^{(1)} + \Phi^{(2)} - \Phi^{(3)} - \Phi^{(4)} \\ \Phi^{(1)} - \Phi^{(2)} + \Phi^{(3)} - \Phi^{(4)} \\ \Phi^{(1)} - \Phi^{(2)} - \Phi^{(3)} + \Phi^{(4)} \end{bmatrix}$$

and

$$\begin{bmatrix} \bar{P}^{(1)} \\ \bar{P}^{(2)} \\ \bar{P}^{(3)} \\ \bar{P}^{(4)} \end{bmatrix} = - \begin{bmatrix} P^{(1)} + P^{(2)} + P^{(3)} + P^{(4)} \\ P^{(1)} + P^{(2)} - P^{(3)} - P^{(4)} \\ P^{(1)} - P^{(2)} + P^{(3)} - P^{(4)} \\ P^{(1)} - P^{(2)} - P^{(3)} + P^{(4)} \end{bmatrix}.$$

Conversely, the system of equations $\bar{G}\bar{\Phi} = \bar{P}$, with the group supermatrix $\bar{G}$ in diagonal form, is transformed into the system of equations $G\Phi = P$, with the group supermatrix G in normal form, by the inverse group supermatrix transformation

$$T^{-1}\bar{G}TT^{-1}\bar{\Phi} = T^{-1}\bar{P}, \quad \text{or}$$

$$\begin{bmatrix} E & E & E & E \\ E & E & -E & -E \\ E & -E & E & -E \\ E & -E & -E & E \end{bmatrix} \begin{bmatrix} A+B+C+D & & & \\ & A+B-C-D & & \\ & & A-B+C-D & \\ & & & A-B-C+D \end{bmatrix}$$

$$\cdot \frac{1}{4} \begin{bmatrix} E & E & E & E \\ E & E & -E & -E \\ E & -E & E & -E \\ E & -E & -E & E \end{bmatrix} \begin{bmatrix} E & E & E & E \\ E & E & -E & -E \\ E & -E & E & -E \\ E & -E & -E & E \end{bmatrix} \frac{1}{4} \begin{bmatrix} \bar{\Phi}^{(1)} + \bar{\Phi}^{(2)} + \bar{\Phi}^{(3)} + \bar{\Phi}^{(4)} \\ \bar{\Phi}^{(1)} + \bar{\Phi}^{(2)} - \bar{\Phi}^{(3)} - \bar{\Phi}^{(4)} \\ \bar{\Phi}^{(1)} - \bar{\Phi}^{(2)} + \bar{\Phi}^{(3)} - \bar{\Phi}^{(4)} \\ \bar{\Phi}^{(1)} - \bar{\Phi}^{(2)} - \bar{\Phi}^{(3)} + \bar{\Phi}^{(4)} \end{bmatrix}$$

$$= \begin{bmatrix} E & E & E & E \\ E & -E & -E & -E \\ E & -E & E & -E \\ E & -E & -E & E \end{bmatrix} \frac{1}{4} \begin{bmatrix} \bar{P}^{(1)} + \bar{P}^{(2)} + \bar{P}^{(3)} + \bar{P}^{(4)} \\ \bar{P}^{(1)} + \bar{P}^{(2)} - \bar{P}^{(3)} - \bar{P}^{(4)} \\ \bar{P}^{(1)} - \bar{P}^{(2)} + \bar{P}^{(3)} - \bar{P}^{(4)} \\ \bar{P}^{(1)} - \bar{P}^{(2)} - \bar{P}^{(3)} + \bar{P}^{(4)} \end{bmatrix},$$

resulting in

$$\begin{bmatrix} A & B & C & D \\ B & A & D & C \\ C & D & A & B \\ D & C & B & A \end{bmatrix} \begin{bmatrix} \Phi^{(1)} \\ \Phi^{(2)} \\ \Phi^{(3)} \\ \Phi^{(4)} \end{bmatrix} = \begin{bmatrix} P^{(1)} \\ P^{(2)} \\ P^{(3)} \\ P^{(4)} \end{bmatrix}.$$

Sec. 3.5] **Group supermatrices of the group C_{2v}** 119

As in the case of the four-node rectangular element, the sequence of symmetry operations E, C_2, σ_1, σ_2, existing in the character table of the group C_{2v}, will be used for definition of the unique nodal numbering of the four-node rhombic element.

The position of the first node, numbered by 1, is on the positive branch of the x axis, as given in Fig. 3.8(a). It is assumed, according to the rule, that application of E, C_2, σ_1, σ_2 (in this order) to the initial nodes 1, 3 provides the nodes 1, 2, 3, 4 (in this sequence). This definition determines the unique nodal numbering in Fig. 3.8(a), stated by the following table

Node	x axis	y axis
1	+	
2	−	
3		+
4		−

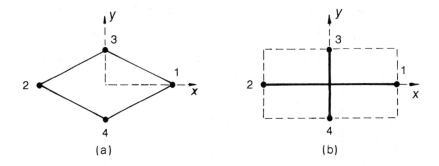

Fig. 3.8. (a) The four-node rhombic element described by the group C_{2v}; (b) the assembly of two two-node linear elements.

Expressed as permutations of the nodes, the actions of symmetry operations on the initial nodal sequence [1 2 3 4] are

$$E = \begin{bmatrix} 1 & 2 & 3 & 4 \\ 1 & 2 & 3 & 4 \end{bmatrix}, \quad C_2 = \begin{bmatrix} 1 & 2 & 3 & 4 \\ 2 & 1 & 4 & 3 \end{bmatrix}, \quad \sigma_1 = \begin{bmatrix} 1 & 2 & 3 & 4 \\ 1 & 2 & 4 & 3 \end{bmatrix},$$

$$\sigma_2 = \begin{bmatrix} 1 & 2 & 3 & 4 \\ 2 & 1 & 3 & 4 \end{bmatrix},$$

or given in form of a permutation table

Node	E	C_2	σ_1	σ_2
1	1	2	1	2
2	2	1	2	1
3	3	4	4	3
4	4	3	3	4

These permutations show that the nodes are grouped into two nodal sets $S_1(1, 2)$ and $S_2(3, 4)$, where the nodes are permuted by action of symmetry operations $E, C_2, \sigma_1, \sigma_2$.

Applying the idempotents $\pi_1, \pi_2, \pi_3, \pi_4$ to the sets of nodal functions (Φ_1, Φ_2), (Φ_3, Φ_4) pertaining to the nodal sets $S_1(1, 2), S_2(3, 4)$, derives the sets of basis vectors of G-invariant subspaces U_i

$U_1:$ $\bar{\Phi}_1 = \pi_1 \Phi_1 = \pi_1 \Phi_2 = \frac{1}{2}(\Phi_1 + \Phi_2)$
$\bar{\Phi}_3 = \pi_1 \Phi_3 = \pi_1 \Phi_4 = \frac{1}{2}(\Phi_3 + \Phi_4)$

$U_2:$ zero subspace, since $\pi_2 \Phi_1 = \pi_2 \Phi_2 = 0, \quad \pi_2 \Phi_3 = \pi_2 \Phi_4 = 0$

$U_3:$ $\bar{\Phi}_2 = \pi_3 \Phi_1 = -\pi_3 \Phi_2 = \frac{1}{2}(\Phi_1 - \Phi_2), \quad \pi_3 \Phi_3 = \pi_3 \Phi_4 = 0$

$U_4:$ $\bar{\Phi}_4 = \pi_3 \Phi_3 = -\pi_3 \Phi_4 = \frac{1}{2}(\Phi_3 - \Phi_4), \quad \pi_4 \Phi_1 = \pi_4 \Phi_2 = 0.$

The relation of the sets of basis vectors $\bar{\Phi}$ to the sets of nodal functions Φ is

$$\begin{bmatrix} \bar{\Phi}_1 \\ \bar{\Phi}_2 \\ \bar{\Phi}_3 \\ \bar{\Phi}_4 \end{bmatrix} = \frac{1}{2} \begin{bmatrix} E & E & & \\ E & -E & & \\ & & E & E \\ & & E & -E \end{bmatrix} \begin{bmatrix} \Phi_1 \\ \Phi_2 \\ \Phi_3 \\ \Phi_4 \end{bmatrix}.$$

Conversely, the relation of the sets of nodal functions Φ to the sets of basis vectors $\bar{\Phi}$ is

$$\begin{bmatrix} \Phi_1 \\ \Phi_2 \\ \Phi_3 \\ \Phi_4 \end{bmatrix} = \begin{bmatrix} E & E & & \\ E & -E & & \\ & & E & E \\ & & E & -E \end{bmatrix} \begin{bmatrix} \bar{\Phi}_1 \\ \bar{\Phi}_2 \\ \bar{\Phi}_3 \\ \bar{\Phi}_4 \end{bmatrix}.$$

The supermatrix

$$T = \frac{1}{2} \begin{bmatrix} E & E \\ E & -E \end{bmatrix}, \quad \text{with} \quad T^{-1} = 2T,$$

is the transformation supermatrix of the two-node linear element pertaining to the group C_2, as derived in section 3.2.

Thus, the four-node rhombic element, as given in Fig. 3.8(a), may be regarded as an assembly of two two-node linear elements shown in Fig. 3.8(b).

3.6 GROUP SUPERMATRICES OF THE GROUP C_{3v}

The character table of the group C_{3v}

C_{3v}	E	$2C_3$	$3\sigma_v$
A_1	1	1	1
A_2	1	1	−1
E	2	−1	0

has the sequence of symmetry operations E, C_3, C_3^{-1}, σ_1, σ_2, σ_3 that is used for definition of the nodal numbering of the six-node hexagonal element shown in Fig. 3.9. The convention is adopted that the sequence of reflection planes of $\sigma_1, \sigma_2, \sigma_3$ is counterclockwise and that the position of the first node, numbered by 1, is in the sector between reflection planes of σ_1, σ_3 and nearer to σ_1.

As previously, it is assumed that application of E, C_3, C_3^{-1}, σ_1, σ_2, σ_3 (in this order) to the initial node 1 provides the nodes 1, 2, 3, 4, 5, 6 (in this sequence).

Expressed as permutations of the nodes, the results of actions of symmetry operations on the nodal sequence [1 2 3 4 5 6] are given in the permutation table as follows:

Node	E	C_3	C_3^{-1}	σ_1	σ_2	σ_3
1	1	2	3	4	5	6
2	2	3	1	6	4	5
3	3	1	2	5	6	4
4	4	5	6	1	2	3
5	5	6	4	3	1	2
6	6	4	5	2	3	1

By replacing the numbers 1, 2, 3, 4, 5, 6 by matrices A, B, C, D, H, I in the above permutation table one obtains the supermatrix

$$G = \begin{bmatrix} A & B & C & D & H & I \\ B & C & A & I & D & H \\ C & A & B & H & I & D \\ D & H & I & A & B & C \\ H & I & D & C & A & B \\ I & D & H & B & C & A \end{bmatrix},$$

which is the group supermatrix in normal form pertaining to the group C_{3v}.

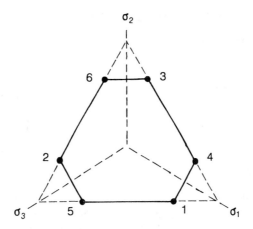

Fig. 3.9. The six-node hexagonal element described by the group C_{3v}.

The idempotents π_i ($i = 1, 2, 3$) of the centre of group algebra are determined by

$$\pi_i = \frac{h_i}{h} \sum_\sigma \chi_i(\sigma^{-1}) \sigma,$$

with explanations concerning this expression given in section 3.2, so that

$$\pi_1 = \tfrac{1}{6}(E + C_3 + C_3^{-1} + \sigma_1 + \sigma_2 + \sigma_3)$$
$$\pi_2 = \tfrac{1}{6}(E + C_3 + C_3^{-1} - \sigma_1 - \sigma_2 - \sigma_3)$$
$$\pi_3 = \tfrac{2}{6}(2E - C_3 - C_3^{-1}).$$

Applying the idempotents π_1, π_2, π_3 to the sets of nodal functions $\Phi^{(1)}, \Phi^{(2)}, \Phi^{(3)}, \Phi^{(4)}, \Phi^{(5)}, \Phi^{(6)}$ derives the basis vectors $\bar{\Phi}^{(1)}, \bar{\Phi}^{(2)}, \bar{\Phi}^{(3)}, \bar{\Phi}^{(4)}, \bar{\Phi}^{(5)}, \bar{\Phi}^{(6)}$ of G-invariant subspaces U_1, U_2, U_3

$U_1:\quad \bar{\Phi}^{(1)} = \pi_1 \Phi^{(1)} = \pi_1 \Phi^{(2)} = \pi_1 \Phi^{(3)} + \pi_1 \Phi^{(4)} = \pi_1 \Phi^{(5)} = \pi_1 \Phi^{(6)}$
$\qquad = \tfrac{1}{6}(\Phi^{(1)} + \Phi^{(2)} + \Phi^{(3)} + \Phi^{(4)} + \Phi^{(5)} + \Phi^{(6)})$

$U_2:\quad \bar{\Phi}^{(2)} = \pi_2 \Phi^{(1)} = \pi_2 \Phi^{(2)} = \pi_2 \Phi^{(3)} = -\pi_2 \Phi^{(4)} = -\pi_2 \Phi^{(5)} = -\pi_2 \Phi^{(6)}$
$\qquad = \tfrac{1}{6}(\Phi^{(1)} + \Phi^{(2)} + \Phi^{(3)} - \Phi^{(4)} - \Phi^{(5)} - \Phi^{(6)})$

$U_3:\quad \bar{\Phi}^{(3)} = \pi_3 \Phi^{(1)} = \tfrac{2}{6}(2\Phi^{(1)} - \Phi^{(2)} - \Phi^{(3)})$
$\qquad \bar{\Phi}^{(4)} = \pi_3 \Phi^{(2)} = \tfrac{2}{6}(-\Phi^{(1)} + 2\Phi^{(2)} - \Phi^{(3)})$
$\qquad [\pi_3 \Phi^{(3)} = \tfrac{2}{6}(-\Phi^{(1)} - \Phi^{(2)} + 2\Phi^{(3)})$ is linearly dependent]
$\qquad \bar{\Phi}^{(5)} = \pi_3 \Phi^{(4)} = \tfrac{2}{6}(2\Phi^{(4)} - \Phi^{(5)} - \Phi^{(6)})$
$\qquad \bar{\Phi}^{(6)} = \pi_3 \Phi^{(5)} = \tfrac{2}{6}(-\Phi^{(4)} + 2\Phi^{(5)} - \Phi^{(6)})$
$\qquad [\pi_3 \Phi^{(6)} = \tfrac{2}{6}(-\Phi^{(4)} - \Phi^{(5)} + 2\Phi^{(6)})$ is linearly dependent].

In supermatrix form

$$\begin{bmatrix} \bar{\Phi}^{(1)} \\ \bar{\Phi}^{(2)} \\ \bar{\Phi}^{(3)} \\ \bar{\Phi}^{(4)} \\ \bar{\Phi}^{(5)} \\ \bar{\Phi}^{(6)} \end{bmatrix} = \frac{1}{3} \begin{bmatrix} E & E & E & E & E & E \\ E & E & E & -E & -E & -E \\ 4E & -2E & -2E & & & \\ -2E & 4E & -2E & & & \\ & & & 4E & -2E & -2E \\ & & & -2E & 4E & -2E \end{bmatrix} \begin{bmatrix} \Phi^{(1)} \\ \Phi^{(2)} \\ \Phi^{(3)} \\ \Phi^{(4)} \\ \Phi^{(5)} \\ \Phi^{(6)} \end{bmatrix}$$

or $\bar{\Phi} = T\Phi$, with

$$T^{-1} = \frac{1}{2} \begin{bmatrix} E & E & E & & & \\ E & E & & E & & \\ E & E & -E & -E & & \\ E & -E & & & E & \\ E & -E & & & & E \\ E & -E & & & -E & -E \end{bmatrix} \quad \text{and} \quad \Phi = T^{-1}\bar{\Phi}.$$

The system of equations $G\Phi = P$ or

$$\begin{bmatrix} A & B & C & D & H & I \\ B & C & A & I & D & H \\ C & A & B & H & I & D \\ D & H & I & A & B & C \\ H & I & D & C & A & B \\ I & D & H & B & C & A \end{bmatrix} \begin{bmatrix} \Phi^{(1)} \\ \Phi^{(2)} \\ \Phi^{(3)} \\ \Phi^{(4)} \\ \Phi^{(5)} \\ \Phi^{(6)} \end{bmatrix} = \begin{bmatrix} P^{(1)} \\ P^{(2)} \\ P^{(3)} \\ P^{(4)} \\ P^{(5)} \\ P^{(6)} \end{bmatrix},$$

with the group supermatrix G in normal form given earlier, is transformed into the system of equations $\bar{G}\bar{\Phi} = \bar{P}$, with the group supermatrix $\bar{G}$ in diagonal form, by the group supermatrix transformation $TGT^{-1}T\Phi = TP$,

$$\frac{1}{3} \begin{bmatrix} E & E & E & E & E & E \\ E & E & E & -E & -E & -E \\ 4E & -2E & -2E & & & \\ -2E & 4E & -2E & & & \\ & & & 4E & -2E & -2E \\ & & & -2E & 4E & -2E \end{bmatrix} \begin{bmatrix} A & B & C & D & H & I \\ B & C & A & I & D & H \\ C & A & B & H & I & D \\ D & H & I & A & B & C \\ H & I & D & C & A & B \\ I & D & H & B & C & A \end{bmatrix}$$

$$\cdot \frac{1}{2} \begin{bmatrix} E & E & E & & & \\ E & E & & & E & \\ E & E & -E & -E & & \\ E & -E & & & E & \\ E & -E & & & & E \\ E & -E & & & -E & -E \end{bmatrix}$$

$$\cdot \frac{1}{3} \begin{bmatrix} E & E & E & E & E & E \\ E & E & E & -E & -E & -E \\ 4E & -2E & -2E & & & \\ -2E & 4E & -2E & & & \\ & & & 4E & -2E & -2E \\ & & & -2E & 4E & -2E \end{bmatrix} \begin{bmatrix} \Phi^{(1)} \\ \Phi^{(2)} \\ \Phi^{(3)} \\ \Phi^{(4)} \\ \Phi^{(5)} \\ \Phi^{(6)} \end{bmatrix}$$

$$= \frac{1}{3} \begin{bmatrix} E & E & E & E & E & E \\ E & E & E & -E & -E & -E \\ 4E & -2E & -2E & & & \\ -2E & 4E & -2E & & & \\ & & & 4E & -2E & -2E \\ & & & -2E & 4E & -2E \end{bmatrix} \begin{bmatrix} P^{(1)} \\ P^{(2)} \\ P^{(3)} \\ P^{(4)} \\ P^{(5)} \\ P^{(6)} \end{bmatrix},$$

resulting in

$$\begin{bmatrix} A+B+C+D+H+I & & & & & \\ & A+B+C-D-H-I & & & & \\ & & A-C & B-C & D-I & H-I \\ & & -A+B & -A+C & -H+I & D-H \\ & & D-I & H-I & A-C & B-C \\ & & -D+H & -D+I & -B+C & A-B \end{bmatrix} \begin{bmatrix} \bar{\Phi}^{(1)} \\ \bar{\Phi}^{(2)} \\ \bar{\Phi}^{(3)} \\ \bar{\Phi}^{(4)} \\ \bar{\Phi}^{(5)} \\ \bar{\Phi}^{(6)} \end{bmatrix}$$

$$= \begin{bmatrix} \bar{p}^{(1)} \\ \bar{p}^{(2)} \\ \bar{p}^{(3)} \\ \bar{p}^{(4)} \\ \bar{p}^{(5)} \\ \bar{p}^{(6)} \end{bmatrix},$$

Sec. 3.6]   **Group supermatrices of the group C_{3v}**   125

with the relation

$$\begin{bmatrix} \Phi^{(1)} \\ \Phi^{(2)} \\ \Phi^{(3)} \\ \Phi^{(4)} \\ \Phi^{(5)} \\ \Phi^{(6)} \end{bmatrix} = \frac{1}{2} \begin{bmatrix} E & E & E & & & \\ E & E & & E & & \\ E & E & -E & -E & & \\ E & -E & & & E & \\ E & -E & & & & E \\ E & -E & & & -E & -E \end{bmatrix} \begin{bmatrix} \bar{\Phi}^{(1)} \\ \bar{\Phi}^{(2)} \\ \bar{\Phi}^{(3)} \\ \bar{\Phi}^{(4)} \\ \bar{\Phi}^{(5)} \\ \bar{\Phi}^{(6)} \end{bmatrix} \quad \text{or} \quad \Phi = T^{-1} \bar{\Phi}.$$

The system of equations $\bar{G}\bar{\Phi} = \bar{P}$ is transformed into the system of equations $G\Phi = P$, with the group supermatrix G in normal form, by the inverse group supermatrix transformation $T^{-1}\bar{G}TT^{-1}\bar{\Phi} = T\bar{P}$:

$$\frac{1}{2} \begin{bmatrix} E & E & E & & & \\ E & E & & E & & \\ E & E & -E & -E & & \\ E & -E & & & E & \\ E & -E & & & & E \\ E & -E & & & -E & -E \end{bmatrix}$$

$$\begin{bmatrix} A+B+C+D+H+I & & & & & \\ & A+B+C-D-H-I & & & & \\ & & A-C & B-C & D-I & H-I \\ & & -A+B & -A+C & -H+I & D-H \\ & & D-I & H-I & A-C & B-C \\ & & -D+H & -D+I & -B+C & A-B \end{bmatrix}$$

$$\cdot \frac{1}{3} \begin{bmatrix} E & E & E & E & E & E \\ E & E & E & -E & -E & -E \\ 4E & -2E & -2E & & & \\ -2E & 4E & -2E & & & \\ & & & 4E & -2E & -2E \\ & & & -2E & 4E & -2E \end{bmatrix}$$

$$\cdot \frac{1}{2} \begin{bmatrix} E & E & E & & & \\ E & E & & E & & \\ E & E & -E & -E & & \\ E & -E & & & E & \\ E & -E & & & & E \\ E & -E & & & -E & -E \end{bmatrix} \begin{bmatrix} \bar{\Phi}^{(1)} \\ \bar{\Phi}^{(2)} \\ \bar{\Phi}^{(3)} \\ \bar{\Phi}^{(4)} \\ \bar{\Phi}^{(5)} \\ \bar{\Phi}^{(6)} \end{bmatrix}$$

$$= \frac{1}{2} \begin{bmatrix} E & E & E & & & \\ E & E & & E & & \\ E & E & -E & -E & & \\ E & -E & & & E & \\ E & -E & & & & E \\ E & -E & & & -E & -E \end{bmatrix} \begin{bmatrix} \bar{P}^{(1)} \\ \bar{P}^{(2)} \\ \bar{P}^{(3)} \\ \bar{P}^{(4)} \\ \bar{P}^{(5)} \\ \bar{P}^{(6)} \end{bmatrix}$$

resulting in

$$\begin{bmatrix} A & B & C & D & H & I \\ B & C & A & I & D & H \\ C & A & B & H & I & D \\ D & H & I & A & B & C \\ H & I & D & C & A & B \\ I & D & H & B & C & A \end{bmatrix} \begin{bmatrix} \Phi^{(1)} \\ \Phi^{(2)} \\ \Phi^{(3)} \\ \Phi^{(4)} \\ \Phi^{(5)} \\ \Phi^{(6)} \end{bmatrix} = \begin{bmatrix} P^{(1)} \\ P^{(2)} \\ P^{(3)} \\ P^{(4)} \\ P^{(5)} \\ P^{(6)} \end{bmatrix}.$$

3.7 GROUP SUPERMATRICES OF THE GROUP C_{4v}

The character table of the group C_{4v}

C_{4v}	E	$2C_4$	C_2	$2\sigma_v$	$2\sigma_d$
A_1	1	1	1	1	1
A_2	1	1	1	−1	−1
B_1	1	−1	1	1	−1
B_2	1	−1	1	−1	1
E	2	0	−2	0	0

Sec. 3.7] Group supermatrices of the group C_{4v} 127

has the sequence of symmetry operations E, C_4, C_4^{-1}, C_2, σ_x, σ_y, σ_1, σ_2, which is used for definition of the nodal numbering of the eight-node octogonal element shown in Fig. 3.10. It is adopted that the position of the first node, numbered by 1, is in the first quadrant of the xy coordinate system and nearer to the x axis.

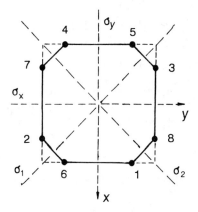

Fig. 3.10. The eight-node octogonal element described by the group C_{4v}.

As previously, it is assumed that application of E, C_4, C_4^{-1}, C_2, σ_x, σ_y, σ_1, σ_2 (in this order) to the initial node 1 provides the nodes 1, 2, 3, 4, 5, 6, 7, 8 (in this sequence).

Expressed as permutations of the nodes, the results of actions of symmetry operations on the initial sequence [1 2 3 4 5 6 7 8] are given in the permutation table as follows:

Node	E	C_4	C_4^{-1}	C_2	σ_x	σ_y	σ_1	σ_2
1	1	2	3	4	5	6	7	8
2	2	4	1	3	7	8	6	5
3	3	1	4	2	8	7	5	6
4	4	3	2	1	6	5	8	7
5	5	8	7	6	1	4	3	2
6	6	7	8	5	4	1	2	3
7	7	5	6	8	2	3	1	4
8	8	6	5	7	3	2	4	1

By replacing the numbers 1, 2, 3, 4, 5, 6, 7, 8 by matrices A, B, C, D, H, I, J, K in the above permutation table one obtains the supermatrix

$$G = \begin{bmatrix} A & B & C & D & H & I & J & K \\ B & D & A & C & J & K & I & H \\ C & A & D & B & K & J & H & I \\ D & C & B & A & I & H & K & J \\ H & K & J & I & A & D & C & B \\ I & J & K & H & D & A & B & C \\ J & H & I & K & B & C & A & D \\ K & I & H & J & C & B & D & A \end{bmatrix},$$

which is the group supermatrix in normal form pertaining to the group C_{4v}.

The idempotents π_i ($i = 1, 2, 3, 4, 5$) of the centre of group algebra are determined by

$$\pi_i = \frac{h_i}{h} \sum_\sigma \chi_i(\sigma^{-1})\sigma,$$

with explanations concerning this expression given in 3.2, so that

$$\pi_1 = \tfrac{1}{8}(E + C_4 + C_4^{-1} + C_2 + \sigma_x + \sigma_y + \sigma_1 + \sigma_2)$$
$$\pi_2 = \tfrac{1}{8}(E + C_4 + C_4^{-1} + C_2 - \sigma_x - \sigma_y - \sigma_1 - \sigma_2)$$
$$\pi_3 = \tfrac{1}{8}(E - C_4 - C_4^{-1} + C_2 + \sigma_x + \sigma_y - \sigma_1 - \sigma_2)$$
$$\pi_4 = \tfrac{1}{8}(E - C_4 - C_4^{-1} + C_2 - \sigma_x - \sigma_y + \sigma_1 + \sigma_2)$$
$$\pi_5 = \tfrac{2}{8}(2E - 2C_2) = \tfrac{1}{2}(E - C_2).$$

Applying the idempotents $\pi_1, \pi_2, \pi_2, \pi_4, \pi_5$ to the sets of nodal functions $\Phi^{(1)}, \Phi^{(2)}, \Phi^{(3)}, \Phi^{(4)}, \Phi^{(5)}, \Phi^{(6)}, \Phi^{(7)}, \Phi^{(8)}$ derives the sets of basis vectors $\bar\Phi^{(1)}, \bar\Phi^{(2)}, \bar\Phi^{(3)}, \bar\Phi^{(4)}, \bar\Phi^{(5)}, \bar\Phi^{(6)}, \bar\Phi^{(7)}, \bar\Phi^{(8)}$ of G-invariant subspaces U_1, U_2, U_3, U_4, U_5

$U_1:$ $\bar\Phi^{(1)} = \pi_1\Phi^{(1)} = \pi_1\Phi^{(2)} = \pi_1\Phi^{(3)} = \pi_1\Phi^{(4)} = \pi_1\Phi^{(5)} = \pi_1\Phi^{(6)} = \pi_1\Phi^{(7)} = \pi_1\Phi^{(8)}$
$= \tfrac{1}{8}(\Phi^{(1)} + \Phi^{(2)} + \Phi^{(3)} + \Phi^{(4)} + \Phi^{(5)} + \Phi^{(6)} + \Phi^{(7)} + \Phi^{(8)})$

$U_2:$ $\bar\Phi^{(2)} = \pi_2\Phi^{(1)} = \pi_2\Phi^{(2)} = \pi_2\Phi^{(3)} = \pi_2\Phi^{(4)} = -\pi_2\Phi^{(5)} = -\pi_2\Phi^{(6)} = -\pi_2\Phi^{(7)}$
$= -\pi_2\Phi^{(8)} = \tfrac{1}{8}(\Phi^{(1)} + \Phi^{(2)} + \Phi^{(3)} + \Phi^{(4)} - \Phi^{(5)} - \Phi^{(6)} - \Phi^{(7)} - \Phi^{(8)})$

$U_3:$ $\bar\Phi^{(3)} = \pi_3\Phi^{(1)} = -\pi_3\Phi^{(2)} = -\pi_3\Phi^{(3)} = \pi_3\Phi^{(4)} = \pi_3\Phi^{(5)} = \pi_3\Phi^{(6)} = -\pi_3\Phi^{(7)}$
$= -\pi_3\Phi^{(8)} = \tfrac{1}{8}(\Phi^{(1)} - \Phi^{(2)} - \Phi^{(3)} + \Phi^{(4)} + \Phi^{(5)} + \Phi^{(6)} - \Phi^{(7)} - \Phi^{(8)})$

$U_4:$ $\bar\Phi^{(4)} = \pi_4\Phi^{(1)} = -\pi_4\Phi^{(2)} = -\pi_4\Phi^{(3)} = \pi_4\Phi^{(4)} = -\pi_4\Phi^{(5)} = -\pi_4\Phi^{(6)} = \pi_4\Phi^{(7)}$
$= \pi_4\Phi^{(8)} = \tfrac{1}{8}(\Phi^{(1)} - \Phi^{(2)} - \Phi^{(3)} + \Phi^{(4)} - \Phi^{(5)} - \Phi^{(6)} + \Phi^{(7)} + \Phi^{(8)})$

$U_5:$ $\bar{\Phi}^{(5)} = \pi_5 \Phi^{(1)} = -\pi_5 \Phi^{(4)} = \frac{1}{2}(\Phi^{(1)} - \Phi^{(4)})$

$\Phi^{(6)} = \pi_5 \Phi^{(2)} = -\pi_5 \Phi^{(3)} = \frac{1}{2}(\Phi^{(2)} - \Phi^{(3)})$

$\Phi^{(7)} = \pi_5 \Phi^{(5)} = -\pi_5 \Phi^{(6)} = \frac{1}{2}(\Phi^{(5)} - \Phi^{(6)})$

$\Phi^{(8)} = \pi_5 \Phi^{(7)} = -\pi_5 \Phi^{(8)} = \frac{1}{8}(\Phi^{(7)} - \Phi^{(8)}).$

In supermatrix form

$$\begin{bmatrix} \bar{\Phi}^{(1)} \\ \bar{\Phi}^{(2)} \\ \bar{\Phi}^{(3)} \\ \bar{\Phi}^{(4)} \\ \bar{\Phi}^{(5)} \\ \bar{\Phi}^{(6)} \\ \bar{\Phi}^{(7)} \\ \bar{\Phi}^{(8)} \end{bmatrix} = \frac{1}{8} \begin{bmatrix} E & E & E & E & E & E & E & E \\ E & E & E & E & -E & -E & -E & -E \\ E & -E & -E & E & E & E & -E & -E \\ E & -E & -E & E & -E & -E & E & E \\ 4E & & & & -4E & & & \\ & 4E & -4E & & & & & \\ & & & & 4E & -4E & & \\ & & & & & & 4E & -4E \end{bmatrix} \begin{bmatrix} \Phi^{(1)} \\ \Phi^{(2)} \\ \Phi^{(3)} \\ \Phi^{(4)} \\ \Phi^{(5)} \\ \Phi^{(6)} \\ \Phi^{(7)} \\ \Phi^{(8)} \end{bmatrix}$$

or $\bar{\Phi} = T\Phi$, with

$$T^{-1} = \begin{bmatrix} E & E & E & E & E & & & \\ E & E & -E & -E & & E & & \\ E & E & -E & -E & & -E & & \\ E & E & E & E & -E & & & \\ E & -E & E & -E & & & E & \\ E & -E & E & -E & & & -E & \\ E & -E & -E & E & & & & E \\ E & -E & -E & E & & & & -E \end{bmatrix}$$ and $\Phi = T^{-1}\bar{\Phi}.$

The system of equations $G\Phi = P$ or

$$\begin{bmatrix} A & B & C & D & H & I & J & K \\ B & D & A & C & J & K & I & H \\ C & A & D & B & K & J & H & I \\ D & C & B & A & I & H & K & J \\ H & K & J & I & A & D & C & B \\ I & J & K & H & D & A & B & C \\ J & H & I & K & B & C & A & D \\ K & I & H & J & C & B & D & A \end{bmatrix} \begin{bmatrix} \Phi^{(1)} \\ \Phi^{(2)} \\ \Phi^{(3)} \\ \Phi^{(4)} \\ \Phi^{(5)} \\ \Phi^{(6)} \\ \Phi^{(7)} \\ \Phi^{(8)} \end{bmatrix} = \begin{bmatrix} P^{(1)} \\ P^{(2)} \\ P^{(3)} \\ P^{(4)} \\ P^{(5)} \\ P^{(6)} \\ P^{(7)} \\ P^{(8)} \end{bmatrix}$$

with the group supermatrix G in normal form given earlier, is transformed into the system of equations $\bar{G}\bar{\Phi} = \bar{P}$, with the group supermatrix $\bar{G}$ in diagonal form, by the group supermatrix transformation $TGT^{-1}T\Phi = TP$, which results in (see p. 131).

The system of equations $\bar{G}\bar{\Phi} = \bar{P}$ is transformed into the system of equations $G\Phi = P$, with the group supermatrix G in normal form, by the inverse group supermatrix transformation $T^{-1}\bar{G}TT^{-1}\bar{\Phi} = T^{-1}\bar{P}$.

3.8 GROUP SUPERMATRICES OF THE GROUP D_{2h}

The character table of the group D_{2h}

D_{2h}	E	C_2^z	C_2^y	C_2^x	i	σ_{xy}	σ_{xz}	σ_{yz}
A_g	1	1	1	1	1	1	1	1
B_{1g}	1	1	-1	-1	1	1	-1	-1
B_{2g}	1	-1	1	-1	1	-1	1	-1
B_{3g}	1	-1	-1	1	1	-1	-1	1
A_u	1	1	1	1	-1	-1	-1	-1
B_{1u}	1	1	-1	-1	-1	-1	1	1
B_{2u}	1	-1	1	-1	-1	1	-1	1
B_{3u}	1	-1	-1	1	-1	1	1	-1

has the character matrix that can be written as

$$\begin{bmatrix} M & M \\ M & -M \end{bmatrix} = \begin{bmatrix} M & \\ & M \end{bmatrix} \begin{bmatrix} E & E \\ E & -E \end{bmatrix},$$

where

$$M = \begin{bmatrix} 1 & 1 & 1 & 1 \\ 1 & 1 & -1 & -1 \\ 1 & -1 & 1 & -1 \\ 1 & -1 & -1 & 1 \end{bmatrix}$$

is the matrix of the character table of the group C_{2v}, while

$$\begin{bmatrix} E & E \\ E & -E \end{bmatrix}$$

is the supermatrix of the character table of the group C_2.

The sequence of symmetry operations $E, C_2^z, C_2^y, C_2^x, i, \sigma_{xy}, \sigma_{xz}, \sigma_{yz}$ at the top of the character table of the group D_{2h} will be used to derive the nodal numbering of the eight-node rectangular hexahedral element.

$$
\begin{bmatrix}
A+B+C+D+H+I+J+K \\
A+B+C+D-H-I-J-K \\
A-B-C+D+H+I-J-K \\
A-B-C+D-H-I+J+K \\
A-D \quad B-C \quad H-D \quad J-K \\
B-C \quad -A+D \quad J-K \quad -H+I \\
H-I \quad -J+K \quad A-D \quad -B+C \\
J-K \quad H-I \quad B-C \quad A-D
\end{bmatrix}
\begin{bmatrix}
\bar{\Phi}^{(1)} \\
\bar{\Phi}^{(2)} \\
\bar{\Phi}^{(3)} \\
\bar{\Phi}^{(4)} \\
\bar{\Phi}^{(5)} \\
\bar{\Phi}^{(6)} \\
\bar{\Phi}^{(7)} \\
\bar{\Phi}^{(8)}
\end{bmatrix}
=
\begin{bmatrix}
\bar{P}^{(1)} \\
\bar{P}^{(2)} \\
\bar{P}^{(3)} \\
\bar{P}^{(4)} \\
\bar{P}^{(5)} \\
\bar{P}^{(6)} \\
\bar{P}^{(7)} \\
\bar{P}^{(8)}
\end{bmatrix}
$$

132   Group supermatrix transformations   [Ch. 3

The position of the first node, numbered by 1, is in the first octant of the rectangular *xyz* coordinate system, where branches of *x*, *y* and *z* axes are positive, as given in Fig. 3.11.

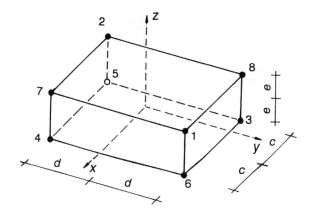

Fig. 3.11. The eight-node rectangular hexahedral element described by the group D_{2h}.

According to the rule applied in previous cases, it is assumed that application of E, C_2^z, C_2^y, C_2^x, i, σ_{xy}, σ_{xz}, σ_{yz} (in this order) to the initial node 1 provides the nodes 1, 2, 3, 4, 5, 6, 7, 8 (in this sequence). This definition determines the unique nodal numbering in Fig. 3.11, which is stated by the following table

Node	1	2	3	4	5	6	7	8
x	+	−	−	+	−	+	+	−
y	+	−	+	−	−	+	−	+
z	+	+	−	−	−	−	+	+

Expressed as permutations of the nodes, the results of actions of symmetry operations on the initial sequence [1 2 3 4 5 6 7 8] are

$$E = \begin{bmatrix} 1 & 2 & 3 & 4 & 5 & 6 & 7 & 8 \\ 1 & 2 & 3 & 4 & 5 & 6 & 7 & 8 \end{bmatrix}, \quad i = \begin{bmatrix} 1 & 2 & 3 & 4 & 5 & 6 & 7 & 8 \\ 5 & 6 & 7 & 8 & 1 & 2 & 3 & 4 \end{bmatrix}$$

$$C_2^z = \begin{bmatrix} 1 & 2 & 3 & 4 & 5 & 6 & 7 & 8 \\ 2 & 1 & 4 & 3 & 6 & 5 & 8 & 7 \end{bmatrix} \quad \sigma_{xy} = \begin{bmatrix} 1 & 2 & 3 & 4 & 5 & 6 & 7 & 8 \\ 6 & 5 & 8 & 7 & 2 & 1 & 4 & 3 \end{bmatrix}$$

$$C_2^y = \begin{bmatrix} 1 & 2 & 3 & 4 & 5 & 6 & 7 & 8 \\ 3 & 4 & 1 & 2 & 7 & 8 & 5 & 6 \end{bmatrix} \quad \sigma_{xz} = \begin{bmatrix} 1 & 2 & 3 & 4 & 5 & 6 & 7 & 8 \\ 7 & 8 & 5 & 6 & 3 & 4 & 1 & 2 \end{bmatrix}$$

$$C_2^x = \begin{bmatrix} 1 & 2 & 3 & 4 & 5 & 6 & 7 & 8 \\ 4 & 3 & 2 & 1 & 8 & 7 & 6 & 5 \end{bmatrix} \qquad \sigma_{yz} = \begin{bmatrix} 1 & 2 & 3 & 4 & 5 & 6 & 7 & 8 \\ 8 & 7 & 6 & 5 & 4 & 3 & 2 & 1 \end{bmatrix},$$

or presented in the form of a permutation table:

Node	E	C_2^z	C_2^y	C_2^x	i	σ_{xy}	σ_{xz}	σ_{yz}
1	1	2	3	4	5	6	7	8
2	2	1	4	3	6	5	8	7
3	3	4	1	2	7	8	5	6
4	4	3	2	1	8	7	6	5
5	5	6	7	8	1	2	3	4
6	6	5	8	7	2	1	4	3
7	7	8	5	6	3	4	1	2
8	8	7	6	5	4	3	2	1

By replacing the numbers 1, 2, 3, 4, 5, 6, 7, 8 by matrices A, B, C, D, H, I, J, K in the above permutation table, one obtains the supermatrix

$$G = \begin{bmatrix} A & B & C & D & H & I & J & K \\ B & A & D & C & I & H & K & J \\ C & D & A & B & J & K & H & I \\ D & C & B & A & K & J & I & H \\ H & I & J & K & A & B & C & D \\ I & H & K & J & B & A & D & C \\ J & K & H & I & C & D & A & B \\ K & J & I & H & D & C & B & A \end{bmatrix},$$

which is the group supermatrix in normal form pertaining to the group D_{2h}. It can be written as the group supermatrix in normal form pertaining to the group C_{2v}

$$G = \begin{bmatrix} G_1 & G_2 & G_3 & G_4 \\ G_2 & G_1 & G_4 & G_3 \\ G_3 & G_4 & G_1 & G_2 \\ G_4 & G_3 & G_2 & G_1 \end{bmatrix},$$

where

$$G_1 = \begin{bmatrix} A & B \\ B & A \end{bmatrix}, \quad G_2 = \begin{bmatrix} C & D \\ D & C \end{bmatrix}, \quad G_3 = \begin{bmatrix} H & I \\ I & H \end{bmatrix}, \quad G_4 = \begin{bmatrix} J & K \\ K & J \end{bmatrix}$$

are group supermatrices in normal form pertaining to the group C_2 derived in 3.2.

The supermatrix G can be also written as

$$G = \begin{bmatrix} G_I & G_{II} \\ G_{II} & G_I \end{bmatrix}$$

where

$$G_I = \begin{bmatrix} A & B & C & D \\ B & A & D & C \\ C & D & A & B \\ D & C & B & A \end{bmatrix}, \quad G_{II} = \begin{bmatrix} H & I & J & K \\ I & H & K & J \\ J & K & H & I \\ K & J & I & H \end{bmatrix}$$

are group supermatrices in normal form pertaining to the group C_{2v} derived in section 3.5.

The idempotents π_i ($i = 1, 2, \ldots, 8$) of the centre of the group algebra are determined by

$$\pi_i = \frac{h_i}{h} \sum_\sigma \chi_i(\sigma^{-1})\sigma,$$

with $h_i = \chi_i(E)$, h the order, χ_i the ith character, σ the element and σ^{-1} the inverse of the element of the group, so that

$$\begin{bmatrix} \pi_1 \\ \pi_2 \\ \pi_3 \\ \pi_4 \\ \pi_5 \\ \pi_6 \\ \pi_7 \\ \pi_8 \end{bmatrix} = \frac{1}{8} \begin{bmatrix} 1 & 1 & 1 & 1 & 1 & 1 & 1 & 1 \\ 1 & 1 & -1 & -1 & 1 & 1 & -1 & -1 \\ 1 & -1 & 1 & -1 & 1 & -1 & 1 & -1 \\ 1 & -1 & -1 & 1 & 1 & -1 & -1 & 1 \\ 1 & 1 & 1 & 1 & -1 & -1 & -1 & -1 \\ 1 & 1 & -1 & -1 & -1 & -1 & 1 & 1 \\ 1 & -1 & 1 & -1 & -1 & 1 & -1 & 1 \\ 1 & -1 & -1 & 1 & -1 & 1 & 1 & -1 \end{bmatrix} \begin{bmatrix} E \\ C_2^z \\ C_2^y \\ C_2^x \\ i \\ \sigma_{xy} \\ \sigma_{xz} \\ \sigma_{yz} \end{bmatrix}$$

or $\Pi = T_c \Sigma$, where T_c is the transformation matrix, with

$$T_c^{-1} = \frac{h}{h_i} T_c = 8 T_c.$$

When in the matrix T_c the unit values are replaced by $l \times l$ unit matrices E, the transformation supermatrix, pertaining to the group D_{2h}, is obtained:

$$T_c = \frac{1}{8} \begin{bmatrix} E & E & E & E & E & E & E & E \\ E & E & -E & -E & E & E & -E & -E \\ E & -E & E & -E & E & -E & E & -E \\ E & -E & -E & E & E & -E & -E & E \\ E & E & E & E & -E & -E & -E & -E \\ E & E & -E & -E & -E & -E & E & E \\ E & -E & E & -E & -E & E & -E & E \\ E & -E & -E & E & -E & E & E & -E \end{bmatrix},$$

with

$$T_c^{-1} = \begin{bmatrix} E & E & E & E & E & E & E & E \\ E & E & -E & -E & E & E & -E & -E \\ E & -E & E & -E & E & -E & E & -E \\ E & -E & -E & E & E & -E & -E & E \\ E & E & E & E & -E & -E & -E & -E \\ E & E & -E & -E & -E & -E & E & E \\ E & -E & E & -E & -E & E & -E & E \\ E & -E & -E & E & -E & E & E & -E \end{bmatrix}$$

since $T_c^{-1} = 8T_c$.

The sets of the nodal functions $\Phi^{(i)}$ ($i = 1, 2, \ldots, 8$), pertaining to the nodes $1, 2, \ldots, 8$, may contain one or more functions at a node. When l is the number of nodal functions at each node, the following numbering of nodal functions φ_j ($j = 1, 2, \ldots, n$; $n = 8l$) is introduced:

for $l = 1$: $\begin{bmatrix} \Phi^{(1)} \\ \Phi^{(2)} \\ \Phi^{(3)} \\ \Phi^{(4)} \\ \Phi^{(5)} \\ \Phi^{(6)} \\ \Phi^{(7)} \\ \Phi^{(8)} \end{bmatrix} = \begin{bmatrix} \varphi_1 \\ \varphi_2 \\ \varphi_3 \\ \varphi_4 \\ \varphi_5 \\ \varphi_6 \\ \varphi_7 \\ \varphi_8 \end{bmatrix}$, for $l = 2$: $\begin{bmatrix} \Phi^{(1)} \\ \Phi^{(2)} \\ \Phi^{(3)} \\ \Phi^{(4)} \\ \Phi^{(5)} \\ \Phi^{(6)} \\ \Phi^{(7)} \\ \Phi^{(8)} \end{bmatrix} = \begin{bmatrix} \varphi_1 & \varphi_9 \\ \varphi_2 & \varphi_{10} \\ \varphi_3 & \varphi_{11} \\ \varphi_4 & \varphi_{12} \\ \varphi_5 & \varphi_{13} \\ \varphi_6 & \varphi_{14} \\ \varphi_7 & \varphi_{15} \\ \varphi_8 & \varphi_{16} \end{bmatrix}$,

and similarly for larger values of l.

Applying the idempotents $\pi_1, \pi_2, \ldots, \pi_8$ to the sets of nodal functions $\Phi^{(1)}, \Phi^{(2)}, \ldots, \Phi^{(8)}$ derives the sets of basis vectors $\bar{\Phi}^{(1)}, \bar{\Phi}^{(2)}, \ldots, \bar{\Phi}^{(8)}$ of G-invariant subspaces $U_1, U_2, \ldots, U_8$:

$$\begin{bmatrix} \bar\Phi^{(1)} \\ \bar\Phi^{(2)} \\ \bar\Phi^{(3)} \\ \bar\Phi^{(4)} \\ \bar\Phi^{(5)} \\ \bar\Phi^{(6)} \\ \bar\Phi^{(7)} \\ \bar\Phi^{(8)} \end{bmatrix} = \frac{1}{8} \begin{bmatrix} E & E & E & E & E & E & E & E \\ E & E & -E & -E & E & E & -E & -E \\ E & -E & E & -E & E & -E & E & -E \\ E & -E & -E & E & E & -E & -E & E \\ E & E & E & E & -E & -E & -E & -E \\ E & E & -E & -E & -E & -E & E & E \\ E & -E & E & -E & -E & E & -E & E \\ E & -E & -E & E & -E & E & E & -E \end{bmatrix} \begin{bmatrix} \Phi^{(1)} \\ \Phi^{(2)} \\ \Phi^{(3)} \\ \Phi^{(4)} \\ \Phi^{(5)} \\ \Phi^{(6)} \\ \Phi^{(7)} \\ \Phi^{(8)} \end{bmatrix}$$

or $\bar\Phi = T\Phi$, where $T = T_c$ is the basis transformation supermatrix.

If l is the number of nodal functions at each node, the numbering of basis vectors $\bar\varphi_j$ ($j = 1, 2, \ldots, n; n = 8l$) corresponds to the numbering of nodal functions φ_j:

for $l = 1$: $\begin{bmatrix} \bar\Phi^{(1)} \\ \bar\Phi^{(2)} \\ \bar\Phi^{(3)} \\ \bar\Phi^{(4)} \\ \bar\Phi^{(5)} \\ \bar\Phi^{(6)} \\ \bar\Phi^{(7)} \\ \bar\Phi^{(8)} \end{bmatrix} = \begin{bmatrix} \bar\varphi_1 \\ \bar\varphi_2 \\ \bar\varphi_3 \\ \bar\varphi_4 \\ \bar\varphi_5 \\ \bar\varphi_6 \\ \bar\varphi_7 \\ \bar\varphi_8 \end{bmatrix}$, for $l = 2$: $\begin{bmatrix} \bar\Phi^{(1)} \\ \bar\Phi^{(2)} \\ \bar\Phi^{(3)} \\ \bar\Phi^{(4)} \\ \bar\Phi^{(5)} \\ \bar\Phi^{(6)} \\ \bar\Phi^{(7)} \\ \bar\Phi^{(8)} \end{bmatrix} = \begin{bmatrix} \bar\varphi_1 & \bar\varphi_9 \\ \bar\varphi_2 & \bar\varphi_{10} \\ \bar\varphi_3 & \bar\varphi_{11} \\ \bar\varphi_4 & \bar\varphi_{12} \\ \bar\varphi_5 & \bar\varphi_{13} \\ \bar\varphi_6 & \bar\varphi_{14} \\ \bar\varphi_7 & \bar\varphi_{15} \\ \bar\varphi_8 & \bar\varphi_{16} \end{bmatrix}$,

and similarly for larger values of l.

Conversely, the relation of the sets of nodal functions $\Phi^{(1)}, \Phi^{(2)}, \ldots, \Phi^{(8)}$ to the sets of basis vectors $\bar\Phi^{(1)}, \bar\Phi^{(2)}, \ldots, \bar\Phi^{(8)}$ is

$$\begin{bmatrix} \Phi^{(1)} \\ \Phi^{(2)} \\ \Phi^{(3)} \\ \Phi^{(4)} \\ \Phi^{(5)} \\ \Phi^{(6)} \\ \Phi^{(7)} \\ \Phi^{(8)} \end{bmatrix} = \begin{bmatrix} E & E & E & E & E & E & E & E \\ E & E & -E & -E & E & E & -E & -E \\ E & -E & E & -E & E & -E & E & -E \\ E & -E & -E & E & E & -E & -E & E \\ E & E & E & E & -E & -E & -E & -E \\ E & E & -E & -E & -E & -E & E & E \\ E & -E & E & -E & -E & E & -E & E \\ E & -E & -E & E & -E & E & E & -E \end{bmatrix} \begin{bmatrix} \bar\Phi^{(1)} \\ \bar\Phi^{(2)} \\ \bar\Phi^{(3)} \\ \bar\Phi^{(4)} \\ \bar\Phi^{(5)} \\ \bar\Phi^{(6)} \\ \bar\Phi^{(7)} \\ \bar\Phi^{(8)} \end{bmatrix}$$

or $\Phi = 8T\bar\Phi$.

The sets of basis vectors $\bar\Phi_1, \bar\Phi_2, \ldots, \bar\Phi_l$, written as are in relation to the sets of nodal functions $\Phi_1, \Phi_2, \ldots, \Phi_l$ expressed by

$$[\bar{\Phi}_1 \ \bar{\Phi}_2 \ \ldots \ \bar{\Phi}_l] = \begin{bmatrix} \bar{\varphi}_1 & \bar{\varphi}_9 & \cdots & \bar{\varphi}_{n-7} \\ \bar{\varphi}_2 & \bar{\varphi}_{10} & \cdots & \bar{\varphi}_{n-6} \\ \bar{\varphi}_3 & \bar{\varphi}_{11} & \cdots & \bar{\varphi}_{n-5} \\ \bar{\varphi}_4 & \bar{\varphi}_{12} & \cdots & \bar{\varphi}_{n-4} \\ \bar{\varphi}_5 & \bar{\varphi}_{13} & \cdots & \bar{\varphi}_{n-3} \\ \bar{\varphi}_6 & \bar{\varphi}_{14} & \cdots & \bar{\varphi}_{n-2} \\ \bar{\varphi}_7 & \bar{\varphi}_{15} & \cdots & \bar{\varphi}_{n-1} \\ \bar{\varphi}_8 & \bar{\varphi}_{16} & \cdots & \bar{\varphi}_n \end{bmatrix}$$

$$\begin{bmatrix} \bar{\Phi}_1 \\ \bar{\Phi}_2 \\ \vdots \\ \bar{\Phi}_l \end{bmatrix} = \begin{bmatrix} T & & & \\ & T & & \\ & & \ddots & \\ & & & T \end{bmatrix} \begin{bmatrix} \Phi_1 \\ \Phi_2 \\ \vdots \\ \Phi_l \end{bmatrix} \quad \text{or} \quad \bar{\Phi} = \bar{T}\Phi,$$

with

$$T = \frac{1}{8} \begin{bmatrix} E & E & E & E & E & E & E & E \\ E & E & -E & -E & E & E & -E & -E \\ E & -E & E & -E & E & -E & E & -E \\ E & -E & -E & E & E & -E & -E & E \\ E & E & E & E & -E & -E & -E & -E \\ E & E & -E & -E & -E & -E & E & E \\ E & -E & E & -E & -E & E & -E & E \\ E & -E & -E & E & -E & E & E & -E \end{bmatrix}$$

and

$$[\Phi_1 \ \Phi_2 \ \ldots \ \Phi_l] = \begin{bmatrix} \varphi_1 & \varphi_9 & \cdots & \varphi_{n-7} \\ \varphi_2 & \varphi_{10} & \cdots & \varphi_{n-6} \\ \varphi_3 & \varphi_{11} & \cdots & \varphi_{n-5} \\ \varphi_4 & \varphi_{12} & \cdots & \varphi_{n-4} \\ \varphi_5 & \varphi_{13} & \cdots & \varphi_{n-3} \\ \varphi_6 & \varphi_{14} & \cdots & \varphi_{n-2} \\ \varphi_7 & \varphi_{15} & \cdots & \varphi_{n-1} \\ \varphi_8 & \varphi_{16} & \cdots & \varphi_n \end{bmatrix}.$$

Conversely, the relation of the sets of nodal functions $\Phi_1, \Phi_2, \ldots, \Phi_l$ to the sets of basis vectors $\bar{\Phi}_1, \bar{\Phi}_2, \ldots, \bar{\Phi}_l$ is

$$\begin{bmatrix} \Phi_1 \\ \Phi_2 \\ \vdots \\ \Phi_l \end{bmatrix} = \begin{bmatrix} T^{-1} & & & \\ & T^{-1} & & \\ & & \ddots & \\ & & & T^{-1} \end{bmatrix} \begin{bmatrix} \bar{\Phi}_1 \\ \bar{\Phi}_2 \\ \vdots \\ \bar{\Phi}_l \end{bmatrix} = 8 \begin{bmatrix} T & & & \\ & T & & \\ & & \ddots & \\ & & & T \end{bmatrix} \begin{bmatrix} \bar{\Phi}_1 \\ \bar{\Phi}_2 \\ \vdots \\ \bar{\Phi}_l \end{bmatrix},$$

or $\Phi = 8\bar{T}\bar{\Phi}$, since $T^{-1} = 8T$.

The system of equations $G\Phi = P$, with the group supermatrix G in normal form derived earlier, expressed as

$$\begin{bmatrix} A & B & C & D & H & I & J & K \\ B & A & D & C & I & H & K & J \\ C & D & D & B & J & K & H & I \\ D & C & B & A & K & J & I & H \\ H & I & J & K & A & B & C & D \\ I & H & K & J & B & A & D & C \\ J & K & H & I & C & D & A & B \\ K & J & I & H & D & C & B & A \end{bmatrix} \begin{bmatrix} \Phi^{(1)} \\ \Phi^{(2)} \\ \Phi^{(3)} \\ \Phi^{(4)} \\ \Phi^{(5)} \\ \Phi^{(6)} \\ \Phi^{(7)} \\ \Phi^{(8)} \end{bmatrix} = \begin{bmatrix} P^{(1)} \\ P^{(2)} \\ P^{(3)} \\ P^{(4)} \\ P^{(5)} \\ P^{(6)} \\ P^{(7)} \\ P^{(8)} \end{bmatrix},$$

can be transformed into the system of equations $\bar{G}\bar{\Phi} = \bar{P}$, with the group supermatrix $\bar{G}$ in diagonal form, by the group supermatrix transformation $TGT^{-1}T\Phi = TP$:

$$\frac{1}{8} \begin{bmatrix} E & E & E & E & E & E & E & E \\ E & E & -E & -E & E & E & -E & -E \\ E & -E & E & -E & E & -E & E & -E \\ E & -E & -E & E & E & -E & -E & E \\ E & E & E & E & -E & -E & -E & -E \\ E & E & -E & -E & -E & -E & E & E \\ E & -E & E & -E & -E & E & -E & E \\ E & -E & -E & E & -E & E & E & -E \end{bmatrix} \begin{bmatrix} A & B & C & D & H & I & J & K \\ B & A & D & C & I & H & K & J \\ C & D & A & B & J & K & H & I \\ D & C & B & A & K & J & I & H \\ H & I & J & K & A & B & C & D \\ I & H & K & J & B & A & D & C \\ J & K & H & I & C & D & A & B \\ K & J & I & H & D & C & B & A \end{bmatrix}$$

Group supermatrices of the group D_{2h}

$$\begin{bmatrix} E & E & E & E & E & E & E & E \\ E & E & -E & -E & E & E & -E & -E \\ E & -E & E & -E & E & -E & E & -E \\ E & -E & -E & E & E & -E & -E & E \\ E & E & E & E & -E & -E & -E & -E \\ E & E & -E & -E & -E & -E & E & E \\ E & -E & E & -E & -E & E & -E & E \\ E & -E & -E & E & -E & E & E & -E \end{bmatrix} \cdot$$

$$\cdot \frac{1}{8} \begin{bmatrix} E & E & E & E & E & E & E & E \\ E & E & -E & -E & E & E & -E & -E \\ E & -E & E & -E & E & -E & E & -E \\ E & -E & -E & E & E & -E & -E & E \\ E & E & E & E & -E & -E & -E & -E \\ E & E & -E & -E & -E & -E & E & E \\ E & -E & E & -E & -E & E & -E & E \\ E & -E & -E & E & -E & E & E & -E \end{bmatrix} \begin{bmatrix} \Phi^{(1)} \\ \Phi^{(2)} \\ \Phi^{(3)} \\ \Phi^{(4)} \\ \Phi^{(5)} \\ \Phi^{(6)} \\ \Phi^{(7)} \\ \Phi^{(8)} \end{bmatrix}$$

$$= \frac{1}{8} \begin{bmatrix} E & E & E & E & E & E & E & E \\ E & E & -E & -E & E & E & -E & -E \\ E & -E & E & -E & E & -E & E & -E \\ E & -E & -E & E & E & -E & -E & E \\ E & E & E & E & -E & -E & -E & -E \\ E & E & -E & -E & -E & -E & E & E \\ E & -E & E & -E & -E & E & -E & E \\ E & -E & -E & E & -E & E & E & -E \end{bmatrix} \begin{bmatrix} P^{(1)} \\ P^{(2)} \\ P^{(3)} \\ P^{(4)} \\ P^{(5)} \\ P^{(6)} \\ P^{(7)} \\ P^{(8)} \end{bmatrix},$$

resulting in

140   Group supermatrix transformations   [Ch. 3

$$\text{Diag} \begin{bmatrix} A+B+C+D+H+I+J+K \\ A+B-C-D+H+I-J-K \\ A-B+C-D+H-I+J-K \\ A-B-C+D+H-I-J+K \\ A+B+C+D-H-I-J-K \\ A+B-C-D-H-I+J+K \\ A-B+C-D-H+I-J+K \\ A-B-C+D-H+I+J-K \end{bmatrix} \begin{bmatrix} \bar{\Phi}^{(1)} \\ \bar{\Phi}^{(2)} \\ \bar{\Phi}^{(3)} \\ \bar{\Phi}^{(4)} \\ \bar{\Phi}^{(5)} \\ \bar{\Phi}^{(6)} \\ \bar{\Phi}^{(7)} \\ \bar{\Phi}^{(8)} \end{bmatrix} = \begin{bmatrix} \bar{P}^{(1)} \\ \bar{P}^{(2)} \\ \bar{P}^{(3)} \\ \bar{P}^{(4)} \\ \bar{P}^{(5)} \\ \bar{P}^{(6)} \\ \bar{P}^{(7)} \\ \bar{P}^{(8)} \end{bmatrix}$$

or $\bar{G}\bar{\Phi} = \bar{P}$.

Conversely, the system of equations $\bar{G}\bar{\Phi} = \bar{P}$, with the group supermatrix $\bar{G}$ in diagonal form, is transformed into the system of equations $G\Phi = P$, with the group supermatrix G in normal form, by the inverse group supermatrix transformation $T^{-1}\bar{G}TT^{-1}\bar{\Phi} = T^{-1}\bar{P}$:

$$\begin{bmatrix} E & E & E & E & E & E & E & E \\ E & E & -E & -E & E & E & -E & -E \\ E & -E & E & -E & E & -E & E & -E \\ E & -E & -E & E & E & -E & -E & E \\ E & E & E & E & -E & -E & -E & -E \\ E & E & -E & -E & -E & -E & E & E \\ E & -E & E & -E & -E & E & -E & E \\ E & -E & -E & E & -E & E & E & -E \end{bmatrix} \text{Diag} \begin{bmatrix} A+B+C+D+H+I+J+K \\ A+B-C-D+H+I-J-K \\ A-B+C-D+H-I+J-K \\ A-B-C+D+H-I-J+K \\ A+B+C+D-H-I-J-K \\ A+B-C-D-H-I+J+K \\ A-B+C-D-H+I-J+K \\ A-B-C+D-H+I+J-K \end{bmatrix}$$

$$\cdot \frac{1}{8} \begin{bmatrix} E & E & E & E & E & E & E & E \\ E & E & -E & -E & E & E & -E & -E \\ E & -E & E & -E & E & -E & E & -E \\ E & -E & -E & E & E & -E & -E & E \\ E & E & E & E & -E & -E & -E & -E \\ E & E & -E & -E & -E & -E & E & E \\ E & -E & E & -E & -E & E & -E & E \\ E & -E & -E & E & -E & E & E & -E \end{bmatrix} \cdot$$

$$\circ \begin{bmatrix} E & E & E & E & E & E & E & E \\ E & E & -E & -E & E & E & -E & -E \\ E & -E & E & -E & E & -E & E & -E \\ E & -E & -E & E & E & -E & -E & E \\ E & E & E & E & -E & -E & -E & -E \\ E & E & -E & -E & -E & -E & E & E \\ E & -E & E & -E & -E & E & -E & E \\ E & -E & -E & E & -E & E & E & -E \end{bmatrix} \begin{bmatrix} \bar{\Phi}^{(1)} \\ \bar{\Phi}^{(2)} \\ \bar{\Phi}^{(3)} \\ \bar{\Phi}^{(4)} \\ \bar{\Phi}^{(5)} \\ \bar{\Phi}^{(6)} \\ \bar{\Phi}^{(7)} \\ \bar{\Phi}^{(8)} \end{bmatrix}$$

$$= \begin{bmatrix} E & E & E & E & E & E & E & E \\ E & E & -E & -E & E & E & -E & -E \\ E & -E & E & -E & E & -E & E & -E \\ E & -E & -E & E & E & -E & -E & E \\ E & E & E & E & -E & -E & -E & -E \\ E & E & -E & -E & -E & -E & E & E \\ E & -E & E & -E & -E & E & -E & E \\ E & -E & -E & E & -E & E & E & -E \end{bmatrix} \begin{bmatrix} \bar{P}^{(1)} \\ \bar{P}^{(2)} \\ \bar{P}^{(3)} \\ \bar{P}^{(4)} \\ \bar{P}^{(5)} \\ \bar{P}^{(6)} \\ \bar{P}^{(7)} \\ \bar{P}^{(8)} \end{bmatrix},$$

resulting in

$$\begin{bmatrix} A & B & C & D & H & I & J & K \\ B & A & D & C & I & H & K & J \\ C & D & A & B & J & K & H & I \\ D & C & B & A & K & J & I & H \\ H & I & J & K & A & B & C & D \\ I & H & K & J & B & A & D & C \\ J & K & H & I & C & D & A & B \\ K & J & I & H & D & C & B & A \end{bmatrix} \begin{bmatrix} \Phi^{(1)} \\ \Phi^{(2)} \\ \Phi^{(3)} \\ \Phi^{(4)} \\ \Phi^{(5)} \\ \Phi^{(6)} \\ \Phi^{(7)} \\ \Phi^{(8)} \end{bmatrix} = \begin{bmatrix} P^{(1)} \\ P^{(2)} \\ P^{(3)} \\ P^{(4)} \\ P^{(5)} \\ P^{(6)} \\ P^{(7)} \\ P^{(8)} \end{bmatrix}.$$

As in the case of the eight-node rectangular hexahedral element, the sequence of symmetry operations E, C_2^z, C_2^y, C_2^x, i, σ_{xy}, σ_{xz}, σ_{yz}, existing in the character table of the group D_{2h}, will be used for definition of the unique nodal numbering of the twelve-node truncated rectangular hexahedral element shown in Fig. 3.12(a).

The nodes can be grouped into three nodal sets, where the nodes are permuted by action of symmetry operations of the group

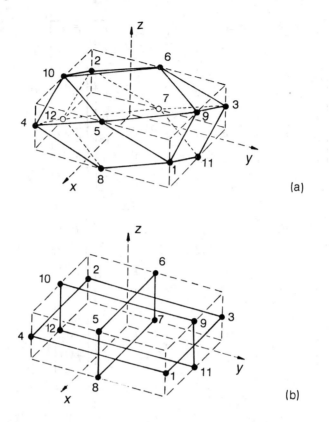

Fig. 3.12. (a) The twelve-node truncated rectangular hexahedral element; (b) the assembly of three four-node perpendicular rectangular elements.

nodes in the xy plane: $S_1(1, 2, 3, 4)$

nodes in the xz plane: $S_2(5, 6, 7, 8)$

nodes in the yz plane: $S_3(9, 10, 11, 12)$,

with the order of xy, xz, yz planes according to the sequence of σ_{xy}, σ_{xz}, σ_{yz} in the character table, as given in Fig. 3.12(b), which shows the assembly of three four-node perpendicular rectangular elements.

The positions of the first nodes in the nodal sets S_1, S_2, S_3, i.e. the nodes 1, 5, 9, are at the borders of the first octant of the xyz coordinate system, where branches of x, y, z axes are positive.

According to the rule, it is assumed that application of E, C_2^z, C_2^y, C_2^x (in this order) to the initial nodes 1, 5, 9 provides the nodes in the following sequences:

to the node 1:   1, 2, 3, 4

to the node 5:   5, 6, 7, 8

to the node 9:   9, 10, 11, 12.

This definition determines the unique nodal numbering in Fig. 3.12(a), (b), which is given in tabular form as

Node	1	2	3	4	5	6	7	8	9	10	11	12
x	+	−	−	+	+	−	−	+				
y	+	−	+	−					+	−	+	−
z					+	+	−	−	+	+	−	−

Expressed as permutations of the nodes, the actions of symmetry operations are presented in the permutation table as follows:

	E	C_2^z	C_2^y	C_2^x	i	σ_{xy}	σ_{xz}	σ_{yz}
1	1	2	3	4	2	1	4	3
2	2	1	4	3	1	2	3	4
3	3	4	1	2	4	3	2	1
4	4	3	2	1	3	4	1	2
5	5	6	7	8	7	8	5	6
6	6	5	8	7	8	7	6	5
7	7	8	5	6	5	6	7	8
8	8	7	6	5	6	5	8	7
9	9	10	11	12	12	11	10	9
10	10	9	12	11	11	12	9	10
11	11	12	9	10	10	9	12	11
12	12	11	10	9	9	10	11	12

This permutation table contains six permutation tables pertaining to the group C_{2v}, as derived in section 3.5 for the four-node rectangular element and expressed by the group supermatrix in normal form:

$$G = \begin{bmatrix} A & B & C & D \\ B & A & D & C \\ C & D & A & B \\ D & C & B & A \end{bmatrix}.$$

Applying the idempotents π_i ($i = 1, 2, \ldots, 8$), derived earlier, to the sets of nodal functions $(\Phi^{(1)}, \Phi^{(2)}, \Phi^{(3)}, \Phi^{(4)})$, $(\Phi^{(5)}, \Phi^{(6)}, \Phi^{(7)}, \Phi^{(8)})$, $(\Phi^{(9)}, \Phi^{(10)}, \Phi^{(11)}, \Phi^{(12)})$, pertaining to the nodal sets $S_1(1, 2, 3, 4), S_2(5, 6, 7, 8), S_3(9, 10, 11, 12)$, derives the sets of basis vectors $\bar{\Phi}$ of G-invariant subspaces U_i ($i = 1, 2, \ldots, 8$)

$U_1:$ $\bar{\Phi}^{(1)} = \frac{1}{4}(\Phi^{(1)} + \Phi^{(2)} + \Phi^{(3)} + \Phi^{(4)})$

$\bar{\Phi}^{(5)} = \frac{1}{4}(\Phi^{(5)} + \Phi^{(6)} + \Phi^{(7)} + \Phi^{(8)})$

$\bar{\Phi}^{(9)} = \frac{1}{4}(\Phi^{(9)} + \Phi^{(10)} + \Phi^{(11)} + \Phi^{(12)})$

$U_2:$ $\bar{\Phi}^{(2)} = \frac{1}{4}(\Phi^{(1)} + \Phi^{(2)} - \Phi^{(3)} - \Phi^{(4)})$

$U_3:$ $\bar{\Phi}^{(7)} = \frac{1}{4}(\Phi^{(5)} - \Phi^{(6)} + \Phi^{(7)} - \Phi^{(8)})$

$U_4:$ $\bar{\Phi}^{(12)} = \frac{1}{4}(\Phi^{(9)} - \Phi^{(10)} - \Phi^{(11)} + \Phi^{(12)})$

$U_5:$ zero subspace

$U_6:$ $\bar{\Phi}^{(6)} = \frac{1}{4}(\Phi^{(5)} + \Phi^{(6)} - \Phi^{(7)} - \Phi^{(8)})$

$\bar{\Phi}^{(10)} = \frac{1}{4}(\Phi^{(9)} + \Phi^{(10)} - \Phi^{(11)} - \Phi^{(12)})$

$U_7:$ $\bar{\Phi}^{(3)} = \frac{1}{4}(\Phi^{(1)} - \Phi^{(2)} + \Phi^{(3)} - \Phi^{(4)})$

$\bar{\Phi}^{(11)} = \frac{1}{4}(\Phi^{(9)} - \Phi^{(10)} + \Phi^{(11)} - \Phi^{(12)})$

$U_8:$ $\bar{\Phi}^{(4)} = \frac{1}{4}(\Phi^{(1)} - \Phi^{(2)} - \Phi^{(3)} + \Phi^{(4)})$

$\bar{\Phi}^{(8)} = \frac{1}{4}(\Phi^{(5)} - \Phi^{(6)} - \Phi^{(7)} + \Phi^{(8)}).$

The relation of the sets of basis vectors $\bar{\Phi}$ to the sets of nodal functions Φ is

$$\begin{bmatrix} \bar{\Phi}^{(1)} \\ \bar{\Phi}^{(2)} \\ \bar{\Phi}^{(3)} \\ \bar{\Phi}^{(4)} \\ \bar{\Phi}^{(5)} \\ \bar{\Phi}^{(6)} \\ \bar{\Phi}^{(7)} \\ \bar{\Phi}^{(8)} \\ \bar{\Phi}^{(9)} \\ \bar{\Phi}^{(10)} \\ \bar{\Phi}^{(11)} \\ \bar{\Phi}^{(12)} \end{bmatrix} = \frac{1}{4} \begin{bmatrix} E & E & E & E & & & & & & & & \\ E & E & -E & -E & & & & & & & & \\ E & -E & E & -E & & & & & & & & \\ E & -E & -E & E & & & & & & & & \\ & & & & E & E & E & E & & & & \\ & & & & E & E & -E & -E & & & & \\ & & & & E & -E & E & -E & & & & \\ & & & & E & -E & -E & E & & & & \\ & & & & & & & & E & E & E & E \\ & & & & & & & & E & E & -E & -E \\ & & & & & & & & E & -E & E & -E \\ & & & & & & & & E & -E & -E & E \end{bmatrix} \begin{bmatrix} \Phi^{(1)} \\ \Phi^{(2)} \\ \Phi^{(3)} \\ \Phi^{(4)} \\ \Phi^{(5)} \\ \Phi^{(6)} \\ \Phi^{(7)} \\ \Phi^{(8)} \\ \Phi^{(9)} \\ \Phi^{(10)} \\ \Phi^{(11)} \\ \Phi^{(12)} \end{bmatrix}$$

Conversely, the relation of the sets of nodal functions Φ to the sets of basis vectors $\bar{\Phi}$ is

$$\begin{bmatrix} \Phi^{(1)} \\ \Phi^{(2)} \\ \Phi^{(3)} \\ \Phi^{(4)} \\ \Phi^{(5)} \\ \Phi^{(6)} \\ \Phi^{(7)} \\ \Phi^{(8)} \\ \Phi^{(9)} \\ \Phi^{(10)} \\ \Phi^{(11)} \\ \Phi^{(12)} \end{bmatrix} = \begin{bmatrix} E & E & E & E & & & & & & & & \\ E & E & -E & -E & & & & & & & & \\ E & -E & E & -E & & & & & & & & \\ E & -E & -E & E & & & & & & & & \\ & & & & E & E & E & E & & & & \\ & & & & E & E & -E & -E & & & & \\ & & & & E & -E & E & -E & & & & \\ & & & & E & -E & -E & E & & & & \\ & & & & & & & & E & E & E & E \\ & & & & & & & & E & E & -E & -E \\ & & & & & & & & E & -E & E & -E \\ & & & & & & & & E & -E & -E & E \end{bmatrix} \begin{bmatrix} \bar{\Phi}^{(1)} \\ \bar{\Phi}^{(2)} \\ \bar{\Phi}^{(3)} \\ \bar{\Phi}^{(4)} \\ \bar{\Phi}^{(5)} \\ \bar{\Phi}^{(6)} \\ \bar{\Phi}^{(7)} \\ \bar{\Phi}^{(8)} \\ \bar{\Phi}^{(9)} \\ \bar{\Phi}^{(10)} \\ \bar{\Phi}^{(11)} \\ \bar{\Phi}^{(12)} \end{bmatrix}.$$

The supermatrix

$$T_c = \frac{1}{4} \begin{bmatrix} E & E & E & E \\ E & E & -E & -E \\ E & -E & E & -E \\ E & -E & -E & E \end{bmatrix}, \quad \text{with} \quad T_c^{-1} = 4T_c,$$

is the transformation supermatrix of the four-node rectangular element pertaining to the group C_{2v}, as derived in section 3.5.

Thus, the twelve-node truncated rectangular hexahedral element, as given in Fig. 3.12(a), may be regarded as an assembly of three four-node perpendicular rectangular elements shown in Fig. 3.12(b).

4

Formulation of shape functions in G-invariant subspaces

4.1 GROUP SUPERMATRIX PROCEDURE FOR DERIVATION OF ELEMENT SHAPE FUNCTIONS IN G-INVARIANT SUBSPACES

When a finite element with its nodal pattern possesses symmetry properties that can be described by a group G, the group supermatrix procedure can provide the system of equations of the displacement field with the supermatrix in diagonal form and formulate element shape functions in G-invariant subspaces.

The group supermatrix procedure for derivation of element shape functions in G-invariant subspaces is systematized in the form of eight consecutive steps.

(1) *Introduction of unique group supermatrix nodal numberings*

The unique nodal numberings, derived in Chapter 3 for elements with symmetry properties described by the groups C_2, C_3, C_4, C_{2v}, C_{3v}, C_{4v}, D_{2h}, provide basis transformation supermatrices in optimum form for formulations in G-vector spaces.

When a symmetry group G corresponds to an element and its nodal pattern, the nodes can be grouped into one or more nonoverlapping nodal sets, each set containing nodes that are permuted by action of symmetry operations of the group

$S_1(1,\ldots,m_1)$

$S_2(m_1+1,\ldots,m_2)$

$\vdots$

$S_l(m_{l-1}+1,\ldots,n)$,

where node numbers run from 1 to n and $m_1, m_2, \ldots, n$ designate the last nodes in the sets $S_1, S_2, \ldots, S_l$.

Sec. 4.1]   **Group supermatrix procedure**   147

The unique nodal numberings are derived for elements with various nodal patterns described by the same group. The same type of nodal numbering may be applied for several nodal sets when they correspond to geometrical figures of the same type that are superposed in an element.

Since the nodes in a nodal set are permuted by action of symmetry operations of the group, it is possible to generate the coordinates of all nodes in the set by applying matrix representatives of symmetry operations to the coordinate vector of a single node. Consequently, in the group supermatrix procedure various relations will be derived by using only the first nodes of the sets $S_1, S_2, \ldots, S_l$, i.e.

$$S_1(1), S_2(m_1 + 1), \ldots, S_l(m_{l-1} + 1).$$

(2) *Derivation of basis vectors of G-invariant subspaces*

The character table of the group G, with Cartesian sets and products used in step (4) for derivation of displacement field functions, provides the characters of irreducible group representations for determination of idempotents of the centre of the group algebra

$$\pi_i = \frac{h_i}{h} \sum_\sigma \chi_i(\sigma^{-1})\sigma,$$

with h_i the dimension of the ith character given by $h_i = \chi_i(E)$, h the order of G, χ_i the ith character of G, σ the element and σ^{-1} the inverse of the element of G, while $i = 1, 2, \ldots, k$ (k being the number of irreducible group representations).

Applying the idempotents π_i to the nodal functions $\Phi = [\varphi_1 \; \varphi_2 \; \ldots \; \varphi_n]$, grouped according to the nodal sets $S_1, S_2, \ldots, S_l$, derives the basis vectors $\bar\Phi = [\bar\varphi_1 \; \bar\varphi_2 \; \ldots \; \bar\varphi_n]$ of G-invariant subspaces U_i of the G-vector space V. This process is given in schematic form in Table 4.1.

By operations given in this table the n-dimensional G-vector space is decomposed into k G-invariant subspaces U_i, where the number of basis vectors that span a subspace determines the dimension of the subspace, so that

$$n = n_1 + n_2 + \ldots + n_k.$$

(3) *Formulation of relations of basis vectors and nodal functions as systems of equations with supermatrices in diagonal form*

The grouping of the basis vectors $\bar\Phi$ according to the nodal numbering in the nodal sets $S_1, S_2, \ldots, S_l$ gives

$$\bar\Phi_1 = [\bar\varphi_1 \; \ldots \; \bar\varphi_{m_1}]$$
$$\bar\Phi_2 = [\bar\varphi_{m_1+1} \; \ldots \; \bar\varphi_{m_2}]$$
$$\vdots$$
$$\bar\Phi_l = [\bar\varphi_{m_{l-1}+1} \; \ldots \; \bar\varphi_n],$$

so that the indices run from 1 to n continuously.

By designating the nodal functions Φ by

148   Formulation of shape functions in G-invariant subspaces   [Ch. 4

Table 4.1. Schematic display of derivation of basis vectors of G-invariant subspaces of a G-vector space with respect to the nodal numbering

Subspace	Application of the idempotents	Numbering of basis vectors	Applied on functions Φ at the nodes contained in nodal sets $S_1, S_2, \ldots, S_l$	Sets of basis vectors
U_1	π_1	S_1	$S_1(1, \ldots, m_1)$	
	π_1	S_2	$S_2(m_1+1, \ldots, m_2)$	
	$\vdots$	$\vdots$	$\vdots$	$\bar{\Phi}^{(1)}$
	π_1	S_l	$S_l(m_{l-1}+1, \ldots, n)$	
U_2	π_2	S_1	$S_1(1, \ldots, m_1)$	
	π_2	S_2	$S_2(m_1+1, \ldots, m_2)$	
	$\vdots$	$\vdots$	$\vdots$	$\bar{\Phi}^{(2)}$
	π_2	S_l	$S_l(m_{l-1}+1, \ldots, n)$	
$\vdots$	$\vdots$		$\vdots$	
U_k	π_k	S_1	$S_1(1, \ldots, m_1)$	
	π_k	S_2	$S_2(m_1+1, \ldots, m_2)$	
	$\vdots$	$\vdots$	$\vdots$	$\bar{\Phi}^{(k)}$
	π_k	S_l	$S_l(m_{l-1}+1, \ldots, n)$	

$$\Phi_1 = [\varphi_1 \quad \cdots \quad \varphi_{m_1}]$$
$$\Phi_2 = [\varphi_{m_1+1} \quad \cdots \quad \varphi_{m_2}]$$
$$\vdots$$
$$\Phi_l = [\varphi_{m_{l-1}+1} \quad \cdots \quad \varphi_n],$$

where $\Phi_1, \Phi_2, \ldots, \Phi_l$ pertain to the nodal sets $S_1, S_2, \ldots, S_l$, the relation of the basis vectors $\bar{\Phi}$ to the nodal functions Φ is obtained as a system of equations with the supermatrix in diagonal form

Sec. 4.1] Group supermatrix procedure 149

$$\begin{bmatrix} \bar{\Phi}_1 \\ \bar{\Phi}_2 \\ \vdots \\ \bar{\Phi}_l \end{bmatrix} = \begin{bmatrix} T_1 & & & \\ & T_2 & & \\ & & \ddots & \\ & & & T_l \end{bmatrix} \begin{bmatrix} \Phi_1 \\ \Phi_2 \\ \vdots \\ \Phi_l \end{bmatrix} \quad \text{or} \quad \bar{\Phi} = \bar{T}\Phi,$$

where $T_1, T_2, \ldots, T_l$ are transformation supermatrices pertaining to the group G or its subgroups.

The relation of the nodal functions Φ to the basis vectors $\bar{\Phi}$ is

$$\begin{bmatrix} \Phi_1 \\ \Phi_2 \\ \vdots \\ \Phi_l \end{bmatrix} = \begin{bmatrix} T_1^{-1} & & & \\ & T_2^{-1} & & \\ & & \ddots & \\ & & & T_l^{-1} \end{bmatrix} \begin{bmatrix} \bar{\Phi}_1 \\ \bar{\Phi}_2 \\ \vdots \\ \bar{\Phi}_l \end{bmatrix} \quad \text{or} \quad \Phi = \bar{T}^{-1}\bar{\Phi}.$$

(4) *Derivation of the function of the displacement field decomposed into G-invariant subspaces*

The u-, v- or w-displacement function is usually assumed as a polynomial expression

$$u(x, y, z) = \sum_{i=1}^{n} \alpha_i f_i,$$

where α_i are the coefficients determined from 'boundary' conditions and f_i the terms chosen from Cartesian sets and products in Pascal's triangle.

The object of the group supermatrix procedure is to classify these terms according to their pertinence to the irreducible representations of the group G, i.e. to the symmetry types of the group, and to obtain the sequence of the terms that provides the most reduced systems of equations with their supermatrices in diagonal form.

This is accomplished by the product

$$\begin{bmatrix} F_1 \\ F_2 \\ \vdots \\ F_k \end{bmatrix} = \begin{bmatrix} g_1 \\ g_2 \\ \vdots \\ g_k \end{bmatrix} [h_1 \; h_2 \; \ldots \; h_l] = \begin{bmatrix} g_1 h_1 & g_1 h_2 & \ldots & g_1 h_l \\ g_2 h_1 & g_2 h_2 & \ldots & g_2 h_l \\ \vdots & \vdots & & \vdots \\ g_k h_1 & g_k h_2 & \ldots & g_k h_l \end{bmatrix}$$

$$= \begin{bmatrix} f_{11} & f_{12} & \ldots & f_{1l} \\ f_{21} & f_{22} & \ldots & f_{2l} \\ \vdots & \vdots & & \vdots \\ f_{k1} & f_{k2} & \ldots & f_{kl} \end{bmatrix} \begin{matrix} \text{subspace} \\ U_1 \\ U_2 \\ \vdots \\ U_k \end{matrix}$$

where $g_1, g_2, \ldots, g_k$ and $h_1, h_2, \ldots, h_l$ are Cartesian sets and products which stand as column and row in addition to the character table of the group G.

The above matrix provides the pertinence of the terms to the subspaces $U_1, U_2, \ldots, U_k$ and the optimum sequence of the terms.

Thus, the polynomial expression for the displacement function

$$u(x, y, z) = [f_{11} \; f_{12} \; \ldots \; f_{1l} \; \vdots \; f_{21} \; f_{22} \; \ldots \; f_{2l} \; \vdots \; \ldots$$
$$\vdots \; f_{k1} \; f_{k2} \; \ldots \; f_{kl}] \, [\alpha_1 \; \alpha_2 \; \ldots \; \alpha_n]^T$$
$$= [F_1 \; \vdots \; F_2 \; \vdots \; \ldots \; \vdots \; F_k] \, [A_1 \; \vdots \; A_2 \; \vdots \; \ldots \; \vdots \; A_k]^T$$

is decomposed into polynomials pertaining to G-invariant subspaces $U_1, U_2, \ldots, U_k$, where the numbers of non-zero terms in $F_1, F_2, \ldots, F_k$ determine the dimensions of the subspaces, which must coincide with dimensions of subspaces obtained in step (2) where basis vectors were derived.

(5) *Determination of relations of displacements $\bar{U}$ to the coefficients A in G-invariant subspaces*

Substitution of the nodal coordinates of the first nodes of the nodal sets $S_1, S_2, \ldots, S_l$ into the sections $F_1, F_2, \ldots, F_k$ of

$$u(x, y, z) = [F_1 \; \vdots \; F_2 \; \vdots \ldots \vdots \; F_k] \, [A_1 \; \vdots \; A_2 \; \vdots \ldots \vdots \; A_k]^T,$$

pertaining to the subspaces $U_1, U_2, \ldots, U_k$, will provide relations of displacements $\bar{U}$ to coefficients A for each subspace separately:

$$\begin{bmatrix} \bar{U}^{(1)} \\ \bar{U}^{(2)} \\ \vdots \\ \bar{U}^{(k)} \end{bmatrix} = \begin{bmatrix} \bar{C}_1 & & & \\ & \bar{C}_2 & & \\ & & \ddots & \\ & & & \bar{C}_k \end{bmatrix} \begin{bmatrix} A_1 \\ A_2 \\ \vdots \\ A_k \end{bmatrix} \quad \text{or} \quad \bar{U} = \bar{C} A,$$

with numberings of displacements u_j in the sets $\bar{U}^{(1)}, \bar{U}^{(2)}, \ldots, \bar{U}^{(k)}$ as applied in step (2) for $\bar{\varphi}_j$ in the sets of basis vectors $\bar{\Phi}^{(1)}, \bar{\Phi}^{(2)}, \ldots, \Phi^{(k)}$.

Thus, the n-dimensional G-vector space is decomposed into k G-invariant subspaces $U_1, U_2, \ldots, U_k$ with $n_1, n_2, \ldots, n_k$ dimensions respectively.

(6) *Determination of the coefficients A in G-invariant subspaces*

The coefficients $A_1, A_2, \ldots, A_k$ are obtained by inverting the matrices $\bar{C}_1, \bar{C}_2, \ldots, \bar{C}_k$ of subspaces $U_1, U_2, \ldots, U_k$ respectively:

$$\begin{bmatrix} A_1 \\ A_2 \\ \vdots \\ A_k \end{bmatrix} = \begin{bmatrix} \bar{C}_1^{-1} & & & \\ & \bar{C}_2^{-1} & & \\ & & \ddots & \\ & & & \bar{C}_k^{-1} \end{bmatrix} \begin{bmatrix} \bar{U}^{(1)} \\ \bar{U}^{(2)} \\ \vdots \\ \bar{U}^{(k)} \end{bmatrix} \quad \text{or} \quad A = \bar{C}^{-1} \bar{U}.$$

Sec. 4.1]   Group supermatrix procedure   151

(7) *Formulation of displacement fields in G-invariant subspaces*

The n-dimensional G-vector space of the displacement field Δ, decomposed into k G-invariant subspaces $U_1, U_2, \ldots, U_k$ with dimensions $n_1, n_2, \ldots, n_k$ respectively, is produced by

$$\Delta = T_D \bar{C}^{-1} \bar{U},$$

with $T_D = \text{Diag}\,[F_1 \,\vdots\, F_2 \,\vdots\, \ldots \,\vdots\, F_k]$, providing

$$\begin{bmatrix} \Delta_1 \\ \Delta_2 \\ \vdots \\ \Delta_k \end{bmatrix} = \begin{bmatrix} \text{Diag}\,F_1 \bar{C}_1^{-1} & & & \\ & \text{Diag}\,F_2 \bar{C}_2^{-1} & & \\ & & \ddots & \\ & & & \text{Diag}\,F_k \bar{C}_k^{-1} \end{bmatrix} \begin{bmatrix} \bar{U}^{(1)} \\ \bar{U}^{(2)} \\ \vdots \\ \bar{U}^{(k)} \end{bmatrix}$$

and with substitutions $\xi = x/c$, $\eta = y/d$, $\zeta = z/e$, the relation of Δ to $\bar{U}$ expressed by ξ, η, ζ is obtained

$$\begin{bmatrix} \Delta_1 \\ \Delta_2 \\ \vdots \\ \Delta_k \end{bmatrix} = \begin{bmatrix} \bar{N}_{\Delta_1} & & & \\ & \bar{N}_{\Delta_2} & & \\ & & \ddots & \\ & & & \bar{N}_{\Delta_k} \end{bmatrix} \begin{bmatrix} \bar{U}^{(1)} \\ \bar{U}^{(2)} \\ \vdots \\ \bar{U}^{(k)} \end{bmatrix} \quad \text{or} \quad \Delta = \bar{N}_\Delta \bar{U}.$$

(8) *Derivation of element shape functions $\bar{N}$ in G-invariant subspaces*

The shape functions $\bar{N}^{(1)}, \bar{N}^{(2)}, \ldots, \bar{N}^{(k)}$ in G-invariant subspaces $U_1, U_2, \ldots, U_k$ are obtained by

$$\bar{N} = S \bar{N}_\Delta,$$

with $S = [1\ 1\ \ldots\ 1]$ (vector with n coordinates equalling 1) and the numbering of $\bar{N}$ according to the numbering of basis vectors $\bar{\Phi}^{(1)}, \bar{\Phi}^{(2)}, \ldots, \bar{\Phi}^{(k)}$ derived in step (2), so that

$$\begin{bmatrix} \bar{N}^{(1)} \\ \bar{N}^{(2)} \\ \vdots \\ \bar{N}^{(k)} \end{bmatrix} = \begin{bmatrix} H_1 & & & \\ & H_2 & & \\ & & \ddots & \\ & & & H_k \end{bmatrix} \begin{bmatrix} F_1 \\ F_2 \\ \vdots \\ F_k \end{bmatrix} \quad \text{or} \quad \bar{N} = HF.$$

The shape functions $N_1, N_2, \ldots, N_n$ at the nodes $1, 2, \ldots, n$ are obtained by

$$N = T\bar{N} = THF \quad \text{or}$$

$$\begin{bmatrix} N^{(1)} \\ N^{(2)} \\ \vdots \\ N^{(k)} \end{bmatrix} = \begin{bmatrix} T_1 H_1 & & & \\ & T_2 H_2 & & \\ & & \ddots & \\ & & & T_k H_k \end{bmatrix} \begin{bmatrix} F_1 \\ F_2 \\ \vdots \\ F_k \end{bmatrix},$$

where $T_1, T_2, \ldots, T_k$ are transformation supermatrices used in step (3).

Element shape functions in G-invariant subspaces can be derived also by means of group supermatrix transformations of the shape functions of serendipity elements.

The group supermatrix procedure for derivation of element shape functions in G-invariant subspaces from shape functions of serendipity elements is systematized in the form of six consecutive steps. Steps (I), (II), (III), (IV) coincide with steps (1), (2), (3), (4) of the previous procedure.

(I) *Introduction of unique group supermatrix nodal numberings.*
(II) *Derivation of basis vectors of G-invariant subspaces.*
(III) *Formulation of relations of basis vectors and nodal functions as systems of equations with supermatrices in diagonal form.*
(IV) *Derivation of the function of the displacement field decomposed into G-invariant subspaces.*
(V) *Formulation of the known shape functions of the serendipity element according to the group supermatrix procedure.*

The known shape functions of the serendipity slement will be expressed here using the unique group supermatrix nodal numbering (from step (I)) and with the optimum sequence of terms classified into their G-invariant subspaces (from step (IV))

$$\begin{bmatrix} N_1 \\ N_2 \\ \vdots \\ N_n \end{bmatrix} = \begin{bmatrix} a_{11} & a_{12} & \ldots & a_{1n} \\ a_{21} & a_{22} & \ldots & a_{2n} \\ \vdots & \vdots & & \vdots \\ a_{n1} & a_{n2} & \ldots & a_{nn} \end{bmatrix} \begin{bmatrix} f_1 \\ f_2 \\ \vdots \\ f_n \end{bmatrix}.$$

After rearranging the order of $N_1, N_2, \ldots, N_n$ according to the numbering of the sets of basis vectors $\bar{\Phi}^{(1)}, \bar{\Phi}^{(2)}, \ldots, \bar{\Phi}^{(k)}$ of subspaces $U_1, U_2, \ldots, U_k$ in step (II), one obtains

Sec. 4.2]   Four-node rectangular element   153

$$\begin{bmatrix} N^{(1)} \\ N^{(2)} \\ \vdots \\ N^{(k)} \end{bmatrix} = \begin{bmatrix} G_{11} & G_{12} & \cdots & G_{1k} \\ G_{21} & G_{22} & \cdots & G_{2k} \\ \vdots & \vdots & & \vdots \\ G_{k1} & G_{k2} & \cdots & G_{kk} \end{bmatrix} \begin{bmatrix} F_1 \\ F_2 \\ \vdots \\ F_k \end{bmatrix} \quad \text{or} \quad N = GF.$$

(VI) *Derivation of shape functions in G-invariant subspaces.*

The shape functions $\bar{N}$ in G-invariant subspaces are obtained by

$$\bar{N} = TN,$$

where T is the transformation supermatrix used in step (III), so that

$$\begin{bmatrix} \bar{N}^{(1)} \\ \bar{N}^{(2)} \\ \vdots \\ \bar{N}^{(k)} \end{bmatrix} = \begin{bmatrix} H_1 & & & \\ & H_2 & & \\ & & \ddots & \\ & & & H_k \end{bmatrix} \begin{bmatrix} F_1 \\ F_2 \\ \vdots \\ F_k \end{bmatrix}.$$

Application of the group supermatrix procedure for derivation of element shape functions in G-invariant subspaces are developed in the following sections. A survey of them is given in Table 4.2.

The group supermatrix procedure for derivation of element shape functions in G-invariant subspaces may be regarded as an implantation of group theoretical formulations into conventional methods for derivation of element shape functions, which were developed

— for the 4-node rectangular element: Argyris and Kelsey (1960),
— for the 8-node and 12-node rectangular elements: Ergatoudis *et al.* (1968b), Taylor (1972),
— for the 16-node rectangular element: Argyris and Fried (1968),
— for the 8-node, 20-node and 32-node rectangular hexahedral elements: Clough (1969), Melosh (1963b), Rigby and McNeice (1972), Zienkiewicz *et al.* (1971), Pawsey and Clough (1971), Ergatoudis *et al.* (1968a),
— for the 64-node rectangular hexahedral element: Argyris and Fried (1968).

4.2 FOUR-NODE RECTANGULAR ELEMENT

The four-node rectangular element, as given in Fig. 4.1, with the unique nodal numbering derived in section 3.5, is described by the group C_{2v}. The nodes of its nodal set $S(1, 2, 3, 4)$ are permuted by actions of group elements, i.e. symmetry operations E (identity), C_2 (rotation through 180° about z axis), σ_1 and σ_2 (reflections in xz and yz planes respectively).

The character table of the group C_{2v} with Cartesian sets and products is

Table 4.2. Survey of derived element shape functions in G-invariant subspaces

Section	Shape of the element	Number of nodes	Derivation of shape functions		
			$\bar{N}$ in G-invariant subspaces — Direct derivation	N at nodes — Derivation from shape functions $\bar{N}$	$\bar{N}$ in G-invariant subspaces — Derivation from shape functions of serendipity elements
4.2		4	+	+	+
4.3	rectangular	8	+	+	+
4.4	element	12	+	+	+
4.5		16	+	+	
4.6		8	+	+	+
4.7	rectangular	20	+		
4.8	hexahedral	32	+	+	
4.9	element	64	+		

C_{2v}	E	C_2	σ_1	σ_2		
A_1	1	1	1	1	z	x^2, y^2, z^2
A_2	1	1	-1	-1		xy
B_1	1	-1	1	-1	x	xz
B_2	1	-1	-1	1	y	yz

while the idempotents of the centre of the group algebra are obtained by

$$\pi_i = \frac{h_i}{h} \sum_\sigma \chi_i(\sigma^{-1})\sigma,$$

where h_i is the dimension of the ith character given by $h_i = \chi_i(E)$, h is the order, χ_i the ith character, σ the element and σ^{-1} the inverse of the element of the group, and $i = 1, 2, 3, 4$.

Thus

Sec. 4.2]	Four-node rectangular element	155

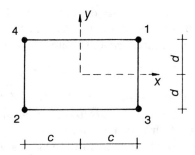

Fig. 4.1. The four-node rectangular element with the nodal numbering of the group supermatrix procedure.

$$\begin{bmatrix} \pi_1 \\ \pi_2 \\ \pi_3 \\ \pi_4 \end{bmatrix} = \frac{1}{4} \begin{bmatrix} 1 & 1 & 1 & 1 \\ 1 & 1 & -1 & -1 \\ 1 & -1 & 1 & -1 \\ 1 & -1 & -1 & 1 \end{bmatrix} \begin{bmatrix} E \\ C_2 \\ \sigma_1 \\ \sigma_2 \end{bmatrix}.$$

By application of the idempotents π_1, π_2, π_3, π_4 to the nodal functions φ_j ($j = 1, 2, 3, 4$) the basis vectors $\bar{\varphi}_j$ of G-invariant subspaces are obtained:

$$U_1: \quad \bar{\varphi}_1 = \pi_1 \varphi_1 = \pi_1 \varphi_2 = \pi_1 \varphi_3 = \pi_1 \varphi_4 = \tfrac{1}{4}(\varphi_1 + \varphi_2 + \varphi_3 + \varphi_4)$$
$$U_2: \quad \bar{\varphi}_2 = \pi_2 \varphi_1 = \pi_2 \varphi_2 = -\pi_2 \varphi_3 = -\pi_2 \varphi_4 = \tfrac{1}{4}(\varphi_1 + \varphi_2 - \varphi_3 - \varphi_4)$$
$$U_3: \quad \bar{\varphi}_3 = \pi_3 \varphi_1 = -\pi_3 \varphi_2 = \pi_3 \varphi_3 = -\pi_3 \varphi_4 = \tfrac{1}{4}(\varphi_1 - \varphi_2 + \varphi_3 - \varphi_4)$$
$$U_4: \quad \bar{\varphi}_4 = \pi_4 \varphi_1 = -\pi_4 \varphi_2 = -\pi_4 \varphi_3 = \pi_4 \varphi_4 = \tfrac{1}{4}(\varphi_1 - \varphi_2 - \varphi_3 + \varphi_4).$$

or, expressed in matrix form,

$$\begin{bmatrix} \bar{\varphi}_1 \\ \bar{\varphi}_2 \\ \bar{\varphi}_3 \\ \bar{\varphi}_4 \end{bmatrix} = \frac{1}{4} \begin{bmatrix} 1 & 1 & 1 & 1 \\ 1 & 1 & -1 & -1 \\ 1 & -1 & 1 & -1 \\ 1 & -1 & -1 & 1 \end{bmatrix} \begin{bmatrix} \varphi_1 \\ \varphi_2 \\ \varphi_3 \\ \varphi_4 \end{bmatrix} \quad \text{or} \quad \bar{\Phi} = T\Phi.$$

Conversely, the relation of the nodal functions Φ to the basis vectors $\bar{\Phi}$ is

$$\begin{bmatrix} \varphi_1 \\ \varphi_2 \\ \varphi_3 \\ \varphi_4 \end{bmatrix} = \begin{bmatrix} 1 & 1 & 1 & 1 \\ 1 & 1 & -1 & -1 \\ 1 & -1 & 1 & -1 \\ 1 & -1 & -1 & 1 \end{bmatrix} \begin{bmatrix} \bar{\varphi}_1 \\ \bar{\varphi}_2 \\ \bar{\varphi}_3 \\ \bar{\varphi}_4 \end{bmatrix}$$

or $\Phi = T^{-1} \bar{\Phi}$, with $T^{-1} = 4T$.

The u- or v-displacement function is usually assumed as

$$u(x, y) = \alpha_1 + \alpha_2 x + \alpha_3 y + \alpha_4 xy.$$

For the group supermatrix procedure the order of terms in this polynomial will be changed according to the pertinence of the terms to the symmetry types of the group C_{2v}, as given in addition to its character table. Therefore, in the polynomial

$$u(x, y) = [1 \;\vdots\; xy \;\vdots\; x \;\vdots\; y][\alpha_1 \;\vdots\; \alpha_2 \;\vdots\; \alpha_3 \;\vdots\; \alpha_4]^T$$

the terms $1, xy, x, y$ belong to the subspaces U_1, U_2, U_3, U_4 respectively. With this order of terms the relation of displacements $\bar{U}$ to coefficients A will be produced directly as a system of equations with the matrix in diagonal form.

Substitution of the nodal coordinates c, d of the node 1, as the first node in the nodal set $S(1, 2, 3, 4)$, into the above polynomial, gives

$$\begin{bmatrix} \bar{u}_1 \\ \bar{u}_2 \\ \bar{u}_3 \\ \bar{u}_4 \end{bmatrix} = \frac{1}{4} \begin{bmatrix} u_1 + u_2 + u_3 + u_4 \\ u_1 + u_2 - u_3 - u_4 \\ u_1 - u_2 + u_3 - u_4 \\ u_1 - u_2 - u_3 + u_4 \end{bmatrix}$$

$$= \frac{1}{4} \begin{bmatrix} 1 & 1 & 1 & 1 \\ 1 & 1 & -1 & -1 \\ 1 & -1 & 1 & -1 \\ 1 & -1 & -1 & 1 \end{bmatrix} \begin{bmatrix} u_1 \\ u_2 \\ u_3 \\ u_4 \end{bmatrix} = \begin{bmatrix} 1 & & & \\ & cd & & \\ & & c & \\ & & & d \end{bmatrix} \begin{bmatrix} \alpha_1 \\ \alpha_2 \\ \alpha_3 \\ \alpha_4 \end{bmatrix}$$

or $\bar{U} = \bar{C}A$.

Thus, the coefficients A are

$$\begin{bmatrix} \alpha_1 \\ \alpha_2 \\ \alpha_3 \\ \alpha_4 \end{bmatrix} = \begin{bmatrix} 1 & & & \\ & \frac{1}{cd} & & \\ & & \frac{1}{c} & \\ & & & \frac{1}{d} \end{bmatrix} \begin{bmatrix} \bar{u}_1 \\ \bar{u}_2 \\ \bar{u}_3 \\ \bar{u}_4 \end{bmatrix} \quad \text{or} \quad A = \bar{C}^{-1}\bar{U}.$$

The four-dimensional G-vector space of the displacement field Δ, decomposed into four one-dimensional subspaces U_1, U_2, U_3, U_4, is produced by

$$\Delta = T_D \bar{C}^{-1} \bar{U},$$

with $T_D = \text{diag}\,[1 \;\vdots\; xy \;\vdots\; x \;\vdots\; y]$, giving

$$\begin{bmatrix}\Delta_1\\\Delta_2\\\Delta_3\\\Delta_4\end{bmatrix}=\begin{bmatrix}1&&&\\&\dfrac{xy}{cd}&&\\&&\dfrac{x}{c}&\\&&&\dfrac{y}{d}\end{bmatrix}\begin{bmatrix}\bar{u}_1\\\bar{u}_2\\\bar{u}_3\\\bar{u}_4\end{bmatrix}$$

and with substitutions $\xi = x/c$, $\eta = y/d$

$$\begin{bmatrix}\Delta_1\\\Delta_2\\\Delta_3\\\Delta_4\end{bmatrix}=\begin{bmatrix}1&&&\\&\xi\eta&&\\&&\xi&\\&&&\eta\end{bmatrix}\begin{bmatrix}\bar{u}_1\\\bar{u}_2\\\bar{u}_3\\\bar{u}_4\end{bmatrix}=\begin{bmatrix}\bar{N}_1&&&\\&\bar{N}_2&&\\&&\bar{N}_3&\\&&&\bar{N}_4\end{bmatrix}\begin{bmatrix}\bar{u}_1\\\bar{u}_2\\\bar{u}_3\\\bar{u}_4\end{bmatrix}$$

with

$$\begin{bmatrix}\bar{N}_1\\\bar{N}_2\\\bar{N}_3\\\bar{N}_4\end{bmatrix}=\begin{bmatrix}N_1+N_2+N_3+N_4\\N_1+N_2-N_3-N_4\\N_1-N_2+N_3-N_4\\N_1-N_2-N_3+N_4\end{bmatrix}=\begin{bmatrix}1&1&1&1\\1&1&-1&-1\\1&-1&1&-1\\1&-1&-1&1\end{bmatrix}\begin{bmatrix}N_1\\N_2\\N_3\\N_4\end{bmatrix}=\begin{bmatrix}1\\\xi\eta\\\xi\\\eta\end{bmatrix}$$

or $\bar{N} = T^{-1}N = F$.

The nodal shape functions N_1, N_2, N_3, N_4, pertaining to the nodes 1, 2, 3, 4, are obtained by

$$N = T\bar{N} = TT^{-1}N = TF \quad \text{or}$$

$$\begin{bmatrix}N_1\\N_2\\N_3\\N_4\end{bmatrix}=\dfrac{1}{4}\begin{bmatrix}1&1&1&1\\1&1&-1&-1\\1&-1&1&-1\\1&-1&-1&1\end{bmatrix}\begin{bmatrix}\bar{N}_1\\\bar{N}_2\\\bar{N}_3\\\bar{N}_4\end{bmatrix}=\dfrac{1}{4}\begin{bmatrix}1&1&1&1\\1&1&-1&-1\\1&-1&1&-1\\1&-1&-1&1\end{bmatrix}\begin{bmatrix}1\\\xi\eta\\\xi\\\eta\end{bmatrix}.$$

The rearrangement of the above equations with respect to the usual nodal numbering according to Fig. 4.2 gives

$$\begin{bmatrix}N_1\\N_2\\N_3\\N_4\end{bmatrix}=\dfrac{1}{4}\begin{bmatrix}1&1&-1&-1\\1&-1&1&-1\\1&1&1&1\\1&-1&-1&1\end{bmatrix}\begin{bmatrix}1\\\xi\eta\\\xi\\\eta\end{bmatrix}$$

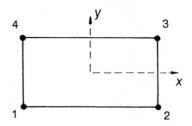

Fig. 4.2. The four-node rectangular element with the usual nodal numbering.

or

$$N_1 = \tfrac{1}{4}(1-\xi)(1-\eta)$$
$$N_2 = \tfrac{1}{4}(1+\xi)(1-\eta)$$
$$N_3 = \tfrac{1}{4}(1+\xi)(1+\eta)$$
$$N_4 = \tfrac{1}{4}(1-\xi)(1+\eta),$$

which is the expression in conventional form for the set of shape functions N_1, N_2, N_3, N_4 of the four-node serendipity rectangle.

Derivation of shape functions in G-invariant subspaces of rectangular elements is also possible by means of group supermatrix transformations of shape functions of serendipity rectangles.

For the four-node serendipity rectangular element with the nodal coordinates $\xi_j = \pm 1$ and $\eta_j = \pm 1$, the shape functions are

$$N_j = \tfrac{1}{4}(1+\xi\xi_j)(1+\eta\eta_j).$$

By adopting the unique group supermatrix nodal numbering, given in Fig. 4.1, and the optimum order of terms derived in this section, the following set of shape functions N is obtained:

$$\begin{bmatrix} N_1 \\ N_2 \\ N_3 \\ N_3 \end{bmatrix} = \frac{1}{4} \begin{bmatrix} 1 & 1 & 1 & 1 \\ 1 & 1 & -1 & -1 \\ 1 & -1 & 1 & -1 \\ 1 & -1 & -1 & 1 \end{bmatrix} \begin{bmatrix} 1 \\ \xi\eta \\ \xi \\ \eta \end{bmatrix} \quad \text{or} \quad N = TF.$$

Multiplying the above equation by T^{-1} produces the shape functions $\bar{N}$ in G-invariant subspaces

$$\bar{N} = T^{-1}N = T^{-1}TF = F,$$

or

Sec. 4.3] Eight-node rectangular element 159

$$\begin{bmatrix} \bar{N}_1 \\ \bar{N}_2 \\ \bar{N}_3 \\ \bar{N}_4 \end{bmatrix} = \begin{bmatrix} N_1 + N_2 + N_3 + N_4 \\ N_1 + N_2 - N_3 - N_4 \\ N_1 - N_2 + N_3 - N_4 \\ N_1 - N_2 - N_3 + N_4 \end{bmatrix} = \begin{bmatrix} 1 & 1 & 1 & 1 \\ 1 & 1 & -1 & -1 \\ 1 & -1 & 1 & -1 \\ 1 & -1 & -1 & 1 \end{bmatrix} \begin{bmatrix} N_1 \\ N_2 \\ N_3 \\ N_4 \end{bmatrix} = \begin{bmatrix} 1 \\ \xi\eta \\ \xi \\ \eta \end{bmatrix}.$$

4.3 EIGHT-NODE RECTANGULAR ELEMENT

The eight-node rectangular element shown in Fig. 4.3 is described by the group C_{2v} and it has three nodal sets

$$S_1(1, 2, 3, 4), \quad S_2(5, 6), \quad S_3(7, 8),$$

each set containing nodes that are permuted by actions of symmetry operations of the group.

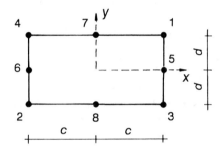

Fig. 4.3. The eight-node rectangular element with the nodal numbering of the group supermatrix procedure.

The unique nodal numberings for the four-node rectangular element and the four-node rhombic element, derived in section 3.5 and used here for the nodal sets S_1 and S_2, S_3 respectively, provide the positions of the nodes 1, 2, 3, 4, 5, 6, 7, 8 shown in Table 4.3.

The character table of the group C_{2v}, with Cartesian sets and products, is

C_{2v}	E	C_2	σ_1	σ_2		
A_1	1	1	1	1	z	x^2, y^2, z^2
A_2	1	1	−1	−1		xy
B_1	1	−1	1	−1	x	xz
B_2	1	−1	−1	1	y	yz

and the idempotents of the centre of the group algebra are

Table 4.3 Positions of nodes

Set	Node	Position
S_1	1	I quadrant
	2	III quadrant
	3	IV quadrant
	4	II quadrant
S_2	5	positive branch of the x axis
	6	negative branch of the x axis
S_3	7	positive branch of the y axis
	8	negative branch of the y axis

$$\begin{bmatrix} \pi_1 \\ \pi_2 \\ \pi_3 \\ \pi_4 \end{bmatrix} = \frac{1}{4} \begin{bmatrix} 1 & 1 & 1 & 1 \\ 1 & 1 & -1 & -1 \\ 1 & -1 & 1 & -1 \\ 1 & -1 & -1 & 1 \end{bmatrix} \begin{bmatrix} E \\ C_2 \\ \sigma_1 \\ \sigma_2 \end{bmatrix}.$$

Applying the idempotents π_i to the nodal functions Φ derives the basis vectors of G-invariant subspaces U_i

U_1: $\bar{\varphi}_1 = \pi_1 \varphi_1 = \pi_1 \varphi_2 = \pi_1 \varphi_3 = \pi_1 \varphi_4 = \frac{1}{4}(\varphi_1 + \varphi_2 + \varphi_3 + \varphi_4)$

$\bar{\varphi}_5 = \pi_1 \varphi_5 = \pi_1 \varphi_6 = \frac{1}{2}(\varphi_5 + \varphi_6)$

$\bar{\varphi}_7 = \pi_1 \varphi_7 = \pi_1 \varphi_8 = \frac{1}{8}(\varphi_7 + \varphi_8)$

U_2: $\bar{\varphi}_2 = \pi_2 \varphi_1 = \pi_2 \varphi_2 = -\pi_2 \varphi_3 = -\pi_2 \varphi_4 = \frac{1}{4}(\varphi_1 + \varphi_2 - \varphi_3 - \varphi_4)$

$(\pi_2 \varphi_5 = \pi_2 \varphi_6 = 0, \quad \pi_2 \varphi_7 = \pi_2 \varphi_8 = 0)$

U_3: $\bar{\varphi}_3 = \pi_3 \varphi_1 = -\pi_3 \varphi_2 = \pi_3 \varphi_3 = -\pi_3 \varphi_4 = \frac{1}{4}(\varphi_1 - \varphi_2 + \varphi_3 - \varphi_4)$

$\bar{\varphi}_6 = \pi_3 \varphi_5 = -\pi_3 \varphi_6 = \frac{1}{2}(\varphi_5 - \varphi_6), \quad (\pi_3 \varphi_7 = \pi_3 \varphi_8 = 0)$

U_4: $\bar{\varphi}_4 = \pi_4 \varphi_1 = -\pi_4 \varphi_2 = -\pi_4 \varphi_3 = \pi_4 \varphi_4 = \frac{1}{4}(\varphi_1 - \varphi_2 - \varphi_3 + \varphi_4)$

$\bar{\varphi}_8 = \pi_4 \varphi_7 = -\pi_4 \varphi_8 = \frac{1}{2}(\varphi_7 - \varphi_8), \quad (\pi_4 \varphi_5 = \pi_4 \varphi_6 = 0).$

Thus, the n-dimensional G-vector space is decomposed into four G-invariant subspaces U_1, U_2, U_3, U_4 with 3, 1, 2, 2 dimensions respectively, spanned by the corresponding sets of basis vectors $\bar{\Phi}^{(1)}$, $\bar{\Phi}^{(2)}$, $\bar{\Phi}^{(3)}$, $\bar{\Phi}^{(4)}$:

Sec. 4.3] **Eight-node rectangular element** 161

$$\begin{bmatrix} \bar{\Phi}^{(1)} \\ \bar{\Phi}^{(2)} \\ \bar{\Phi}^{(3)} \\ \bar{\Phi}^{(4)} \end{bmatrix} = \begin{bmatrix} \bar{\varphi}_1 & \bar{\varphi}_5 & \bar{\varphi}_7 \\ \bar{\varphi}_2 & & \\ \bar{\varphi}_3 & \bar{\varphi}_6 & \\ \bar{\varphi}_4 & & \bar{\varphi}_8 \end{bmatrix}.$$

The sets $\bar{\Phi}_1, \bar{\Phi}_2, \bar{\Phi}_3$ of basis vectors φ_j ($j = 1, 2, \ldots, 8$), ordered according to the numbering of the nodes in the nodal sets S_1, S_2, S_3, are given by

$$[\bar{\Phi}_1 \ \bar{\Phi}_2 \ \bar{\Phi}_3] = \begin{bmatrix} \bar{\varphi}_1 & & \\ \bar{\varphi}_2 & \bar{\varphi}_5 & \bar{\varphi}_7 \\ \bar{\varphi}_3 & \bar{\varphi}_6 & \bar{\varphi}_8 \\ \bar{\varphi}_4 & & \end{bmatrix}.$$

The relation of the sets of basis vectors $\bar{\Phi}_l$ to the sets of nodal functions Φ_l ($l = 1, 2, 3$) is expressed by the following system of equations with the supermatrix in diagonal form, containing one transformation matrix of the group C_{2v} and two transformation matrices of the group C_2:

$$\begin{bmatrix} \bar{\Phi}_1 \\ \bar{\Phi}_2 \\ \bar{\Phi}_3 \end{bmatrix} = \begin{bmatrix} T_{C_{2v}} & & \\ & T_{C_2} & \\ & & T_{C_2} \end{bmatrix} \begin{bmatrix} \Phi_1 \\ \Phi_2 \\ \Phi_3 \end{bmatrix}$$

with

$$[\Phi_1 \ \Phi_2 \ \Phi_3] = \begin{bmatrix} \varphi_1 & & \\ \varphi_2 & \varphi_5 & \varphi_7 \\ \varphi_3 & \varphi_6 & \varphi_8 \\ \varphi_4 & & \end{bmatrix},$$

or, in expanded form, this relation is

$$\begin{bmatrix} \bar{\varphi}_1 \\ \bar{\varphi}_2 \\ \bar{\varphi}_3 \\ \bar{\varphi}_4 \\ \bar{\varphi}_5 \\ \bar{\varphi}_6 \\ \bar{\varphi}_7 \\ \bar{\varphi}_8 \end{bmatrix} = \begin{bmatrix} \frac{1}{4}\begin{bmatrix} 1 & 1 & 1 & 1 \\ 1 & 1 & -1 & -1 \\ 1 & -1 & 1 & -1 \\ 1 & -1 & -1 & 1 \end{bmatrix} & & \\ & \frac{1}{2}\begin{bmatrix} 1 & 1 \\ 1 & -1 \end{bmatrix} & \\ & & \frac{1}{2}\begin{bmatrix} 1 & 1 \\ 1 & -1 \end{bmatrix} \end{bmatrix} \begin{bmatrix} \varphi_1 \\ \varphi_2 \\ \varphi_3 \\ \varphi_4 \\ \varphi_5 \\ \varphi_6 \\ \varphi_7 \\ \varphi_8 \end{bmatrix}$$

or $\bar{\Phi} = \bar{T}\Phi$.

Conversely, the relation of the sets of nodal functions Φ_l to the sets of basis vectors $\bar{\Phi}_l$ is

$$\begin{bmatrix} \Phi_1 \\ \Phi_2 \\ \Phi_3 \end{bmatrix} = \begin{bmatrix} T_{C_{2v}}^{-1} & & \\ & T_{C_2}^{-1} & \\ & & T_{C_2}^{-1} \end{bmatrix} \begin{bmatrix} \bar{\Phi}_1 \\ \bar{\Phi}_2 \\ \bar{\Phi}_3 \end{bmatrix}$$

with $T_{C_{2v}}^{-1} = 4T_{C_{2v}}$ and $T_{C_2}^{-1} = 2T_{C_2}$.

In expanded form this relation is

$$\begin{bmatrix} \varphi_1 \\ \varphi_2 \\ \varphi_3 \\ \varphi_4 \\ \varphi_5 \\ \varphi_6 \\ \varphi_7 \\ \varphi_8 \end{bmatrix} = \begin{bmatrix} 1 & 1 & 1 & 1 & & & & \\ 1 & 1 & -1 & -1 & & & & \\ 1 & -1 & 1 & -1 & & & & \\ 1 & -1 & -1 & 1 & & & & \\ & & & & 1 & 1 & & \\ & & & & 1 & -1 & & \\ & & & & & & 1 & 1 \\ & & & & & & 1 & -1 \end{bmatrix} \begin{bmatrix} \bar{\varphi}_1 \\ \bar{\varphi}_2 \\ \bar{\varphi}_3 \\ \bar{\varphi}_4 \\ \bar{\varphi}_5 \\ \bar{\varphi}_6 \\ \bar{\varphi}_7 \\ \bar{\varphi}_8 \end{bmatrix}$$

or $\Phi = \bar{T}^{-1} \bar{\Phi}$.

The u- or v-displacement function is usually assumed as

$$u(x, y) = \alpha_1 + \alpha_2 x + \alpha_3 y + \alpha_4 x^2 + \alpha_5 xy + \alpha_6 y^2 + \alpha_7 x^2 y + \alpha_8 xy^2.$$

For the group supermatrix procedure the terms in the above polynomial will be allocated by the following expression into respective G-invariant subspaces where they belong and with the unique order of terms that provides the diagonal form of the supermatrix of equations relating displacements U to coefficients A:

$$\begin{bmatrix} F_1 \\ F_2 \\ F_3 \\ F_4 \end{bmatrix} = \begin{bmatrix} 1 \\ xy \\ x \\ y \end{bmatrix} \begin{bmatrix} 1 & x^2 & y^2 \end{bmatrix} = \begin{bmatrix} 1 & x^2 & y^2 \\ xy & x^3 y & xy^3 \\ x & x^3 & xy^2 \\ y & x^2 y & y^3 \end{bmatrix}.$$

The column vector and the row vector are taken from the Cartesian sets and products as they stand in addition to the character table of the group C_{2v} according to the pertinence of the terms to the symmetry types of the group representations A_1, A_2, B_1, B_2. By omitting the terms $x^3 y, xy^3, x^3, y^3$, which do not appear in the polynomial $u(x, y)$, one obtains

Sec. 4.3]   Eight-node rectangular element   163

$$\begin{bmatrix} F'_1 \\ F'_2 \\ F'_3 \\ F'_4 \end{bmatrix} = \begin{bmatrix} 1 & x^2 & & y^2 \\ & xy & & \\ x & xy^2 & & \\ y & x^2y & & \end{bmatrix} \begin{matrix} \text{subspace} \\ U_1 \\ U_2 \\ U_3 \\ U_4 \end{matrix}$$

with the pertinence of the terms to the subspaces U_1, U_2, U_3, U_4 and the optimum order of the terms.

Thus, the displacement function

$$u(x, y) = \begin{bmatrix} 1 & x^2 & y^2 & \vdots & xy & \vdots & x & xy^2 & \vdots & y & x^2y \end{bmatrix}$$
$$\quad\quad\quad [\alpha_1 \; \alpha_2 \; \alpha_3 \; \vdots \; \alpha_4 \; \vdots \; \alpha_5 \; \alpha_6 \; \vdots \; \alpha_7 \; \alpha_8]^T$$
$$= [F'_1 \; \vdots \; F'_2 \; \vdots \; F'_3 \; \vdots \; F'_4][A_1 \; \vdots \; A_2 \; \vdots \; A_3 \; \vdots \; A_4]^T$$

is decomposed into polynomials pertaining to G-invariant subspaces U_1, U_2, U_3, U_4 with 3, 1, 2, 2 dimensions respectively, which coincides with dimensions of subspaces spanned by basis vectors derived earlier.

Substitution of the nodal coordinates of the first nodes in the nodal sets $S_1(1, 2, 3, 4)$, $S_2(5, 6)$, $S_3(7, 8)$

	1	5	7
x	c	c	0
y	d	0	d

into the sections F'_1, F'_2, F'_3, F'_4 of $u(x, y)$, pertaining to the subspaces U_1, U_2, U_3, U_4, provides relations of displacements $\bar{U}$ to the coefficients A for each subspace separately.

$$\begin{bmatrix} \bar{u}_1 \\ \bar{u}_5 \\ \bar{u}_7 \\ \hdashline \bar{u}_2 \\ \hdashline \bar{u}_3 \\ \bar{u}_6 \\ \hdashline \bar{u}_4 \\ \bar{u}_8 \end{bmatrix} = \frac{1}{4} \begin{bmatrix} u_1 + u_2 + u_3 + u_4 \\ 2u_5 + 2u_6 \\ 2u_7 + 2u_8 \\ \hdashline u_1 + u_2 - u_3 - u_4 \\ \hdashline u_1 - u_2 + u_3 - u_4 \\ 2u_5 - 2u_6 \\ \hdashline u_1 - u_2 - u_3 + u_4 \\ 2u_7 - 2u_8 \end{bmatrix} = \begin{bmatrix} 1 & c^2 & d^2 & & & & & \\ 1 & c^2 & 0 & & & & & \\ 1 & 0 & d^2 & & & & & \\ \hdashline & & & cd & & & & \\ \hdashline & & & & c & cd^2 & & \\ & & & & c & 0 & & \\ \hdashline & & & & & & d & c^2d \\ & & & & & & d & 0 \end{bmatrix} \begin{bmatrix} \alpha_1 \\ \alpha_2 \\ \alpha_3 \\ \hdashline \alpha_4 \\ \hdashline \alpha_5 \\ \alpha_6 \\ \hdashline \alpha_7 \\ \alpha_8 \end{bmatrix}$$

or $\bar{U} = \bar{C}A$, where $\bar{U}$ is the set of displacements with the numbering according to the numbering of basis vectors derived earlier. Thus, the eight-dimensional G-vector space of the problem is decomposed into four G-invariant subspaces U_1, U_2, U_3, U_4 with 3, 1, 2, 2 dimensions respectively.

The coefficients A_1, A_2, A_3, A_4 are found by inverting the matrices $\bar{C}_1$, $\bar{C}_2$, $\bar{C}_3$, $\bar{C}_4$ of the subspaces U_1, U_2, U_3, U_4.

$$\begin{bmatrix} \alpha_1 \\ \alpha_2 \\ \alpha_3 \\ \alpha_4 \\ \alpha_5 \\ \alpha_6 \\ \alpha_7 \\ \alpha_8 \end{bmatrix} = \text{Diag}\left[1 \quad \frac{1}{c^2} \quad \frac{1}{d^2} \; \vdots \; \frac{1}{cd} \; \vdots \; \frac{1}{c} \quad \frac{1}{cd^2} \; \vdots \; \frac{1}{d} \quad \frac{1}{c^2 d}\right]$$

$$= \begin{bmatrix} -1 & 1 & 1 & & & & & \\ 1 & 0 & -1 & & & & & \\ 1 & -1 & 1 & & & & & \\ \hdashline & & & 1 & & & & \\ \hdashline & & & & 0 & 1 & & \\ & & & & 1 & -1 & & \\ \hdashline & & & & & & 0 & 1 \\ & & & & & & 1 & -1 \end{bmatrix} \begin{bmatrix} \bar{u}_1 \\ \bar{u}_5 \\ \bar{u}_7 \\ \bar{u}_2 \\ \bar{u}_3 \\ \bar{u}_6 \\ \bar{u}_4 \\ \bar{u}_8 \end{bmatrix}$$

or $A = \bar{C}^{-1}\bar{U}$.

The eight-dimensional G-vector space of the displacement field Δ, decomposed into subspaces U_1, U_2, U_3, U_4 with 3, 1, 2, 2 dimensions respectively, is produced by

$$\Delta = T_D \bar{C}^{-1} \bar{U},$$

with

$$T_D = \text{Diag}\begin{bmatrix} 1 & x^2 & y^2 \; \vdots \; xy \; \vdots \; x & xy^2 \; \vdots \; y & x^2 y \end{bmatrix} = \text{Diag}\begin{bmatrix} F'_1 \; \vdots \; F'_2 \; \vdots \; F'_3 \; \vdots \; F'_4 \end{bmatrix}$$

and $\xi = x/c$, $\eta = y/d$, giving

Sec. 4.3] Eight-node rectangular element

$$\begin{bmatrix} \Delta_1 \\ \cdots \\ \Delta_2 \\ \cdots \\ \Delta_3 \\ \cdots \\ \Delta_4 \end{bmatrix} = \text{Diag}\begin{bmatrix} 1 & \dfrac{x^2}{c^2} & \dfrac{y^2}{d^2} & \vdots & \dfrac{xy}{cd} & \vdots & \dfrac{x}{c} & \dfrac{xy^2}{cd^2} & \vdots & \dfrac{y}{d} & \dfrac{x^2y}{c^2d} \end{bmatrix}$$

$$= \begin{bmatrix} -1 & 1 & 1 & & & & & & \\ 1 & 0 & -1 & & & & & & \\ 1 & -1 & 0 & & & & & & \\ \hline & & & 1 & & & & & \\ \hline & & & & 0 & 1 & & & \\ & & & & 1 & -1 & & & \\ \hline & & & & & & 0 & 1 \\ & & & & & & 1 & -1 \end{bmatrix} \begin{bmatrix} \bar{u}_1 \\ \bar{u}_5 \\ \bar{u}_7 \\ \cdots \\ \bar{u}_2 \\ \cdots \\ \bar{u}_3 \\ \bar{u}_6 \\ \cdots \\ \bar{u}_4 \\ \bar{u}_8 \end{bmatrix}$$

$$= \begin{bmatrix} -1 & 1 & 1 & & & & & & \\ \xi^2 & 0 & -\xi^2 & & & & & & \\ \eta^2 & -\eta^2 & 0 & & & & & & \\ \hline & & & \xi\eta & & & & & \\ \hline & & & & 0 & \xi & & & \\ & & & & \xi\eta^2 & -\xi\eta^2 & & & \\ \hline & & & & & & 0 & \eta \\ & & & & & & \xi^2\eta & -\xi^2\eta \end{bmatrix} \begin{bmatrix} \bar{u}_1 \\ \bar{u}_5 \\ \bar{u}_7 \\ \cdots \\ \bar{u}_2 \\ \cdots \\ \bar{u}_3 \\ \bar{u}_6 \\ \cdots \\ \bar{u}_4 \\ \bar{u}_8 \end{bmatrix}$$

or $\Delta = \bar{N}_\Delta \bar{U}$.

The shape functions $\bar{N}$ in G-invariant subspaces are obtained by

$$\bar{N} = S\bar{N}_\Delta,$$

with

$$S = [1 \quad 1 \quad 1 \quad 1 \quad 1 \quad 1 \quad 1 \quad 1],$$

or

166   Formulation of shape functions in G-invariant subspaces   [Ch. 4

$$\begin{bmatrix} \bar{N}_1 \\ \bar{N}_5 \\ \bar{N}_7 \\ \hdashline \bar{N}_2 \\ \hdashline \bar{N}_3 \\ \bar{N}_6 \\ \hdashline \bar{N}_4 \\ \bar{N}_8 \end{bmatrix} = \begin{bmatrix} N_1 + N_2 + N_3 + N_4 \\ N_5 + N_6 \\ N_7 + N_8 \\ \hdashline N_1 + N_2 - N_3 - N_4 \\ \hdashline N_1 - N_2 + N_3 - N_4 \\ N_5 - N_6 \\ \hdashline N_1 - N_2 - N_3 + N_4 \\ N_7 - N_8 \end{bmatrix}$$

$$= \begin{bmatrix} 1 \\ 1 \\ 1 \\ \hdashline 1 \\ \hdashline 1 \\ 1 \\ \hdashline 1 \\ 1 \end{bmatrix}^T \begin{bmatrix} -1 & 1 & 1 & & & & & \\ \xi^2 & 0 & -\xi^2 & & & & & \\ \eta^2 & -\eta^2 & 0 & & & & & \\ & & & \xi\eta & & & & \\ & & & & 0 & \xi & & \\ & & & & \xi\eta^2 & -\xi\eta^2 & & \\ & & & & & & 0 & \eta \\ & & & & & & \xi^2\eta & -\xi^2\eta \end{bmatrix}$$

$$= \begin{bmatrix} -1 + \xi^2 + \eta^2 \\ 1 - \eta^2 \\ 1 - \xi^2 \\ \hdashline \xi\eta \\ \hdashline \xi\eta^2 \\ \xi - \xi\eta^2 \\ \hdashline \xi^2\eta \\ \eta - \xi^2\eta \end{bmatrix}$$

or

$$\begin{bmatrix} \bar N_1 \\ \bar N_5 \\ \bar N_7 \\ \cdots \\ \bar N_2 \\ \cdots \\ \bar N_3 \\ \bar N_6 \\ \cdots \\ \bar N_4 \\ \bar N_8 \end{bmatrix} = \begin{bmatrix} -1 & 1 & 1 & & & & & \\ 1 & 0 & -1 & & & & & \\ 1 & -1 & 0 & & & & & \\ \hline & & & 1 & & & & \\ \hline & & & & 0 & 1 & & \\ & & & & 1 & -1 & & \\ \hline & & & & & & 0 & 1 \\ & & & & & & 1 & -1 \end{bmatrix} \begin{bmatrix} 1 \\ \xi^2 \\ \eta^2 \\ \cdots \\ \xi\eta \\ \cdots \\ \xi \\ \xi\eta^2 \\ \cdots \\ \eta \\ \xi^2\eta \end{bmatrix} \quad \text{or} \quad \bar N = \bar N_\Delta F.$$

The shape functions $N_1, N_2, \ldots, N_8$ pertaining to the nodes $1, 2, \ldots, 8$ respectively, are obtained by

$$N = T\bar N = T\bar N_\Delta F,$$

which, after rearrangement of the sequence of the equations, gives

$$\begin{bmatrix} N_1 \\ N_2 \\ N_3 \\ N_4 \\ N_5 \\ N_6 \\ N_7 \\ N_8 \end{bmatrix} = T \begin{bmatrix} \bar N_1 \\ \bar N_2 \\ \bar N_3 \\ \bar N_4 \\ \bar N_5 \\ \bar N_6 \\ \bar N_7 \\ \bar N_8 \end{bmatrix}$$

$$= \begin{bmatrix} \dfrac{1}{4}\begin{bmatrix} 1 & 1 & 1 & 1 \\ 1 & 1 & -1 & -1 \\ 1 & -1 & 1 & -1 \\ 1 & -1 & -1 & 1 \end{bmatrix} & & \\ & \dfrac{1}{2}\begin{bmatrix} 1 & 1 \\ 1 & -1 \end{bmatrix} & \\ & & \dfrac{1}{2}\begin{bmatrix} 1 & 1 \\ 1 & -1 \end{bmatrix} \end{bmatrix}$$

$$\begin{bmatrix} -1 & 1 & 1 & & & & & \\ & & & 1 & & & & \\ & & & & 0 & 1 & & \\ & & & & & & 0 & 1 \\ \hdashline 1 & 0 & -1 & & & & & \\ & & & 1 & -1 & & & \\ \hdashline 1 & -1 & 0 & & & & & \\ & & & & & & 1 & -1 \end{bmatrix} \begin{bmatrix} 1 \\ \xi^2 \\ \eta^2 \\ \xi\eta \\ \xi \\ \xi\eta^2 \\ \eta \\ \xi^2\eta \end{bmatrix}$$

$$= \frac{1}{4} \begin{bmatrix} -1 & 1 & 1 & 1 & 0 & 1 & 0 & 1 \\ -1 & 1 & 1 & 1 & 0 & -1 & 0 & -1 \\ -1 & 1 & 1 & -1 & 0 & 1 & 0 & -1 \\ -1 & 1 & 1 & -1 & 0 & -1 & 0 & 1 \\ 2 & 0 & -2 & 0 & 2 & -2 & 0 & 0 \\ 2 & 0 & -2 & 0 & -2 & 2 & 0 & 0 \\ 2 & -2 & 0 & 0 & 0 & 0 & 2 & -2 \\ 2 & -2 & 0 & 0 & 0 & 0 & -2 & 2 \end{bmatrix} \begin{bmatrix} 1 \\ \xi^2 \\ \eta^2 \\ \xi\eta \\ \xi \\ \xi\eta^2 \\ \eta \\ \xi^2\eta \end{bmatrix}.$$

As in the case of the four-node rectangular element, shape functions in G-invariant subspaces for the eight-node rectangular element will be derived also by means of group supermatrix transformation of the shape functions for the eight-node serendipity rectangular element. These shape functions are

for corner nodes at $\xi_i = \pm 1$, $\eta_i = \pm 1$:

$$\tfrac{1}{4}(1+\xi\xi_i)(1+\eta\eta_i) - \tfrac{1}{4}(1-\xi^2)(1+\eta\eta_i) - \tfrac{1}{4}(1+\xi\xi_i)(1-\eta^2);$$

for side nodes at $\xi_i = \pm 1$, $\eta_i = 0$:

$$\tfrac{1}{2}(1-\eta^2)(1+\xi\xi_i);$$

for side nodes at $\xi_i = 0$, $\eta_i = \pm 1$:

$$\tfrac{1}{2}(1-\xi^2)(1+\eta\eta_i).$$

By adopting the unique group supermatrix nodal numbering given in Fig. 4.3, and the optimum order of terms derived in this section, after expansion of expressions for the above shape functions the following set of shape functions is obtained:

Sec. 4.3] Eight-node rectangular element

$$\begin{bmatrix} N_1 \\ N_2 \\ N_3 \\ N_4 \\ N_5 \\ N_6 \\ N_7 \\ N_8 \end{bmatrix} = \frac{1}{4} \begin{bmatrix} -1 & 1 & 1 & 1 & 0 & 1 & 0 & 1 \\ -1 & 1 & 1 & 1 & 0 & -1 & 0 & -1 \\ -1 & 1 & 1 & -1 & 0 & 1 & 0 & -1 \\ -1 & 1 & 1 & -1 & 0 & -1 & 0 & 1 \\ 2 & 0 & -2 & 0 & 2 & -2 & 0 & 0 \\ 2 & 0 & -2 & 0 & -2 & 2 & 0 & 0 \\ 2 & -2 & 0 & 0 & 0 & 0 & 2 & -2 \\ 2 & -2 & 0 & 0 & 0 & 0 & -2 & 2 \end{bmatrix} \begin{bmatrix} 1 \\ \xi^2 \\ \eta^2 \\ \xi\eta \\ \xi \\ \xi\eta^2 \\ \eta \\ \xi^2\eta \end{bmatrix}$$

By applying $T^{-1}N$, with $T^{-1} = 4T$ and

$$T^{-1} = \begin{bmatrix} 1 & 1 & 1 & 1 & & & & \\ 1 & 1 & -1 & -1 & & & & \\ 1 & -1 & 1 & -1 & & & & \\ 1 & -1 & -1 & 1 & & & & \\ & & & & 1 & 1 & & \\ & & & & 1 & -1 & & \\ & & & & & & 1 & 1 \\ & & & & & & 1 & -1 \end{bmatrix},$$

the shape functions $\bar{N}$ appearing in G-invariant subspaces are produced:

$$\begin{bmatrix} \bar{N}_1 \\ \bar{N}_2 \\ \bar{N}_3 \\ \bar{N}_4 \\ \bar{N}_5 \\ \bar{N}_6 \\ \bar{N}_7 \\ \bar{N}_8 \end{bmatrix} = \begin{bmatrix} N_1 + N_2 + N_3 + N_4 \\ N_1 + N_2 - N_3 - N_4 \\ N_1 - N_2 + N_3 - N_4 \\ N_1 - N_2 - N_3 + N_4 \\ N_5 + N_6 \\ N_7 + N_8 \\ N_5 - N_6 \\ N_7 - N_8 \end{bmatrix} = \begin{bmatrix} -1 & 1 & 1 & & & & & \\ & & & 1 & & & & \\ & & & & & 0 & 1 & \\ & & & & & & 0 & 1 \\ 1 & 0 & -1 & & & & & \\ & & & & & 1 & -1 & \\ 1 & -1 & 0 & & & & & \\ & & & & & & 1 & -1 \end{bmatrix} \begin{bmatrix} 1 \\ \xi^2 \\ \eta^2 \\ \xi\eta \\ \xi \\ \xi\eta^2 \\ \eta \\ \xi^2\eta \end{bmatrix}.$$

Rearranging $\bar{N}_j$ according to the numbering of basis vectors (derived earlier) makes the matrix of shape functions $\bar{N}$ appear in block diagonal form

$$
\begin{bmatrix}
\bar{N}_1 \\ \bar{N}_5 \\ \bar{N}_7 \\ \text{---} \\ \bar{N}_2 \\ \text{---} \\ \bar{N}_3 \\ \bar{N}_6 \\ \text{---} \\ \bar{N}_4 \\ \bar{N}_8
\end{bmatrix}
\begin{bmatrix}
-1 & 1 & 1 & & & & & \\
1 & 0 & -1 & & & & & \\
1 & -1 & 0 & & & & & \\
\hline
 & & & 1 & & & & \\
\hline
 & & & & 0 & 1 & & \\
 & & & & 1 & -1 & & \\
\hline
 & & & & & & 0 & 1 \\
 & & & & & & 1 & -1
\end{bmatrix}
\begin{matrix}
1 \\ \xi^2 \\ \eta^2 \\ \text{---} \\ \xi\eta \\ \text{---} \\ \xi \\ \xi\eta^2 \\ \text{---} \\ \eta \\ \xi^2\eta
\end{matrix}
\quad
\begin{matrix}
\text{Subspace} \\ \\ U_1 \\ \\ \text{-----} \\ U_2 \\ \text{-----} \\ \\ U_3 \\ \\ \text{-----} \\ \\ U_4 \\
\end{matrix}
$$

Thus, the G-vector space of element shape functions is decomposed into G-invariant subspaces U_1, U_2, U_3, U_4 with 3, 1, 2, 2 dimensions and spanned by shape functions $[\bar{N}_1, \bar{N}_5, \bar{N}_7]$, $[\bar{N}_2]$, $[\bar{N}_3, \bar{N}_6]$, $[\bar{N}_4, \bar{N}_8]$ respectively.

4.4 TWELVE-NODE RECTANGULAR ELEMENT

The twelve-node rectangular element, given in Fig. 4.4, is described by the group C_{2v} and it has three nodal sets

$$S_1(1, 2, 3, 4), \quad S_2(5, 6, 7, 8), \quad S_3(9, 10, 11, 12),$$

each set containing nodes that are permuted by action of symmetry operations of the group.

The unique nodal numbering for the four-node rectangular element, derived in section 3.5, is used here for the nodal sets S_1, S_2, S_3, giving that the first nodes in these sets lie in

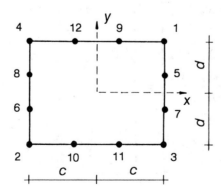

Fig. 4.4. The twelve-node rectangular element with the nodal numbering of the group supermatrix procedure.

Sec. 4.4]   Twelve-node rectangular element   171

the first quadrant, the second nodes in the third quadrant, the third nodes in the fourth quadrant and the fourth nodes in the second quadrant.

The character table of the group C_{2v} (with Cartesian sets and products) and the idempotents of the centre of the group algebra are given in section 4.3. By application of the idempotents π_i to the sets of nodal functions $\Phi^{(1)}$, $\Phi^{(2)}$, $\Phi^{(3)}$, $\Phi^{(4)}$, the sets of basis vectors $\bar{\Phi}^{(1)}$, $\bar{\Phi}^{(2)}$, $\bar{\Phi}^{(3)}$, $\bar{\Phi}^{(4)}$ of G-invariant subspaces U_i ($i = 1, 2, 3, 4$) are derived:

$$\begin{bmatrix} \bar{\Phi}^{(1)} \\ \bar{\Phi}^{(2)} \\ \bar{\Phi}^{(3)} \\ \bar{\Phi}^{(4)} \end{bmatrix} = \begin{bmatrix} \bar{\varphi}_1 & \bar{\varphi}_5 & \bar{\varphi}_9 \\ \bar{\varphi}_2 & \bar{\varphi}_6 & \bar{\varphi}_{10} \\ \bar{\varphi}_3 & \bar{\varphi}_7 & \bar{\varphi}_{11} \\ \bar{\varphi}_4 & \bar{\varphi}_8 & \bar{\varphi}_{12} \end{bmatrix} = \frac{1}{4} \begin{bmatrix} 1 & 1 & 1 & 1 \\ 1 & 1 & -1 & -1 \\ 1 & -1 & 1 & -1 \\ 1 & -1 & -1 & 1 \end{bmatrix} \begin{bmatrix} \varphi_1 & \varphi_5 & \varphi_9 \\ \varphi_2 & \varphi_6 & \varphi_{10} \\ \varphi_3 & \varphi_7 & \varphi_{11} \\ \varphi_4 & \varphi_8 & \varphi_{12} \end{bmatrix}$$

$$= \frac{1}{4} \begin{bmatrix} \varphi_1 + \varphi_2 + \varphi_3 + \varphi_4 & \varphi_5 + \varphi_6 + \varphi_7 + \varphi_8 & \varphi_9 + \varphi_{10} + \varphi_{11} + \varphi_{12} \\ \varphi_1 + \varphi_2 - \varphi_3 - \varphi_4 & \varphi_5 + \varphi_6 - \varphi_7 - \varphi_8 & \varphi_9 + \varphi_{10} - \varphi_{11} - \varphi_{12} \\ \varphi_1 - \varphi_2 + \varphi_3 - \varphi_4 & \varphi_5 - \varphi_6 + \varphi_7 - \varphi_8 & \varphi_9 - \varphi_{10} + \varphi_{11} - \varphi_{12} \\ \varphi_1 - \varphi_2 - \varphi_3 + \varphi_4 & \varphi_5 - \varphi_6 - \varphi_7 + \varphi_8 & \varphi_9 - \varphi_{10} - \varphi_{11} + \varphi_{12} \end{bmatrix}$$

or $\bar{\Phi} = T\Phi$, with $T^{-1} = 4T$.

Conversely, the relation of the sets of nodal functions $\Phi^{(1)}$, $\Phi^{(2)}$, $\Phi^{(3)}$, $\Phi^{(4)}$ to the sets of basis vectors $\bar{\Phi}^{(1)}$, $\bar{\Phi}^{(2)}$, $\bar{\Phi}^{(3)}$, $\bar{\Phi}^{(4)}$ is

$$\begin{bmatrix} \Phi^{(1)} \\ \Phi^{(2)} \\ \Phi^{(3)} \\ \Phi^{(4)} \end{bmatrix} = \begin{bmatrix} \varphi_1 & \varphi_5 & \varphi_9 \\ \varphi_2 & \varphi_6 & \varphi_{10} \\ \varphi_3 & \varphi_7 & \varphi_{11} \\ \varphi_4 & \varphi_8 & \varphi_{12} \end{bmatrix} = \begin{bmatrix} 1 & 1 & 1 & 1 \\ 1 & 1 & -1 & -1 \\ 1 & -1 & 1 & -1 \\ 1 & -1 & -1 & 1 \end{bmatrix} \begin{bmatrix} \bar{\Phi}^{(1)} \\ \bar{\Phi}^{(2)} \\ \bar{\Phi}^{(3)} \\ \bar{\Phi}^{(4)} \end{bmatrix}$$

$$= \begin{bmatrix} \bar{\varphi}_1 + \bar{\varphi}_2 + \bar{\varphi}_3 + \bar{\varphi}_4 & \bar{\varphi}_5 + \bar{\varphi}_6 + \bar{\varphi}_7 + \bar{\varphi}_8 & \bar{\varphi}_9 + \bar{\varphi}_{10} + \bar{\varphi}_{11} + \bar{\varphi}_{12} \\ \bar{\varphi}_1 + \bar{\varphi}_2 - \bar{\varphi}_3 - \bar{\varphi}_4 & \bar{\varphi}_5 + \bar{\varphi}_6 - \bar{\varphi}_7 - \bar{\varphi}_8 & \bar{\varphi}_9 + \bar{\varphi}_{10} - \bar{\varphi}_{11} - \bar{\varphi}_{12} \\ \bar{\varphi}_1 - \bar{\varphi}_2 + \bar{\varphi}_3 - \bar{\varphi}_4 & \bar{\varphi}_5 - \bar{\varphi}_6 + \bar{\varphi}_7 - \bar{\varphi}_8 & \bar{\varphi}_9 - \bar{\varphi}_{10} + \bar{\varphi}_{11} - \bar{\varphi}_{12} \\ \bar{\varphi}_1 - \bar{\varphi}_2 - \bar{\varphi}_3 + \bar{\varphi}_4 & \bar{\varphi}_5 - \bar{\varphi}_6 - \bar{\varphi}_7 + \bar{\varphi}_8 & \bar{\varphi}_9 - \bar{\varphi}_{10} - \bar{\varphi}_{11} + \bar{\varphi}_{12} \end{bmatrix}$$

or $\Phi = T^{-1} \bar{\Phi} = 4T\bar{\Phi}$.

The sets $\bar{\Phi}_1$, $\bar{\Phi}_2$, $\bar{\Phi}_3$ of basis vectors $\bar{\varphi}_j$ ($j = 1, 2, \ldots, 12$), ordered according to the numbering of the nodes in the nodal sets S_1, S_2, S_3, are given by

$$[\bar{\Phi}_1 \quad \bar{\Phi}_2 \quad \bar{\Phi}_3] = \begin{bmatrix} \bar{\varphi}_1 & \bar{\varphi}_5 & \bar{\varphi}_9 \\ \bar{\varphi}_2 & \bar{\varphi}_6 & \bar{\varphi}_{10} \\ \bar{\varphi}_3 & \bar{\varphi}_7 & \bar{\varphi}_{11} \\ \bar{\varphi}_4 & \bar{\varphi}_8 & \bar{\varphi}_{12} \end{bmatrix}.$$

The relation of the sets of basis vectors $\bar{\Phi}_l$ to the sets of nodal functions Φ_l ($l = 1, 2, 3$) is expressed by the following system of equations with the supermatrix in diagonal form, containing three transformation matrices T of the group C_{2v}:

$$\begin{bmatrix} \bar{\Phi}_1 \\ \bar{\Phi}_2 \\ \bar{\Phi}_3 \end{bmatrix} = \begin{bmatrix} T & & \\ & T & \\ & & T \end{bmatrix} \begin{bmatrix} \Phi_1 \\ \Phi_2 \\ \Phi_3 \end{bmatrix} \quad \text{or} \quad \bar{\Phi} = \bar{T}\Phi,$$

with

$$T = \frac{1}{4} \begin{bmatrix} 1 & 1 & 1 & 1 \\ 1 & 1 & -1 & -1 \\ 1 & -1 & 1 & -1 \\ 1 & -1 & -1 & 1 \end{bmatrix} \quad \text{and} \quad [\Phi_1 \; \Phi_2 \; \Phi_3] = \begin{bmatrix} \varphi_1 & \varphi_5 & \varphi_9 \\ \varphi_2 & \varphi_6 & \varphi_{10} \\ \varphi_3 & \varphi_7 & \varphi_{11} \\ \varphi_4 & \varphi_8 & \varphi_{12} \end{bmatrix}.$$

Conversely, the relation of the sets of nodal functions Φ_l to the sets of basis vectors $\bar{\Phi}_l$ is

$$\begin{bmatrix} \Phi_1 \\ \Phi_2 \\ \Phi_3 \end{bmatrix} = 4 \begin{bmatrix} T & & \\ & T & \\ & & T \end{bmatrix} \begin{bmatrix} \bar{\Phi}_1 \\ \bar{\Phi}_2 \\ \bar{\Phi}_3 \end{bmatrix} \quad \text{or} \quad \Phi = 4\bar{T}\bar{\Phi},$$

since $T^{-1} = 4T$.

The u- or v-displacement function is usually assumed as

$$u(x, y) = \alpha_1 + \alpha_2 x + \alpha_3 y + \alpha_4 x^2 + \alpha_5 xy + \alpha_6 y^2 + \alpha_7 x^3 + \alpha_8 x^2 y + \alpha_9 xy^2 +$$
$$+ \alpha_{10} y^3 + \alpha_{11} x^3 y + \alpha_{12} xy^3.$$

For the group supermatrix procedure, by the following expression the terms of the above polynomial will be allocated to respective G-invariant subspaces where they belong, and with the unique order of terms that provides the block diagonal form of the matrix of equations relating displacements $\bar{U}$ to coefficients A:

$$\begin{bmatrix} F_1 \\ F_2 \\ F_3 \\ F_4 \end{bmatrix} = \begin{bmatrix} 1 \\ xy \\ x \\ y \end{bmatrix} [1 \; x^2 \; y^2] = \begin{bmatrix} 1 & x^2 & y^2 \\ xy & x^3 y & xy^3 \\ x & x^3 & xy^2 \\ y & x^2 y & y^3 \end{bmatrix} \quad \begin{matrix} \text{Subspace} \\ U_1 \\ U_2 \\ U_3 \\ U_4 \end{matrix}$$

The column vector and the row vector are taken from the Cartesian sets and products as they stand, in addition to the character table of the group C_{2v} according to the pertinence of the terms to the symmetry types of group representations A_1, A_2, B_1, B_2.

Thus, the displacement function

Sec. 4.4]   Twelve-node rectangular element   173

$$u(x, y) = [1 \quad x^2 \quad y^2 \mid xy \quad x^3y \quad xy^3 \mid x \quad x^3 \quad xy^2$$
$$\mid y \quad x^2y \quad y^3][\alpha_1 \quad \alpha_2 \quad \ldots \quad \alpha_{12}]^T$$
$$= [F_1 \mid F_2 \mid F_3 \mid F_4][A_1 \mid A_2 \mid A_3 \mid A_4]^T$$

is decomposed into four three-dimensional G-invariant subspaces U_1, U_2, U_3, U_4.

Substitution of nodal coordinates of the first nodes in the nodal sets $S_1(1, 2, 3, 4)$, $S_2(5, 6, 7, 8), S_3(9, 10, 11, 12)$,

Node	1	5	9
x	c	c	$c/3$
y	d	$d/3$	d

into the sections F_1, F_2, F_3, F_4 of $u(x, y)$, pertaining to subspaces U_1, U_2, U_3, U_4, will provide the relation of the displacements $\bar{U}$ to the coefficients A for each subspace separately (see p. 174) or $\bar{U} = \bar{C}A$, where $\bar{U}$ is the set of displacements $\bar{U}^{(1)}, \bar{U}^{(2)}, \bar{U}^{(3)}$, $\bar{U}^{(4)}$ with the numbering according to the numbering of the sets of basis vectors $\bar{\Phi}^{(1)}$, $\bar{\Phi}^{(2)}, \bar{\Phi}^{(3)}, \bar{\Phi}^{(4)}$ derived earlier. Thus, the twelve-dimensional G-vector space of the problem is decomposed into four three-dimensional G-invariant subspaces U_1, U_2, U_3, U_4.

The coefficients A are determined by inverting the matrices $\bar{C}_1, \bar{C}_2, \bar{C}_3, \bar{C}_4$ of subspaces U_1, U_2, U_3, U_4, (see p. 175).

The twelve dimensional G-vector space of the displacement field Δ, decomposed into four three-dimensional G-invariant subspaces U_1, U_2, U_3, U_4, is given by

$$\Delta = T_D \bar{C}^{-1} \bar{U},$$

with

$$T_D = \text{Diag } [1 \quad x^2 \quad y^2 \mid xy \quad x^3y \quad xy^3 \mid x \quad x^3 \quad xy^2 \mid y \quad x^2y \quad y^3]$$
$$= \text{Diag } [F_1 \mid F_2 \mid F_3 \mid F_4]$$

and $\xi = x/c, \eta = y/d$, giving (see pp. 176–177) or $\Delta = \bar{N}_\Delta \bar{U}$.

The shape functions $\bar{N}$ of G-invariant subspaces are obtained by

$$\bar{N} = S\bar{N}_\Delta \quad \text{with} \quad S = [1 \quad 1 \quad 1 \mid 1 \quad 1 \quad 1 \mid 1 \quad 1 \quad 1 \mid 1 \quad 1 \quad 1]$$

(text follows on p. 178)

174 Formulation of shape functions in G-invariant subspaces [Ch. 4

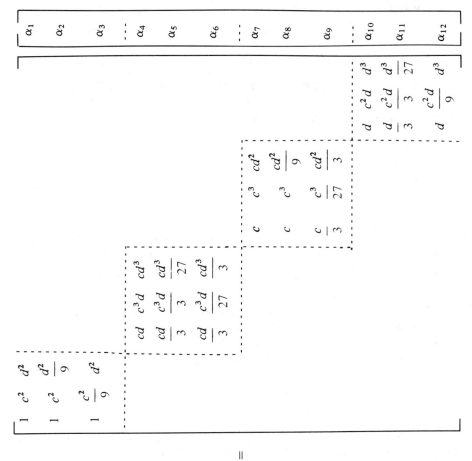

Sec. 4.4] Twelve-node rectangular element 175

$$
\begin{bmatrix} \bar{u}_1 \\ \bar{u}_5 \\ \bar{u}_9 \\ \hline \bar{u}_2 \\ \bar{u}_6 \\ \bar{u}_{10} \\ \hline \bar{u}_3 \\ \bar{u}_7 \\ \bar{u}_{11} \\ \hline \bar{u}_4 \\ \bar{u}_8 \\ \bar{u}_{12} \end{bmatrix}
= \frac{1}{8}
\begin{bmatrix}
-10 & 9 & 9 & & & & & & & & & \\
9 & 0 & -9 & & & & & & & & & \\
9 & -9 & 0 & & & & & & & & & \\
\hline
& & & -10 & 27 & 27 & & & & & & \\
& & & 9 & 0 & -27 & & & & & & \\
& & & 9 & -27 & 0 & & & & & & \\
\hline
& & & & & & -10 & 9 & 27 & & & \\
& & & & & & 9 & 0 & -27 & & & \\
& & & & & & 9 & -9 & 0 & & & \\
\hline
& & & & & & & & & -10 & 27 & 9 \\
& & & & & & & & & 9 & 0 & -9 \\
& & & & & & & & & 9 & -27 & 0
\end{bmatrix}
\cdot \mathrm{Diag}\begin{bmatrix} 1 \\ \dfrac{1}{c^2} \\ \dfrac{1}{d^2} \\ \hline \dfrac{1}{cd} \\ \dfrac{1}{c^3 d} \\ \dfrac{1}{cd^3} \\ \hline \dfrac{1}{c} \\ \dfrac{1}{c^3} \\ \dfrac{1}{cd^2} \\ \hline \dfrac{1}{d} \\ \dfrac{1}{c^2 d} \\ \dfrac{1}{d^3} \end{bmatrix}
\begin{bmatrix} \alpha_1 \\ \alpha_2 \\ \alpha_3 \\ \hline \alpha_4 \\ \alpha_5 \\ \alpha_6 \\ \hline \alpha_7 \\ \alpha_8 \\ \alpha_9 \\ \hline \alpha_{10} \\ \alpha_{11} \\ \alpha_{12} \end{bmatrix}
$$

176 Formulation of shape functions in G-invariant subspaces [Ch. 4

$$
\begin{bmatrix} 1 \\ \dfrac{x^2}{c^2} \\ \dfrac{y^2}{d^2} \\ \hline \dfrac{xy}{cd} \\ \dfrac{x^3y}{c^3d} \\ \dfrac{xy^3}{cd^3} \\ \hline \dfrac{x}{c} \\ \dfrac{x^3}{c^3} \\ \dfrac{xy^3}{cd^3}... \\ \hline \dfrac{y}{d} \\ \dfrac{x^2y}{c^2d} \\ \dfrac{y^3}{d^3} \end{bmatrix}
= \frac{1}{8}
\begin{bmatrix}
-10 & 9 & 9 & & & & & & & & & \\
9 & 0 & 9 & & & & & & & & & \\
9 & -9 & 0 & & & & & & & & & \\
\hline
 & & & -10 & 27 & 27 & & & & & & \\
 & & & 9 & 0 & -27 & & & & & & \\
 & & & 9 & -27 & 0 & & & & & & \\
\hline
 & & & & & & -10 & 9 & 27 & & & \\
 & & & & & & 9 & 0 & -27 & & & \\
 & & & & & & 9 & -9 & 0 & & & \\
\hline
 & & & & & & & & & -10 & 27 & 9 \\
 & & & & & & & & & 9 & 0 & -9 \\
 & & & & & & & & & 9 & -27 & 0 \\
\end{bmatrix}
\begin{bmatrix} \bar{u}_1 \\ \bar{u}_5 \\ \bar{u}_9 \\ \hline \bar{u}_2 \\ \bar{u}_6 \\ \bar{u}_{10} \\ \hline \bar{u}_3 \\ \bar{u}_7 \\ \bar{u}_{11} \\ \hline \bar{u}_4 \\ \bar{u}_8 \\ \bar{u}_{12} \end{bmatrix}
$$

$$= \operatorname{Diag}\begin{bmatrix} \Delta_1 & \Delta_2 & \Delta_3 & \Delta_4 \end{bmatrix}$$

Sec. 4.4] Twelve-node rectangular element

$$\mathbf{N} = \frac{1}{8}\begin{bmatrix} -10 & 9 & 9 & \vdots & -10\xi\eta & 27\xi\eta & 27\xi\eta & \vdots & -10\xi & 9\xi & 27\xi & \vdots & -10\eta & 27\eta & 9\eta \\ 9\xi^2 & 0 & -9\xi^2 & \vdots & 9\xi^3\eta & 0 & -27\xi^3\eta & \vdots & 9\xi^3 & 0 & -27\xi^3 & \vdots & 9\xi^2\eta & 0 & -9\xi^2\eta \\ 9\eta^2 & -9\eta^2 & 0 & \vdots & 9\xi\eta^3 & -27\xi\eta^3 & 0 & \vdots & 9\xi\eta^2 & -9\xi\eta^2 & 0 & \vdots & 9\eta^3 & -27\eta^3 & 0 \end{bmatrix} \begin{bmatrix} \bar{u}_1 \\ \bar{u}_5 \\ \bar{u}_9 \\ \bar{u}_2 \\ \bar{u}_6 \\ \bar{u}_{10} \\ \bar{u}_3 \\ \bar{u}_7 \\ \bar{u}_{11} \\ \bar{u}_4 \\ \bar{u}_8 \\ \bar{u}_{12} \end{bmatrix}$$

178   Formulation of shape functions in G-invariant subspaces   [Ch. 4

$$\begin{bmatrix} \bar{N}_1 \\ \bar{N}_5 \\ \bar{N}_9 \\ \hdashline \bar{N}_2 \\ \bar{N}_6 \\ \bar{N}_{10} \\ \hdashline \bar{N}_3 \\ \bar{N}_7 \\ \bar{N}_{11} \\ \hdashline \bar{N}_4 \\ \bar{N}_8 \\ \bar{N}_{12} \end{bmatrix} = \begin{bmatrix} N_1 + N_2 + N_3 + N_4 \\ N_5 + N_6 + N_7 + N_8 \\ N_9 + N_{10} + N_{11} + N_{12} \\ \hdashline N_1 + N_2 - N_3 - N_4 \\ N_5 + N_6 - N_7 - N_8 \\ N_9 + N_{10} - N_{11} - N_{12} \\ \hdashline N_1 - N_2 + N_3 - N_4 \\ N_5 - N_6 + N_7 - N_8 \\ N_9 - N_{10} + N_{11} - N_{12} \\ \hdashline N_1 - N_2 - N_3 + N_4 \\ N_5 - N_6 - N_7 + N_8 \\ N_9 - N_{10} - N_{11} + N_{12} \end{bmatrix} = \begin{bmatrix} 1 \\ 1 \\ 1 \\ 1 \\ 1 \\ 1 \\ 1 \\ 1 \\ 1 \\ 1 \\ 1 \\ 1 \end{bmatrix}^T \frac{1}{8} \text{Diag} \begin{bmatrix} -10 & 9 & 9 \\ 9\xi^2 & 0 & -9\xi^2 \\ 9\eta^2 & -9\eta^2 & 0 \\ \hdashline -10\xi\eta & 27\xi\eta & 27\xi\eta \\ 9\xi^3\eta & 0 & -27\xi^3\eta \\ 9\xi\eta^3 & -27\xi\eta^3 & 0 \\ \hdashline -10\xi & 9\xi & 27\xi \\ 9\xi^2 & 0 & -27\xi^3 \\ 9\xi\eta^2 & -9\xi\eta^2 & 0 \\ \hdashline -10\eta & 27\eta & 9\eta \\ 9\xi^2\eta & 0 & -9\xi^2\eta \\ 9\eta^3 & -27\eta^3 & 0 \end{bmatrix}$$

$$= \frac{1}{8} \begin{bmatrix} -10 + 9\xi^2 + 9\eta^2 \\ 9 - 9\eta^2 \\ 9 - 9\xi^2 \\ \hdashline -10\xi\eta + 9\xi^3\eta + 9\xi\eta^3 \\ 27\xi\eta - 27\xi\eta^3 \\ 27\xi\eta - 27\xi^3\eta \\ \hdashline -10\xi + 9\xi^3 + 9\xi\eta^2 \\ 9\xi - 9\xi\eta^2 \\ 27\xi - 27\xi^3 \\ \hdashline -10\eta + 9\xi^2\eta + 9\eta^3 \\ 27\eta - 27\eta^3 \\ 9\eta - 9\xi^2\eta \end{bmatrix}$$

or

Sec. 4.4] Twelve-node rectangular element

$$\begin{bmatrix} \bar{N}_1 \\ \bar{N}_5 \\ \bar{N}_9 \\ \hdashline \bar{N}_2 \\ \bar{N}_6 \\ \bar{N}_{10} \\ \hdashline \bar{N}_3 \\ \bar{N}_7 \\ \bar{N}_{11} \\ \hdashline \bar{N}_4 \\ \bar{N}_8 \\ \bar{N}_{12} \end{bmatrix} = \frac{1}{8} \begin{bmatrix} -10 & 9 & 9 & & & & & & & & & \\ 9 & 0 & -9 & & & & & & & & & \\ 9 & -9 & 0 & & & & & & & & & \\ & & & -10 & 9 & -27 & & & & & & \\ & & & 27 & 0 & -27 & & & & & & \\ & & & 27 & -27 & 0 & & & & & & \\ & & & & & & -10 & 9 & 9 & & & \\ & & & & & & 9 & 0 & -9 & & & \\ & & & & & & 27 & -27 & 0 & & & \\ & & & & & & & & & -10 & 9 & 9 \\ & & & & & & & & & 27 & 0 & -9 \\ & & & & & & & & & 9 & -27 & 0 \end{bmatrix} \begin{bmatrix} 1 \\ \xi^2 \\ \eta^2 \\ \hdashline \xi\eta \\ \xi^3\eta \\ \xi\eta^3 \\ \hdashline \xi \\ \xi^3 \\ \xi\eta^2 \\ \hdashline \eta \\ \xi^2\eta \\ \eta^3 \end{bmatrix}$$

The shape functions $\bar{N}$ can be written in supermatrix form

$$\begin{bmatrix} \bar{N}^{(1)} \\ \bar{N}^{(2)} \\ \bar{N}^{(3)} \\ \bar{N}^{(4)} \end{bmatrix} = \frac{1}{8} \begin{bmatrix} A & & & \\ & B & & \\ & & C & \\ & & & D \end{bmatrix} \begin{bmatrix} F_1 \\ F_2 \\ F_3 \\ F_4 \end{bmatrix},$$

where the shape functions at the nodes

$$N^{(1)} = [N_1 \; N_5 \; N_9]^T,$$
$$N^{(2)} = [N_2 \; N_6 \; N_{10}]^T,$$
$$N^{(3)} = [N_3 \; N_7 \; N_{11}]^T,$$
$$N^{(4)} = [N_4 \; N_8 \; N_{12}]^T,$$

are obtained by $N = T\bar{N}$ or

$$\begin{bmatrix} N^{(1)} \\ N^{(2)} \\ N^{(3)} \\ N^{(4)} \end{bmatrix} = \frac{1}{4} \begin{bmatrix} E & E & E & E \\ E & E & -E & -E \\ E & -E & E & -E \\ E & -E & -E & E \end{bmatrix} \frac{1}{8} \begin{bmatrix} A & & & \\ & B & & \\ & & C & \\ & & & D \end{bmatrix} \begin{bmatrix} F_1 \\ F_2 \\ F_3 \\ F_4 \end{bmatrix}$$

$$= \frac{1}{32} \begin{bmatrix} A & B & C & D \\ A & B & -C & -D \\ A & -B & C & -D \\ A & -B & -C & D \end{bmatrix} \begin{bmatrix} F_1 \\ F_2 \\ F_3 \\ F_4 \end{bmatrix},$$

or explicitly (see p. 181).

Similar to the case of the four-node rectangular element in section 4.2, shape functions in G-invariant subspaces for the twelve-node rectangular element will be derived also by means of group supermatrix transformation of shape functions for the twelve-node serendipity rectangular element. These shape functions are

for corner nodes at $\xi_i = \pm 1, \quad \eta_i = \pm 1,$

$$\tfrac{1}{32} (1 + \xi\xi_i)(1 + \eta\eta_i)[-10 + 9(\xi^2 + \eta^2)],$$

for nodes at $\xi_i = \pm 1, \quad \eta_i = \pm \tfrac{1}{3},$

$$\tfrac{9}{32} (1 + \xi\xi_i)(1 + 9\eta\eta_i)(1 - \eta^2),$$

for nodes at $\xi_i = \pm \tfrac{1}{3}, \quad \eta_i = \pm 1,$

$$\tfrac{9}{32} (1 + 9\xi\xi_i)(1 + \eta\eta_i)(1 - \xi^2).$$

Sec. 4.4] Twelve-node rectangular element 181

$$\begin{bmatrix} N_1 \\ N_5 \\ N_9 \\ \hdashline N_2 \\ N_6 \\ N_{10} \\ \hdashline N_3 \\ N_7 \\ N_{11} \\ \hdashline N_4 \\ N_8 \\ N_{12} \end{bmatrix} = \frac{1}{32} \begin{bmatrix} -10 & 9 & 9 & -10 & 9 & 9 \\ 9 & 0 & -9 & 27 & 0 & -27 \\ 9 & -9 & 0 & 27 & -27 & 0 \\ \hdashline -10 & 9 & 9 & -10 & 9 & 9 \\ 9 & 0 & -9 & 27 & 0 & -27 \\ 9 & -9 & 0 & 27 & -27 & 0 \\ \hdashline -10 & 9 & 9 & 10 & -9 & -9 \\ 9 & 0 & -9 & -27 & 0 & 27 \\ 9 & -9 & 9 & -27 & 27 & 0 \\ \hdashline -10 & 9 & 9 & 10 & -9 & -9 \\ 9 & 0 & -9 & -27 & 0 & 27 \\ 9 & -9 & 0 & -27 & 27 & 0 \end{bmatrix}$$

$$\begin{bmatrix} -10 & 9 & 9 & -10 & 9 & 9 \\ 9 & 0 & -9 & 27 & 0 & -27 \\ 27 & -27 & 0 & 9 & -9 & 0 \\ \hdashline 10 & -9 & -9 & 10 & -9 & -9 \\ -9 & 0 & 9 & -27 & 0 & 27 \\ -27 & 27 & 0 & -9 & 9 & 0 \\ \hdashline -10 & 9 & 9 & 10 & -9 & -9 \\ 9 & 0 & -9 & -27 & 0 & 27 \\ 27 & -27 & 0 & -9 & 9 & 0 \\ \hdashline 10 & -9 & -9 & -10 & 9 & 9 \\ -9 & 0 & 9 & 27 & 0 & -27 \\ -27 & 27 & 0 & 9 & -9 & 0 \end{bmatrix} \begin{bmatrix} 1 \\ \xi^2 \\ \eta^2 \\ \hdashline \xi\eta \\ \xi^3\eta \\ \xi\eta^3 \\ \hdashline \xi \\ \xi^3 \\ \xi\eta^2 \\ \hdashline \eta \\ \xi^2\eta \\ \eta^3 \end{bmatrix}$$

With the unique group supermatrix nodal numbering in Fig. 4.4, the order of the terms derived in this section and expansion of the expression for the above shape functions, the following set of shape functions N is obtained: (see p. 182).

After rearranging the order of N according to the numbering of basis vectors $\bar{\Phi}^{(1)}$, $\bar{\Phi}^{(2)}$, $\bar{\Phi}^{(3)}$, $\bar{\Phi}^{(4)}$, one obtains (see p. 183)

$$
\begin{bmatrix} N_1 \\ N_2 \\ N_3 \\ N_4 \\ \hdashline N_5 \\ N_6 \\ N_7 \\ N_8 \\ \hdashline N_9 \\ N_{10} \\ N_{11} \\ N_{12} \end{bmatrix} = \frac{1}{32} \begin{bmatrix} -10 & 9 & 9 & -10 & 9 & 9 \\ -10 & 9 & 9 & -10 & 9 & 9 \\ -10 & 9 & 9 & 10 & -9 & -9 \\ -10 & 9 & 9 & 10 & -9 & -9 \\ \hdashline 9 & 0 & -9 & 27 & 0 & -27 \\ 9 & 0 & -9 & 27 & 0 & -27 \\ 9 & 0 & -9 & -27 & 0 & 27 \\ 9 & 0 & -9 & -27 & 0 & 27 \\ \hdashline 9 & -9 & 0 & 27 & -27 & 0 \\ 9 & -9 & 0 & 27 & -27 & 0 \\ 9 & -9 & 0 & -27 & 27 & 0 \\ 9 & -9 & 0 & -27 & 27 & 0 \end{bmatrix}
$$

$$
\begin{bmatrix} -10 & 9 & 9 & -10 & 9 & 9 \\ 10 & -9 & -9 & 10 & -9 & -9 \\ -10 & 9 & 9 & 10 & -9 & -9 \\ 10 & -9 & -9 & -10 & 9 & 9 \\ \hdashline 9 & 0 & -9 & 27 & 0 & -27 \\ -9 & 0 & 9 & -27 & 0 & 27 \\ 9 & 0 & -9 & -27 & 0 & 27 \\ -9 & 0 & 9 & 27 & 0 & -27 \\ \hdashline 27 & -27 & 0 & 9 & -9 & 0 \\ -27 & 27 & 0 & -9 & 9 & 0 \\ 27 & -27 & 0 & -9 & 9 & 0 \\ -27 & 27 & 0 & 9 & -9 & 0 \end{bmatrix} \begin{bmatrix} 1 \\ \xi^2 \\ \eta^2 \\ \hdashline \xi\eta \\ \xi^3\eta \\ \xi\eta^3 \\ \hdashline \xi \\ \xi^3 \\ \xi\eta^2 \\ \hdashline \eta \\ \xi^2\eta \\ \eta^3 \end{bmatrix}
$$

Sec. 4.4] Twelve-node rectangular element

$$
\begin{bmatrix} N_1 \\ N_5 \\ N_9 \\ \hline N_2 \\ N_6 \\ N_{10} \\ \hline N_3 \\ N_7 \\ N_{11} \\ \hline N_4 \\ N_8 \\ N_{12} \end{bmatrix} = \frac{1}{32} \begin{bmatrix} -10 & 9 & 9 & -10 & 9 & 9 \\ 9 & 0 & -9 & 27 & 0 & -27 \\ 9 & -9 & 0 & 27 & -27 & 0 \\ \hline -10 & 9 & 9 & -10 & 9 & 9 \\ 9 & 0 & -9 & 27 & 0 & -27 \\ 9 & -9 & 0 & 27 & -27 & 0 \\ \hline -10 & 9 & 9 & 10 & -9 & -9 \\ 9 & 0 & -9 & -27 & 0 & 27 \\ 9 & -9 & 0 & -27 & 27 & 0 \\ \hline -10 & 9 & 9 & 10 & -9 & -9 \\ 9 & 0 & -9 & -27 & 0 & 27 \\ 9 & -9 & 0 & -27 & 27 & 0 \end{bmatrix}
$$

$$
\begin{bmatrix} -10 & 9 & 9 & -10 & 9 & 9 \\ 9 & 0 & -9 & 27 & 0 & -27 \\ 27 & -27 & 0 & 9 & -9 & 0 \\ \hline 10 & -9 & -9 & 10 & -9 & -9 \\ -9 & 0 & 9 & -27 & 0 & 27 \\ -27 & 27 & 0 & -9 & 9 & 0 \\ \hline -10 & 9 & 9 & 10 & -9 & -9 \\ 9 & 0 & -9 & -27 & 0 & 27 \\ 27 & -27 & 0 & -9 & 9 & 0 \\ \hline 10 & -9 & -9 & -10 & 9 & 9 \\ -9 & 0 & 9 & 27 & 0 & -27 \\ -27 & 27 & 0 & 9 & -9 & 0 \end{bmatrix} \begin{bmatrix} 1 \\ \xi^2 \\ \eta^2 \\ \hline \xi\eta \\ \xi^3\eta \\ \xi\eta^3 \\ \hline \xi \\ \xi^3 \\ \xi\eta^2 \\ \hline \eta \\ \xi^2\eta \\ \eta^3 \end{bmatrix}
$$

This relation can be written in the form

$$\begin{bmatrix} N^{(1)} \\ N^{(2)} \\ N^{(3)} \\ N^{(4)} \end{bmatrix} = \frac{1}{32} \begin{bmatrix} A & B & C & D \\ A & B & -C & -D \\ A & -B & C & -D \\ A & -B & -C & D \end{bmatrix} \begin{bmatrix} F_1 \\ F_2 \\ F_3 \\ F_4 \end{bmatrix}$$

$$= \frac{1}{32} \begin{bmatrix} E & E & E & E \\ E & E & -E & -E \\ E & -E & E & -E \\ E & -E & -E & E \end{bmatrix} \begin{bmatrix} A & & & \\ & B & & \\ & & C & \\ & & & D \end{bmatrix} \begin{bmatrix} F_1 \\ F_2 \\ F_3 \\ F_4 \end{bmatrix}$$

or $N = \frac{1}{32} T^{-1}$ Diag $[A\ B\ C\ D] F$.

Since $T^{-1} = 4T$, as given earlier in this section, the shape functions $\bar{N}$ in G-invariant subspaces are obtained by

$$\bar{N} = TN = \frac{1}{32} T T^{-1} \text{ Diag } [A\ B\ C\ D] F = \frac{1}{8} \text{ Diag } [A\ B\ C\ D] F,$$

or

$$\frac{1}{4} \begin{bmatrix} \bar{N}^{(1)} \\ \bar{N}^{(2)} \\ \bar{N}^{(3)} \\ \bar{N}^{(4)} \end{bmatrix} = \frac{1}{32} \begin{bmatrix} A & & & \\ & B & & \\ & & C & \\ & & & D \end{bmatrix} \begin{bmatrix} F_1 \\ F_2 \\ F_3 \\ F_4 \end{bmatrix}$$

or

$$\begin{bmatrix} \bar{N}_1 \\ \bar{N}_5 \\ \bar{N}_9 \\ \cdots \\ \bar{N}_2 \\ \bar{N}_4 \\ \bar{N}_{10} \\ \cdots \\ \bar{N}_3 \\ \bar{N}_7 \\ \bar{N}_{11} \\ \cdots \\ \bar{N}_4 \\ \bar{N}_8 \\ \bar{N}_{12} \end{bmatrix} = \begin{bmatrix} N_1 + N_2 + N_3 + N_4 \\ N_5 + N_6 + N_7 + N_8 \\ N_9 + N_{10} + N_{11} + N_{12} \\ \cdots \\ N_1 + N_2 - N_3 - N_4 \\ N_5 + N_6 - N_7 - N_8 \\ N_9 + N_{10} - N_{11} - N_{12} \\ \cdots \\ N_1 - N_2 + N_3 - N_4 \\ N_5 - N_6 + N_7 - N_8 \\ N_9 - N_{10} + N_{11} - N_{12} \\ \cdots \\ N_1 - N_2 - N_3 + N_4 \\ N_5 - N_6 - N_7 + N_8 \\ N_9 - N_{10} - N_{11} + N_{12} \end{bmatrix}$$

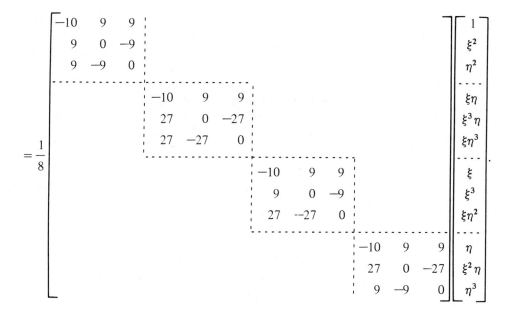

4.5 SIXTEEN-NODE RECTANGULAR ELEMENT

The sixteen-node rectangular element, shown in Fig. 4.5, is described by the group C_{2v} and it has four nodal sets

$S_1(1, 2, 3, 4)$ $S_2(5, 6, 7, 8)$

$S_3(9, 10, 11, 12)$ $S_4(13, 14, 15, 16)$,

each set containing nodes that are permuted by action of symmetry operations of the group.

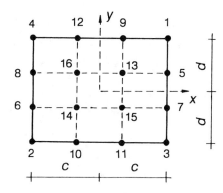

Fig. 4.5. The sixteen-node rectangular element with the nodal numbering of the group supermatrix procedure.

As in the case of the twelve-node rectangular element, the unqiue nodal numbering for the four-node rectangular element, derived in section 3.5, is applied here for the nodal sets S_1, S_2, S_3, S_4, giving that the first nodes in these sets lie in the first quadrant, the second nodes in the third quadrant, the third nodes in the fourth quadrant and the fourth nodes in the second quadrant.

The character table of the group C_{2v} (with Cartesian sets and products) and the idempotents of the centre of the group algebra are given in section 4.3. Applying the idempotents to the sets of nodal functions $\Phi^{(1)}, \Phi^{(2)}, \Phi^{(3)}, \Phi^{(4)}$ derives the sets of basis vectors $\bar{\Phi}^{(1)}, \bar{\Phi}^{(2)}, \bar{\Phi}^{(3)}, \bar{\Phi}^{(4)}$ of G-invariant subspaces U_i ($i = 1, 2, 3, 4$):

$$\begin{bmatrix} \bar{\Phi}^{(1)} \\ \bar{\Phi}^{(2)} \\ \bar{\Phi}^{(3)} \\ \bar{\Phi}^{(4)} \end{bmatrix} = \begin{bmatrix} \bar{\varphi}_1 & \bar{\varphi}_5 & \bar{\varphi}_9 & \bar{\varphi}_{13} \\ \bar{\varphi}_2 & \bar{\varphi}_6 & \bar{\varphi}_{10} & \bar{\varphi}_{14} \\ \bar{\varphi}_3 & \bar{\varphi}_7 & \bar{\varphi}_{11} & \bar{\varphi}_{15} \\ \bar{\varphi}_4 & \bar{\varphi}_8 & \bar{\varphi}_{12} & \bar{\varphi}_{16} \end{bmatrix}$$

$$= \frac{1}{4} \begin{bmatrix} 1 & 1 & 1 & 1 \\ 1 & 1 & -1 & -1 \\ 1 & -1 & 1 & -1 \\ 1 & -1 & -1 & 1 \end{bmatrix} \begin{bmatrix} \varphi_1 & \varphi_5 & \varphi_9 & \varphi_{13} \\ \varphi_2 & \varphi_6 & \varphi_{10} & \varphi_{14} \\ \varphi_3 & \varphi_7 & \varphi_{11} & \varphi_{15} \\ \varphi_4 & \varphi_8 & \varphi_{12} & \varphi_{16} \end{bmatrix}$$

$$= \frac{1}{4} \begin{bmatrix} \varphi_1 + \varphi_2 + \varphi_3 + \varphi_4 & \varphi_5 + \varphi_6 + \varphi_7 + \varphi_8 \\ \varphi_1 + \varphi_2 - \varphi_3 - \varphi_4 & \varphi_5 + \varphi_6 - \varphi_7 - \varphi_8 \\ \varphi_1 - \varphi_2 + \varphi_3 - \varphi_4 & \varphi_5 - \varphi_6 + \varphi_7 - \varphi_8 \\ \varphi_1 - \varphi_2 - \varphi_3 + \varphi_4 & \varphi_5 - \varphi_6 - \varphi_7 + \varphi_8 \end{bmatrix}$$

$$\begin{bmatrix} \varphi_9 + \varphi_{10} + \varphi_{11} + \varphi_{12} & \varphi_{13} + \varphi_{14} + \varphi_{15} + \varphi_{16} \\ \varphi_9 + \varphi_{10} - \varphi_{11} - \varphi_{12} & \varphi_{13} + \varphi_{14} - \varphi_{15} - \varphi_{16} \\ \varphi_9 - \varphi_{10} + \varphi_{11} - \varphi_{12} & \varphi_{13} - \varphi_{14} + \varphi_{15} - \varphi_{16} \\ \varphi_9 - \varphi_{10} - \varphi_{11} + \varphi_{12} & \varphi_{13} - \varphi_{14} - \varphi_{15} + \varphi_{16} \end{bmatrix}$$

or $\bar{\Phi} = T\Phi$, with $T^{-1} = 4T$.

Conversely, the relation of the nodal functions $\Phi^{(i)}$ to the basis vectors $\bar{\Phi}^{(i)}$ is

$$\begin{bmatrix} \Phi^{(1)} \\ \Phi^{(2)} \\ \Phi^{(3)} \\ \Phi^{(4)} \end{bmatrix} = \begin{bmatrix} \varphi_1 & \varphi_5 & \varphi_9 & \varphi_{13} \\ \varphi_2 & \varphi_6 & \varphi_{10} & \varphi_{14} \\ \varphi_3 & \varphi_7 & \varphi_{11} & \varphi_{15} \\ \varphi_4 & \varphi_8 & \varphi_{12} & \varphi_{16} \end{bmatrix} = \begin{bmatrix} 1 & 1 & 1 & 1 \\ 1 & 1 & -1 & -1 \\ 1 & -1 & 1 & -1 \\ 1 & -1 & -1 & 1 \end{bmatrix} \begin{bmatrix} \bar{\Phi}^{(1)} \\ \bar{\Phi}^{(2)} \\ \bar{\Phi}^{(3)} \\ \bar{\Phi}^{(4)} \end{bmatrix}$$

Sec. 4.5] Sixteen-node rectangular element 187

$$= \begin{bmatrix} \bar{\varphi}_1+\bar{\varphi}_2+\bar{\varphi}_3+\bar{\varphi}_4 & \bar{\varphi}_5+\bar{\varphi}_6+\bar{\varphi}_7+\bar{\varphi}_8 \\ \bar{\varphi}_1+\bar{\varphi}_2-\bar{\varphi}_3-\bar{\varphi}_4 & \bar{\varphi}_5+\bar{\varphi}_6-\bar{\varphi}_7-\bar{\varphi}_8 \\ \bar{\varphi}_1-\bar{\varphi}_2+\bar{\varphi}_3-\bar{\varphi}_4 & \bar{\varphi}_5-\bar{\varphi}_6+\bar{\varphi}_7-\bar{\varphi}_8 \\ \bar{\varphi}_1-\bar{\varphi}_2-\bar{\varphi}_3+\bar{\varphi}_4 & \bar{\varphi}_5-\bar{\varphi}_6-\bar{\varphi}_7+\bar{\varphi}_8 \end{bmatrix}$$

$$\begin{bmatrix} \bar{\varphi}_9+\bar{\varphi}_{10}+\bar{\varphi}_{11}+\bar{\varphi}_{12} & \bar{\varphi}_{13}+\bar{\varphi}_{14}+\bar{\varphi}_{15}+\bar{\varphi}_{16} \\ \bar{\varphi}_9+\bar{\varphi}_{10}-\bar{\varphi}_{11}-\bar{\varphi}_{12} & \bar{\varphi}_{13}+\bar{\varphi}_{14}-\bar{\varphi}_{15}-\bar{\varphi}_{16} \\ \bar{\varphi}_9-\bar{\varphi}_{10}+\bar{\varphi}_{11}-\bar{\varphi}_{12} & \bar{\varphi}_{13}-\bar{\varphi}_{14}+\bar{\varphi}_{15}-\bar{\varphi}_{16} \\ \bar{\varphi}_9-\bar{\varphi}_{10}-\bar{\varphi}_{11}+\bar{\varphi}_{12} & \bar{\varphi}_{13}-\bar{\varphi}_{14}-\bar{\varphi}_{15}+\bar{\varphi}_{16} \end{bmatrix}.$$

The sets $\bar{\Phi}_1, \bar{\Phi}_2, \bar{\Phi}_3, \bar{\Phi}_4$ of basis vectors $\bar{\varphi}_j$ ($j = 1, 2, \ldots, 16$), ordered according to the numbering of the nodes in the nodal sets S_1, S_2, S_3, S_4, are given by

$$[\bar{\Phi}_1 \ \bar{\Phi}_2 \ \bar{\Phi}_3 \ \bar{\Phi}_4] = \begin{bmatrix} \bar{\varphi}_1 & \bar{\varphi}_5 & \bar{\varphi}_9 & \bar{\varphi}_{13} \\ \bar{\varphi}_2 & \bar{\varphi}_6 & \bar{\varphi}_{10} & \bar{\varphi}_{14} \\ \bar{\varphi}_3 & \bar{\varphi}_7 & \bar{\varphi}_{11} & \bar{\varphi}_{15} \\ \bar{\varphi}_4 & \bar{\varphi}_8 & \bar{\varphi}_{12} & \bar{\varphi}_{16} \end{bmatrix}.$$

The relation of the sets of basis vectors $\bar{\Phi}_l$ to the sets of nodal functions Φ_l ($l = 1, 2, 3, 4$) is expressed by the following system of equations, with the supermatrix in diagonal form containing four transformation matrices T of the group C_{2v}:

$$\begin{bmatrix} \bar{\Phi}_1 \\ \bar{\Phi}_2 \\ \bar{\Phi}_3 \\ \bar{\Phi}_4 \end{bmatrix} = \begin{bmatrix} T & & & \\ & T & & \\ & & T & \\ & & & T \end{bmatrix} \begin{bmatrix} \Phi_1 \\ \Phi_2 \\ \Phi_3 \\ \Phi_4 \end{bmatrix} \quad \text{or} \quad \bar{\Phi} = T\Phi,$$

where

$$T = \frac{1}{4} \begin{bmatrix} 1 & 1 & 1 & 1 \\ 1 & 1 & -1 & -1 \\ 1 & -1 & 1 & -1 \\ 1 & -1 & -1 & 1 \end{bmatrix}$$

and

$$[\Phi_1 \ \Phi_2 \ \Phi_3 \ \Phi_4] = \begin{bmatrix} \varphi_1 & \varphi_5 & \varphi_9 & \varphi_{13} \\ \varphi_2 & \varphi_6 & \varphi_{10} & \varphi_{14} \\ \varphi_3 & \varphi_7 & \varphi_{11} & \varphi_{15} \\ \varphi_4 & \varphi_8 & \varphi_{12} & \varphi_{16} \end{bmatrix}.$$

Conversely, the relation of the set of nodal functions Φ_I to the sets of basis vectors $\bar{\Phi}_I$ is

$$\begin{bmatrix} \Phi_1 \\ \Phi_2 \\ \Phi_3 \\ \Phi_4 \end{bmatrix} = 4 \begin{bmatrix} T & & & \\ & T & & \\ & & T & \\ & & & T \end{bmatrix} \begin{bmatrix} \bar{\Phi}_1 \\ \bar{\Phi}_2 \\ \bar{\Phi}_3 \\ \bar{\Phi}_4 \end{bmatrix} \quad \text{or} \quad \Phi = 4\bar{T}\bar{\Phi},$$

since $T^{-1} = 4T$.

The u- or v-displacement function is usually assumed to be

$$u(x, y) = \alpha_1 + \alpha_2 x + \alpha_3 y + \alpha_4 x^2 + \alpha_5 xy + \alpha_6 y^2 + \alpha_7 x^3 + \alpha_8 x^2 y + \alpha_9 xy^2$$
$$+ \alpha_{10} y^3 + \alpha_{11} x^3 y + \alpha_{12} x^2 y^2 + \alpha_{13} xy^3 + \alpha_{14} x^3 y^2 + \alpha_{15} x^2 y^3$$
$$+ \alpha_{16} x^3 y^3.$$

For the group supermatrix procedure, the terms of the above polynomial will be allocated by the following expression to their respective G-invariant subspaces and with the unique order of terms that gives the diagonal form of the supermatrix of the system of equations relating displacements $\bar{U}$ to coefficients A:

$$\begin{bmatrix} F_1 \\ F_2 \\ F_3 \\ F_4 \end{bmatrix} = \begin{bmatrix} 1 \\ xy \\ x \\ y \end{bmatrix} \begin{bmatrix} 1 & x^2 & y^2 & x^2 y^2 \end{bmatrix} = \begin{bmatrix} 1 & x^2 & y^2 & x^2 y^2 \\ xy & x^3 y & xy^3 & x^3 y^3 \\ x & x^3 & xy^2 & x^3 y^2 \\ y & x^2 y & y^3 & x^2 y^3 \end{bmatrix} \quad \begin{matrix} \text{Subspace} \\ U_1 \\ U_2 \\ U_3 \\ U_4 \end{matrix}$$

The terms in the column and row vectors correspond to Cartesian sets and products pertaining to the symmetry types of group representations A_1, A_2, B_1, B_2, as given in addition to the character table of the group C_{2v}.

Thus, the displacement function

$$u(x,y) = \begin{bmatrix} 1 & x^2 & y^2 & x^2 y^2 & \vdots & xy & x^3 y & xy^3 & x^3 y^3 \\ & x & x^3 & xy^2 & x^3 y^2 & \vdots & y & x^2 y & y^3 & x^2 y^3 \end{bmatrix}$$
$$[\alpha_1 \quad \alpha_2 \quad \alpha_3 \quad \alpha_4 \vdots \alpha_5 \quad \alpha_6 \quad \alpha_7 \quad \alpha_8 \vdots \alpha_9 \quad \alpha_{10} \quad \alpha_{11} \quad \alpha_{12}$$
$$\vdots \alpha_{13} \quad \alpha_{14} \quad \alpha_{15} \quad \alpha_{16}]^T$$
$$= [F_1 \vdots F_2 \vdots F_3 \vdots F_4][A_1 \vdots A_2 \vdots A_3 \vdots A_4]^T$$

is decomposed into four four-dimensional subspaces U_1, U_2, U_3, U_4.

Substitution of nodal coordinates of the first nodes of the nodal sets $S_1(1, 2, 3, 4)$, $S_2(5, 6, 7, 8), S_3(9, 10, 11, 12), S_4(13, 14, 15, 16)$

	1	5	9	13
x	c	c	$c/3$	$c/3$
y	d	$d/3$	d	$d/3$

into the sections F_1, F_2, F_3, F_4 of $u(x, y)$ pertaining to subspaces U_1, U_2, U_3, U_4 will give the relation of the displacements $\bar{U}$ to the coefficients A for each subspace separately: (see pp. 190–191).

This relation can be written as $\bar{U} = \bar{C}A$, where $\bar{U}$ is the set of displacements $\bar{U}^{(1)}$, $\bar{U}^{(2)}$, $\bar{U}^{(3)}$, $\bar{U}^{(4)}$ with the numbering according to the numbering of basis vectors $\bar{\Phi}^{(1)}$, $\bar{\Phi}^{(2)}$, $\bar{\Phi}^{(3)}$, $\bar{\Phi}^{(4)}$ derived earlier. Thus, the sixteen-dimensional G-vector space of the problem is decomposed into four four-dimensional G-invariant subspaces U_1, U_2, U_3, U_4.

The coefficients A are determined by inverting the matrices $\bar{C}_1$, $\bar{C}_2$, $\bar{C}_3$, $\bar{C}_4$ of subspaces U_1, U_2, U_3, U_4: (see page 192–193).

The sixteen-dimensional G-vector space of the displacement field Δ, decomposed into four four-dimensional G-invariant subspaces U_1, U_2, U_3, U_4, is given by

$$\Delta = T_D \bar{C}^{-1} \bar{U},$$

with

$$T_D = \text{Diag } [1 \quad x^2 \quad y^2 \quad x^2 y^2 \mid xy \quad x^3 y \quad xy^3 \quad x^3 y^3 \mid x \quad x^3 \quad xy^2 \quad x^3 y^2$$
$$\mid y \quad x^2 y \quad y^3 \quad x^2 y^3] = \text{Diag } [F_1 \mid F_2 \mid F_3 \mid F_4]$$

and $\xi = x/c$, $\eta = y/d$, giving (see page 194–195).

$$
\begin{bmatrix}
\bar{u}_1 \\
\bar{u}_5 \\
\bar{u}_9 \\
\bar{u}_{13} \\
\hdashline
\bar{u}_2 \\
\bar{u}_6 \\
\bar{u}_{10} \\
\bar{u}_{14} \\
\hdashline
\bar{u}_3 \\
\bar{u}_7 \\
\bar{u}_{11} \\
\bar{u}_{15} \\
\hdashline
\bar{u}_4 \\
\bar{u}_8 \\
\bar{u}_{12} \\
\bar{u}_{16}
\end{bmatrix}
= \frac{1}{4}
\begin{bmatrix}
u_1 + u_2 + u_3 + u_4 \\
u_5 + u_6 + u_7 + u_8 \\
u_9 + u_{10} + u_{11} + u_{12} \\
u_{13} + u_{14} + u_{15} + u_{16} \\
\hdashline
u_1 + u_2 - u_3 - u_4 \\
u_5 + u_6 - u_7 - u_8 \\
u_9 + u_{10} - u_{11} - u_{12} \\
u_{13} + u_{14} - u_{15} - u_{16} \\
\hdashline
u_1 - u_2 + u_3 - u_4 \\
u_5 - u_6 + u_7 - u_8 \\
u_9 - u_{10} + u_{11} - u_{12} \\
u_{13} - u_{14} + u_{15} - u_{16} \\
\hdashline
u_1 - u_2 - u_3 + u_4 \\
u_5 - u_6 - u_7 + u_8 \\
u_9 - u_{10} - u_{11} + u_{12} \\
u_{13} - u_{14} - u_{15} + u_{16}
\end{bmatrix}
=
\begin{bmatrix}
1 & c^2 & d^2 & c^2 d^2 \\
1 & c^2 & \dfrac{d^2}{9} & \dfrac{c^2 d^2}{9} \\
1 & \dfrac{c^2}{9} & d^2 & \dfrac{c^2 d^2}{9} \\
1 & \dfrac{c^2}{9} & \dfrac{d^2}{9} & \dfrac{c^2 d^2}{81} \\
\hdashline
& & & \\
& & & \\
& & & \\
& & & \\
\hdashline
\begin{array}{cccc} cd & c^3 d & cd^3 & c^3 d^3 \\ \dfrac{cd}{3} & \dfrac{c^3 d}{3} & \dfrac{cd^3}{27} & \dfrac{c^3 d^3}{27} \\ \dfrac{cd}{3} & \dfrac{c^3 d}{27} & \dfrac{cd^3}{3} & \dfrac{c^3 d^3}{27} \\ \dfrac{cd}{9} & \dfrac{c^3 d}{81} & \dfrac{cd^3}{81} & \dfrac{c^3 d^3}{729} \end{array}
\end{bmatrix}
$$

Sec. 4.5] Sixteen-node rectangular element

$$
\begin{bmatrix}
c & c^3 & cd^2 & c^3d^2 & & & & & & & & & & & & \\
c & c^3 & \dfrac{cd^2}{9} & \dfrac{c^3d^2}{9} & & & & & & & & & & & & \\
\dfrac{c}{3} & \dfrac{c^3}{27} & \dfrac{cd^2}{3} & \dfrac{c^3d^2}{27} & & & & & & & & & & & & \\
\dfrac{c}{3} & \dfrac{c^3}{27} & \dfrac{cd^2}{27} & \dfrac{c^3d^2}{243} & & & & & & & & & & & & \\
& & & & d & c^2d & d^3 & c^2d^3 & & & & & & & & \\
& & & & \dfrac{d}{3} & \dfrac{c^2d}{3} & \dfrac{d^3}{27} & \dfrac{c^2d^3}{27} & & & & & & & & \\
& & & & d & \dfrac{c^2d}{9} & d^3 & \dfrac{c^2d^3}{9} & & & & & & & & \\
& & & & \dfrac{d}{3} & \dfrac{c^2d}{27} & \dfrac{d^3}{27} & \dfrac{c^2d^3}{243} & & & & & & & &
\end{bmatrix}
\begin{bmatrix}
\alpha_1 \\ \alpha_2 \\ \alpha_3 \\ \alpha_4 \\ \alpha_5 \\ \alpha_6 \\ \alpha_7 \\ \alpha_8 \\ \alpha_9 \\ \alpha_{10} \\ \alpha_{11} \\ \alpha_{12} \\ \alpha_{13} \\ \alpha_{14} \\ \alpha_{15} \\ \alpha_{16}
\end{bmatrix}
$$

$$
\begin{bmatrix} \alpha_1 \\ \alpha_2 \\ \alpha_3 \\ \alpha_4 \\ \hdashline \alpha_5 \\ \alpha_6 \\ \alpha_7 \\ \alpha_8 \\ \hdashline \alpha_9 \\ \alpha_{10} \\ \alpha_{11} \\ \alpha_{12} \\ \hdashline \alpha_{13} \\ \alpha_{14} \\ \alpha_{15} \\ \alpha_{16} \end{bmatrix} = \frac{1}{64}\, \mathrm{Diag} \begin{bmatrix} 1 \\ \dfrac{1}{c^2} \\ \dfrac{1}{d^2} \\ \dfrac{1}{c^2 d^2} \\ \hdashline \dfrac{1}{cd} \\ \dfrac{1}{c^3 d} \\ \dfrac{1}{c d^3} \\ \dfrac{1}{c^3 d^3} \\ \hdashline \dfrac{1}{c} \\ \dfrac{1}{c^3} \\ \dfrac{1}{c d^2} \\ \dfrac{1}{c^3 d^2} \\ \hdashline \dfrac{1}{d} \\ \dfrac{1}{c^2 d} \\ \dfrac{1}{d^3} \\ \dfrac{1}{c^2 d^3} \end{bmatrix}
\left[\begin{array}{cccc:cccc}
1 & -9 & -9 & 81 & & & & \\
-9 & 81 & 9 & -81 & & & & \\
-9 & 9 & 81 & -81 & & & & \\
81 & -81 & -81 & 81 & & & & \\
\hdashline
 & & & & 1 & -27 & -27 & 729 \\
 & & & & -9 & 243 & 27 & -729 \\
 & & & & -9 & 27 & 243 & -729 \\
 & & & & 81 & -243 & -243 & 729 \\
\end{array}\right]
$$

$$\begin{bmatrix} 1 & -9 & -27 & 243 & & & & & & & & & & & & \\ -9 & 81 & 27 & -243 & & & & & & & & & & & & \\ -9 & 9 & 243 & -243 & & & & & & & & & & & & \\ 81 & -81 & -243 & 243 & & & & & & & & & & & & \\ & & & & & & & & & & & & & & & \\ & & & & & & & & & & & & & & & \\ & & & & & & & & & & & & & & & \\ & & & & & & & & & & & & & & & \\ & & & & & & & & & & & & & & & \\ & & & & & & & & & & & & & & & \\ & & & & & & & & & & & & & & & \\ & & & & & & & & & & & & & & & \\ & & & & & & & & & 1 & -27 & -9 & 243 \\ & & & & & & & & & -9 & 243 & 9 & -243 \\ & & & & & & & & & -9 & 27 & 81 & -243 \\ & & & & & & & & & 81 & -243 & -81 & 243 \end{bmatrix} \begin{bmatrix} \bar{u}_1 \\ \bar{u}_5 \\ \bar{u}_9 \\ \bar{u}_{13} \\ \cdots \\ \bar{u}_2 \\ \bar{u}_6 \\ \bar{u}_{10} \\ \bar{u}_{14} \\ \cdots \\ \bar{u}_3 \\ \bar{u}_7 \\ \bar{u}_{11} \\ \bar{u}_{15} \\ \cdots \\ \bar{u}_4 \\ \bar{u}_8 \\ \bar{u}_{12} \\ \bar{u}_{16} \end{bmatrix}$$

$$\begin{bmatrix} \Delta_1 \\ \cdots \\ \Delta_2 \\ \cdots \\ \Delta_3 \\ \cdots \\ \Delta_4 \end{bmatrix} = \frac{1}{64} \text{Diag} \begin{bmatrix} 1 \\ \frac{x^2}{c^2} \\ \frac{y^2}{d^2} \\ \frac{x^2 y^2}{c^2 d^2} \\ \cdots \\ \frac{xy}{cd} \\ \frac{x^3 y}{c^3 d} \\ \frac{xy^3}{cd^3} \\ \frac{x^3 y^3}{c^3 d^3} \\ \cdots \\ \frac{x}{c} \\ \frac{x^3}{c^3} \\ \frac{xy^2}{cd^2} \\ \frac{x^3 y^2}{c^3 d^2} \\ \cdots \\ \frac{y}{d} \\ \frac{x^2 y}{c^2 d} \\ \frac{y^3}{d^3} \\ \frac{x^2 y^3}{c^2 d^3} \end{bmatrix} \begin{bmatrix} 1 & -9 & -9 & 81 \\ -9 & 81 & 9 & -81 \\ -9 & 9 & 81 & -81 \\ 81 & -81 & -81 & 81 \\ & & & & 1 & -27 & -27 & 729 \\ & & & & -9 & 243 & 27 & -729 \\ & & & & -9 & 27 & 243 & -729 \\ & & & & 81 & -243 & -243 & 729 \end{bmatrix}$$

Sec. 4.5] **Sixteen-node rectangular element** 195

$$
\begin{bmatrix}
\begin{array}{cccc|cccc|cccc|cccc}
 & & & & & & & & & & & & & & & \\
 & & & & & & & & & & & & & & & \\
 & & & & & & & & & & & & & & & \\
 & & & & & & & & & & & & & & & \\
\hline
 & & & & & & & & & & & & & & & \\
 & & & & & & & & & & & & & & & \\
 & & & & & & & & & & & & & & & \\
 & & & & & & & & & & & & & & & \\
\hline
1 & -9 & -27 & 243 & & & & & & & & & & & & \\
-9 & 81 & 27 & -243 & & & & & & & & & & & & \\
-9 & 9 & 243 & -243 & & & & & & & & & & & & \\
81 & -81 & -243 & 243 & & & & & & & & & & & & \\
\hline
 & & & & & & & & & & & & 1 & -27 & -9 & 243 \\
 & & & & & & & & & & & & -9 & 243 & 9 & -243 \\
 & & & & & & & & & & & & -9 & 27 & 81 & -243 \\
 & & & & & & & & & & & & 81 & -243 & -81 & 243 \\
\end{array}
\end{bmatrix}
\begin{bmatrix}
\bar{u}_1 \\ \bar{u}_5 \\ \bar{u}_9 \\ \bar{u}_{13} \\
\bar{u}_2 \\ \bar{u}_6 \\ \bar{u}_{10} \\ \bar{u}_{14} \\
\bar{u}_3 \\ \bar{u}_7 \\ \bar{u}_{11} \\ \bar{u}_{15} \\
\bar{u}_4 \\ \bar{u}_8 \\ \bar{u}_{12} \\ \bar{u}_{16}
\end{bmatrix}
$$

or $\Delta = \bar{N}_\Delta \bar{U}$, or

$$\begin{bmatrix} \Delta_1 \\ \Delta_2 \\ \Delta_3 \\ \Delta_4 \end{bmatrix} = \begin{bmatrix} \bar{N}_{\Delta_1} & & & \\ & \bar{N}_{\Delta_2} & & \\ & & \bar{N}_{\Delta_3} & \\ & & & \bar{N}_{\Delta_4} \end{bmatrix} \begin{bmatrix} \bar{U}^{(1)} \\ \bar{U}^{(2)} \\ \bar{U}^{(3)} \\ \bar{U}^{(4)} \end{bmatrix}$$

with

$$\bar{N}_{\Delta_1} = \frac{1}{64} \begin{bmatrix} 1 & -9 & -9 & 81 \\ -9\xi^2 & 81\xi^2 & 9\xi^2 & -81\xi^2 \\ -9\eta^2 & 9\eta^2 & 81\eta^2 & -81\eta^2 \\ 81\xi^2\eta^2 & -81\xi^2\eta^2 & -81\xi^2\eta^2 & 81\xi^2\eta^2 \end{bmatrix}$$

$$\bar{N}_{\Delta_2} = \frac{1}{64} \begin{bmatrix} \xi\eta & -27\xi\eta & -27\xi\eta & 729\xi\eta \\ -9\xi^3\eta & 243\xi^3\eta & 27\xi^3\eta & -729\xi^3\eta \\ -9\xi\eta^3 & 27\xi\eta^3 & 243\xi\eta^3 & -729\xi\eta^3 \\ 81\xi^3\eta^3 & -243\xi^3\eta^3 & -243\xi^3\eta^3 & 729\xi^3\eta^3 \end{bmatrix}$$

$$\bar{N}_{\Delta_3} = \frac{1}{64} \begin{bmatrix} \xi & -9\xi & -27\xi & 243\xi \\ -9\xi^3 & 81\xi^3 & 27\xi^3 & -243\xi^3 \\ -9\xi\eta^2 & 9\xi\eta^2 & 243\xi\eta^2 & -243\xi\eta^2 \\ 81\xi^3\eta^2 & -81\xi^3\eta^2 & -243\xi^3\eta^2 & 243\xi^3\eta^2 \end{bmatrix}$$

$$\bar{N}_{\Delta_4} = \frac{1}{64} \begin{bmatrix} \eta & -27\eta & -9\eta & 243\eta \\ -9\xi^2\eta & 243\xi^2\eta & 9\xi^2\eta & -243\xi^2\eta \\ -9\eta^3 & 27\eta^3 & 81\eta^3 & -243\eta^3 \\ 81\xi^2\eta^3 & -243\xi^2\eta^3 & -81\xi^2\eta^3 & 243\xi^2\eta^3 \end{bmatrix}.$$

The shape functions $\bar{N}$ of G-invariant subspaces $\bar{U}_1, \bar{U}_2, \bar{U}_3, \bar{U}_4$ are obtained by

$$\bar{N} = S\bar{N}_\Delta,$$

with

$$S = [1 \quad 1 \quad 1 \quad 1 \vdots 1 \quad 1 \quad 1 \quad 1 \vdots 1 \quad 1 \quad 1 \quad 1 \vdots 1 \quad 1 \quad 1 \quad 1]$$

U_1:
$$\begin{bmatrix} \bar{N}_1 \\ \bar{N}_5 \\ \bar{N}_9 \\ \bar{N}_{13} \end{bmatrix} = \begin{bmatrix} N_1 + N_2 + N_3 + N_4 \\ N_5 + N_6 + N_7 + N_8 \\ N_9 + N_{10} + N_{11} + N_{12} \\ N_{13} + N_{14} + N_{15} + N_{16} \end{bmatrix}$$

$$= \frac{1}{64} \begin{bmatrix} 1 & -9 & -9 & 81 \\ -9 & 81 & 9 & -81 \\ -9 & 9 & 81 & -81 \\ 81 & -81 & -81 & 81 \end{bmatrix} \begin{bmatrix} 1 \\ \xi^2 \\ \eta^2 \\ \xi^2 \eta^2 \end{bmatrix}$$

U_2:
$$\begin{bmatrix} \bar{N}_2 \\ \bar{N}_6 \\ \bar{N}_{10} \\ \bar{N}_{14} \end{bmatrix} = \begin{bmatrix} N_1 + N_2 - N_3 - N_4 \\ N_5 + N_6 - N_7 - N_8 \\ N_9 + N_{10} - N_{11} - N_{12} \\ N_{13} + N_{14} - N_{15} - N_{16} \end{bmatrix}$$

$$= \frac{1}{64} \begin{bmatrix} 1 & -9 & -9 & 81 \\ -27 & 243 & 27 & -243 \\ -27 & 27 & 243 & -243 \\ 729 & -729 & -729 & 729 \end{bmatrix} \begin{bmatrix} \xi\eta \\ \xi^3\eta \\ \xi\eta^3 \\ \xi^3\eta^3 \end{bmatrix}$$

U_3:
$$\begin{bmatrix} \bar{N}_3 \\ \bar{N}_7 \\ \bar{N}_{11} \\ \bar{N}_{15} \end{bmatrix} = \begin{bmatrix} N_1 - N_2 + N_3 - N_4 \\ N_5 - N_6 + N_7 - N_8 \\ N_9 - N_{10} + N_{11} - N_{12} \\ N_{13} - N_{14} + N_{15} - N_{16} \end{bmatrix}$$

$$= \frac{1}{64} \begin{bmatrix} 1 & -9 & -9 & 81 \\ -9 & 81 & 9 & -81 \\ -27 & 27 & 243 & -243 \\ 243 & -243 & -243 & 243 \end{bmatrix} \begin{bmatrix} \xi \\ \xi^3 \\ \xi\eta^3 \\ \xi^3\eta^2 \end{bmatrix}$$

$$U_4: \begin{bmatrix} \bar{N}_4 \\ \bar{N}_8 \\ \bar{N}_{12} \\ \bar{N}_{16} \end{bmatrix} = \begin{bmatrix} N_1 - N_2 - N_3 + N_4 \\ N_5 - N_6 - N_7 + N_8 \\ N_9 - N_{10} - N_{11} + N_{12} \\ N_{13} - N_{14} - N_{15} + N_{16} \end{bmatrix}$$

$$= \frac{1}{64} \begin{bmatrix} 1 & -9 & -9 & 81 \\ -27 & 243 & 27 & -243 \\ -9 & 9 & 81 & -81 \\ 243 & -243 & -243 & 243 \end{bmatrix} \begin{bmatrix} \eta \\ \xi^2 \eta \\ \eta^3 \\ \xi^2 \eta^3 \end{bmatrix}.$$

The shape functions $\bar{N}$ can be written in supermatrix form

$$\begin{bmatrix} \bar{N}^{(1)} \\ \bar{N}^{(2)} \\ \bar{N}^{(3)} \\ \bar{N}^{(4)} \end{bmatrix} = \frac{1}{8} \begin{bmatrix} A & & & \\ & B & & \\ & & C & \\ & & & D \end{bmatrix} \begin{bmatrix} F_1 \\ F_2 \\ F_3 \\ F_4 \end{bmatrix} \quad \text{or} \quad \bar{N} = \bar{N}_\Delta F,$$

while the shape functions

$$N^{(1)} = [N_1 \quad N_5 \quad N_9 \quad N_{13}]^T$$
$$N^{(2)} = [N_2 \quad N_6 \quad N_{10} \quad N_{14}]^T$$
$$N^{(3)} = [N_3 \quad N_7 \quad N_{11} \quad N_{15}]^T$$
$$N^{(4)} = [N_4 \quad N_8 \quad N_{12} \quad N_{16}]^T$$

can be obtained by

$$N = T\bar{N} = T\bar{N}_\Delta F,$$

or

$$\begin{bmatrix} N^{(1)} \\ N^{(2)} \\ N^{(3)} \\ N^{(4)} \end{bmatrix} = \frac{1}{4} \begin{bmatrix} E & E & E & E \\ E & E & -E & -E \\ E & -E & E & -E \\ E & -E & -E & E \end{bmatrix} \frac{1}{8} \begin{bmatrix} A & & & \\ & B & & \\ & & C & \\ & & & D \end{bmatrix} \begin{bmatrix} F_1 \\ F_2 \\ F_3 \\ F_4 \end{bmatrix}$$

$$= \frac{1}{32} \begin{bmatrix} A & B & C & D \\ A & B & -C & -D \\ A & -B & C & -D \\ A & -B & -C & D \end{bmatrix} \begin{bmatrix} F_1 \\ F_2 \\ F_3 \\ F_4 \end{bmatrix},$$

giving $N_1, N_2, \ldots, N_{16}$, as in the case of the twelve-node rectangular element in section 4.4.

4.6 EIGHT-NODE RECTANGULAR HEXAHEDRAL ELEMENT

The eight-node rectangular hexahedral element, as shown in Fig. 4.6 with the unique nodal numbering derived in section 3.8, is described by the group D_{2h}. The nodes of the nodal set $S(1,2,3,4,5,6,7,8)$ are permuted by the action of group elements, i.e. symmetry operations E (identity), C_2^z, C_2^y, C_2^x (rotations through $180°$ about the z, y, x axes respectively), i (inversion), σ_{xy}, σ_{xz}, σ_{yz} (reflections in xy, xz, yz planes respectively).

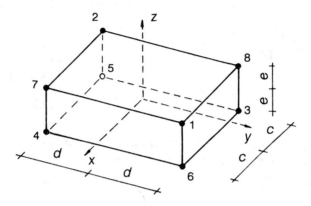

Fig. 4.6. The eight-node rectangular hexahedral element with the nodal numbering of the group supermatrix procedure.

The character table of the group D_{2h} (with Cartesian sets and products) is

D_{2h}	E	C_2^z	C_2^y	C_2^x	i	σ_{xy}	σ_{xz}	σ_{yz}		
A_g	1	1	1	1	1	1	1	1		x^2, y^2, z^2
B_{1g}	1	1	−1	−1	1	1	−1	−1		xy
B_{2g}	1	−1	1	−1	1	−1	1	−1		xz
B_{3g}	1	−1	−1	1	1	−1	−1	1		yz
A_u	1	1	1	1	−1	−1	−1	−1	xyz	
B_{1u}	1	1	−1	−1	−1	−1	1	1	z	
B_{2u}	1	−1	1	−1	−1	1	−1	1	y	
B_{3u}	1	−1	−1	1	−1	1	1	−1	x	

The idempotents π_i of the centre of the group algebra are obtained by

$$\pi_i = \frac{h_i}{h} \sum_\sigma \chi_i(\sigma^{-1}) \sigma,$$

where h_i is the dimension of the ith character given by $h_i = \chi_i(E)$, h is the order, χ_i is the ith character, σ is the element and σ^{-1} its inverse, and $i = 1, 2, \ldots, 8$.

Thus

$$\begin{bmatrix} \pi_1 \\ \pi_2 \\ \pi_3 \\ \pi_4 \\ \pi_5 \\ \pi_6 \\ \pi_7 \\ \pi_8 \end{bmatrix} = \frac{1}{8} \begin{bmatrix} 1 & 1 & 1 & 1 & 1 & 1 & 1 & 1 \\ 1 & 1 & -1 & -1 & 1 & 1 & -1 & -1 \\ 1 & -1 & 1 & -1 & 1 & -1 & 1 & -1 \\ 1 & -1 & -1 & 1 & 1 & -1 & -1 & 1 \\ 1 & 1 & 1 & 1 & -1 & -1 & -1 & -1 \\ 1 & 1 & -1 & -1 & -1 & -1 & 1 & 1 \\ 1 & -1 & 1 & -1 & -1 & 1 & -1 & 1 \\ 1 & -1 & -1 & 1 & -1 & 1 & 1 & -1 \end{bmatrix} \begin{bmatrix} E \\ C_2^z \\ C_2^y \\ C_2^x \\ i \\ \sigma_{xy} \\ \sigma_{xz} \\ \sigma_{yz} \end{bmatrix}$$

or $\Pi = T\Sigma$.

By application of the idempotents π_i to the nodal functions φ_j ($j = 1, 2, \ldots, 8$) the basis vectors $\bar{\varphi}_j$ of G-invariant subspaces are derived:

$$\begin{bmatrix} \bar{\varphi}_1 \\ \bar{\varphi}_2 \\ \bar{\varphi}_3 \\ \bar{\varphi}_4 \\ \bar{\varphi}_5 \\ \bar{\varphi}_6 \\ \bar{\varphi}_7 \\ \bar{\varphi}_8 \end{bmatrix} = \frac{1}{8} \begin{bmatrix} 1 & 1 & 1 & 1 & 1 & 1 & 1 & 1 \\ 1 & 1 & -1 & -1 & 1 & 1 & -1 & -1 \\ 1 & -1 & 1 & -1 & 1 & -1 & 1 & -1 \\ 1 & -1 & -1 & 1 & 1 & -1 & -1 & 1 \\ 1 & 1 & 1 & 1 & -1 & -1 & -1 & -1 \\ 1 & 1 & -1 & -1 & -1 & -1 & 1 & 1 \\ 1 & -1 & 1 & -1 & -1 & 1 & -1 & 1 \\ 1 & -1 & -1 & 1 & -1 & 1 & 1 & -1 \end{bmatrix} \begin{bmatrix} \varphi_1 \\ \varphi_2 \\ \varphi_3 \\ \varphi_4 \\ \varphi_5 \\ \varphi_6 \\ \varphi_7 \\ \varphi_8 \end{bmatrix} \quad \text{or} \quad \bar{\Phi} = T\Phi,$$

with $T^{-1} = 8T$, and conversely $\Phi = T^{-1}\bar{\Phi} = 8T\bar{\Phi}$.

The u-, v- or w-displacement function is usually assumed to be

$$u(x, y, z) = \alpha_1 + \alpha_2 x + \alpha_3 y + \alpha_4 z + \alpha_5 xy + \alpha_6 yz + \alpha_7 xz + \alpha_8 xyz.$$

As in the cases of rectangular elements in the group supermatrix procedure, the order of the terms in the above polynomial will be changed according to the pertinence of the terms to the symmetry types of the group D_{2h}, as given in addition to its character table. Thus, in the polynomial

$$u(x, y, z) = [1 \mid xy \mid xz \mid yz \mid xyz \mid z \mid y \mid x]$$
$$[\alpha_1 \quad \alpha_2 \quad \alpha_3 \quad \alpha_4 \quad \alpha_5 \quad \alpha_6 \quad \alpha_7 \quad \alpha_8]^T$$
$$= [F_1 \mid F_2 \mid F_3 \mid F_4 \mid F_5 \mid F_6 \mid F_7 \mid F_8]$$
$$[\alpha_1 \quad \alpha_2 \quad \alpha_3 \quad \alpha_4 \quad \alpha_5 \quad \alpha_6 \quad \alpha_7 \quad \alpha_8]^T$$

Sec. 4.6] Eight-node rectangular hexahedral element 201

the terms $1, xy, xz, yz, xyz, z, y, x$ belong to the subspaces $U_1, U_2, U_3, U_4, U_5, U_6, U_7, U_8$ respectively. With this order of the terms the relationship of displacement $\bar{U}$ to coefficients A will be obtained as a system of equations with the matrix in diagonal form.

Substitution of the nodal coordinates c, d, e of the node 1, the first node in the nodal set $S(1, 2, 3, 4, 5, 6, 7, 8)$, into the above polynomial gives

$$\begin{bmatrix} \bar{u}_1 \\ \bar{u}_2 \\ \bar{u}_3 \\ \bar{u}_4 \\ \bar{u}_5 \\ \bar{u}_6 \\ \bar{u}_7 \\ \bar{u}_8 \end{bmatrix} = \frac{1}{8} \begin{bmatrix} 1 & 1 & 1 & 1 & 1 & 1 & 1 & 1 \\ 1 & 1 & -1 & -1 & 1 & 1 & -1 & -1 \\ 1 & -1 & 1 & -1 & 1 & -1 & 1 & -1 \\ 1 & -1 & -1 & 1 & 1 & -1 & -1 & 1 \\ 1 & 1 & 1 & 1 & -1 & -1 & -1 & -1 \\ 1 & 1 & -1 & -1 & -1 & -1 & 1 & 1 \\ 1 & -1 & 1 & -1 & -1 & 1 & -1 & 1 \\ 1 & -1 & -1 & 1 & -1 & 1 & 1 & -1 \end{bmatrix} \begin{bmatrix} u_1 \\ u_2 \\ u_3 \\ u_4 \\ u_5 \\ u_6 \\ u_7 \\ u_8 \end{bmatrix}$$

$$= \begin{bmatrix} 1 & & & & & & & \\ & cd & & & & & & \\ & & ce & & & & & \\ & & & de & & & & \\ & & & & cde & & & \\ & & & & & e & & \\ & & & & & & d & \\ & & & & & & & c \end{bmatrix} \begin{bmatrix} \alpha_1 \\ \alpha_2 \\ \alpha_3 \\ \alpha_4 \\ \alpha_5 \\ \alpha_6 \\ \alpha_7 \\ \alpha_8 \end{bmatrix} \quad \text{or} \quad \bar{U} = \bar{C}A.$$

Thus, the coefficients A are (see p. 202 (1)).

The eight-dimensional G-vector space of the displacement field Δ, decomposed into eight one-dimensional subspaces $U_1, U_2, U_3, U_4, U_5, U_6, U_7, U_8$, is produced by

$$\Delta = T_D \bar{C}^{-1} \bar{U},$$

with

$$T_D = \text{Diag } [1 \mid xy \mid xz \mid yz \mid xyz \mid z \mid y \mid x]$$
$$= \text{Diag } [F_1 \mid F_2 \mid F_3 \mid F_4 \mid F_5 \mid F_6 \mid F_7 \mid F_8]$$

producing (see p. 202 (2)).

(1)
$$\begin{bmatrix} \alpha_1 \\ \alpha_2 \\ \alpha_3 \\ \alpha_4 \\ \alpha_5 \\ \alpha_6 \\ \alpha_7 \\ \alpha_8 \end{bmatrix} = \begin{bmatrix} 1 & & & & & & & \\ & \frac{1}{cd} & & & & & & \\ & & \frac{1}{ce} & & & & & \\ & & & \frac{1}{de} & & & & \\ & & & & \frac{1}{cde} & & & \\ & & & & & \frac{1}{e} & & \\ & & & & & & \frac{1}{d} & \\ & & & & & & & \frac{1}{c} \end{bmatrix} \begin{bmatrix} \bar{u}_1 \\ \bar{u}_2 \\ \bar{u}_3 \\ \bar{u}_4 \\ \bar{u}_5 \\ \bar{u}_6 \\ \bar{u}_7 \\ \bar{u}_8 \end{bmatrix} \quad \text{or} \quad A = \bar{C}^{-1} \bar{U}.$$

(2)
$$\begin{bmatrix} \Delta_1 \\ \Delta_2 \\ \Delta_3 \\ \Delta_4 \\ \Delta_5 \\ \Delta_6 \\ \Delta_7 \\ \Delta_8 \end{bmatrix} \begin{bmatrix} 1 & & & & & & & \\ & \frac{xy}{cd} & & & & & & \\ & & \frac{xz}{ce} & & & & & \\ & & & \frac{yz}{de} & & & & \\ & & & & \frac{xyz}{cde} & & & \\ & & & & & \frac{z}{e} & & \\ & & & & & & \frac{y}{d} & \\ & & & & & & & \frac{x}{c} \end{bmatrix} \begin{bmatrix} \bar{u}_1 \\ \bar{u}_2 \\ \bar{u}_3 \\ \bar{u}_4 \\ \bar{u}_5 \\ \bar{u}_6 \\ \bar{u}_7 \\ \bar{u}_8 \end{bmatrix}$$

and with substitutions $\xi = x/c$, $\eta = y/d$, $\zeta = z/e$

$$\begin{bmatrix} \Delta_1 \\ \Delta_2 \\ \Delta_3 \\ \Delta_4 \\ \Delta_5 \\ \Delta_6 \\ \Delta_7 \\ \Delta_8 \end{bmatrix} = \begin{bmatrix} 1 & & & & & & & \\ & \xi\eta & & & & & & \\ & & \xi\zeta & & & & & \\ & & & \eta\zeta & & & & \\ & & & & \xi\eta\zeta & & & \\ & & & & & \zeta & & \\ & & & & & & \eta & \\ & & & & & & & \xi \end{bmatrix} \begin{bmatrix} \bar{u}_1 \\ \bar{u}_2 \\ \bar{u}_3 \\ \bar{u}_4 \\ \bar{u}_5 \\ \bar{u}_6 \\ \bar{u}_7 \\ \bar{u}_8 \end{bmatrix}$$

$$= \begin{bmatrix} \bar{N}_1 & & & & & & & \\ & \bar{N}_2 & & & & & & \\ & & \bar{N}_3 & & & & & \\ & & & \bar{N}_4 & & & & \\ & & & & \bar{N}_5 & & & \\ & & & & & \bar{N}_6 & & \\ & & & & & & \bar{N}_7 & \\ & & & & & & & \bar{N}_8 \end{bmatrix} \begin{bmatrix} \bar{u}_1 \\ \bar{u}_2 \\ \bar{u}_3 \\ \bar{u}_4 \\ \bar{u}_5 \\ \bar{u}_6 \\ \bar{u}_7 \\ \bar{u}_8 \end{bmatrix}$$

and

$$\begin{bmatrix} \bar{N}_1 \\ \bar{N}_2 \\ \bar{N}_3 \\ \bar{N}_4 \\ \bar{N}_5 \\ \bar{N}_6 \\ \bar{N}_7 \\ \bar{N}_8 \end{bmatrix} = \begin{bmatrix} 1 & 1 & 1 & 1 & 1 & 1 & 1 & 1 \\ 1 & 1 & -1 & -1 & 1 & 1 & -1 & -1 \\ 1 & -1 & 1 & -1 & 1 & -1 & 1 & -1 \\ 1 & -1 & -1 & 1 & 1 & -1 & -1 & 1 \\ 1 & 1 & 1 & 1 & -1 & -1 & -1 & -1 \\ 1 & 1 & -1 & -1 & -1 & -1 & 1 & 1 \\ 1 & -1 & 1 & -1 & -1 & 1 & -1 & 1 \\ 1 & -1 & -1 & 1 & -1 & 1 & 1 & -1 \end{bmatrix} \begin{bmatrix} N_1 \\ N_2 \\ N_3 \\ N_4 \\ N_5 \\ N_6 \\ N_7 \\ N_8 \end{bmatrix}$$

$$= \begin{bmatrix} 1 & & & & & & & \\ & 1 & & & & & & \\ & & 1 & & & & & \\ & & & 1 & & & & \\ & & & & 1 & & & \\ & & & & & 1 & & \\ & & & & & & 1 & \\ & & & & & & & 1 \end{bmatrix} \begin{bmatrix} 1 \\ \xi\eta \\ \xi\zeta \\ \eta\zeta \\ \xi\eta\zeta \\ \zeta \\ \eta \\ \xi \end{bmatrix}$$

or $\bar{N} = F$.

The nodal shape functions N_1, N_2, N_3, N_4, N_5, N_6, N_7, N_8, pertaining to nodes 1, 2, 3, 4, 5, 6, 7, 8, are derived by

$$N = T\bar{N} = TF$$

or

$$\begin{bmatrix} N_1 \\ N_2 \\ N_3 \\ N_4 \\ N_5 \\ N_6 \\ N_7 \\ N_8 \end{bmatrix} = \frac{1}{8} \begin{bmatrix} 1 & 1 & 1 & 1 & 1 & 1 & 1 & 1 \\ 1 & 1 & -1 & -1 & 1 & 1 & -1 & -1 \\ 1 & -1 & 1 & -1 & 1 & -1 & 1 & -1 \\ 1 & -1 & -1 & 1 & 1 & -1 & -1 & 1 \\ 1 & 1 & 1 & 1 & -1 & -1 & -1 & -1 \\ 1 & 1 & -1 & -1 & -1 & -1 & 1 & 1 \\ 1 & -1 & 1 & -1 & -1 & 1 & -1 & 1 \\ 1 & -1 & -1 & 1 & -1 & 1 & 1 & -1 \end{bmatrix} \begin{bmatrix} 1 \\ \xi\eta \\ \xi\zeta \\ \eta\zeta \\ \xi\eta\zeta \\ \zeta \\ \eta \\ \xi \end{bmatrix}$$

or

$$N_1 = \tfrac{1}{8}(1+\xi)(1+\eta)(1+\zeta)$$
$$N_2 = \tfrac{1}{8}(1-\xi)(1-\eta)(1+\zeta)$$
$$N_3 = \tfrac{1}{8}(1-\xi)(1+\eta)(1-\zeta)$$
$$N_4 = \tfrac{1}{8}(1+\xi)(1-\eta)(1-\zeta)$$
$$N_5 = \tfrac{1}{8}(1-\xi)(1-\eta)(1-\zeta)$$
$$N_6 = \tfrac{1}{8}(1+\xi)(1+\eta)(1-\zeta)$$
$$N_7 = \tfrac{1}{8}(1+\xi)(1-\eta)(1+\zeta)$$
$$N_8 = \tfrac{1}{8}(1-\xi)(1+\eta)(1+\xi),$$

which is the expression in conventional form for the set of shape functions N_1, N_2, N_3, N_4, N_5, N_6, N_7, N_8 for the serendipity rectangular hexahedral element, but with respect to the unique group supermatrix nodal numbering given in Fig. 4.6.

Derivation of shape functions in G-invariant subspaces for the rectangular hexahedral element is also possible by means of group supermatrix transformation of the known shape functions of serendipity rectangular hexahedra.

For the eight-node serendipity rectangular hexahedron, with $\xi_i = \pm 1$, $\eta_i = \pm 1$, $\zeta_i = \pm 1$, the shape functions are

$$N_i = \tfrac{1}{8}(1 + \xi\xi_i)(1 + \eta\eta_i)(1 + \zeta\zeta_i).$$

With the unique group supermatrix nodal numbering given in Fig. 4.6, and with the unique order of the terms derived in this section, the following set of shape functions N is obtained:

$$\begin{bmatrix} N_1 \\ N_2 \\ N_3 \\ N_4 \\ N_5 \\ N_6 \\ N_7 \\ N_8 \end{bmatrix} = \frac{1}{8} \begin{bmatrix} 1 & 1 & 1 & 1 & 1 & 1 & 1 & 1 \\ 1 & 1 & -1 & -1 & 1 & 1 & -1 & -1 \\ 1 & -1 & 1 & -1 & 1 & -1 & 1 & -1 \\ 1 & -1 & -1 & 1 & 1 & -1 & -1 & 1 \\ 1 & 1 & 1 & 1 & -1 & -1 & -1 & -1 \\ 1 & 1 & -1 & -1 & -1 & -1 & 1 & 1 \\ 1 & -1 & 1 & -1 & -1 & 1 & -1 & 1 \\ 1 & -1 & -1 & 1 & -1 & 1 & 1 & -1 \end{bmatrix} \begin{bmatrix} 1 \\ \xi\eta \\ \xi\zeta \\ \eta\zeta \\ \xi\eta\zeta \\ \zeta \\ \eta \\ \xi \end{bmatrix}$$

or $N = TF$.

Multiplying both sides of the above equation by T^{-1} produces the shape functions $\bar{N}$ in G-invariant subspaces

$$\bar{N} = T^{-1}N = T^{-1}TF = F$$

or

$$\begin{bmatrix} \bar{N}_1 \\ \bar{N}_2 \\ \bar{N}_3 \\ \bar{N}_4 \\ \bar{N}_5 \\ \bar{N}_6 \\ \bar{N}_7 \\ \bar{N}_8 \end{bmatrix} = \begin{bmatrix} 1 & 1 & 1 & 1 & 1 & 1 & 1 & 1 \\ 1 & 1 & -1 & -1 & 1 & 1 & -1 & -1 \\ 1 & -1 & 1 & -1 & 1 & -1 & 1 & -1 \\ 1 & -1 & -1 & 1 & 1 & -1 & -1 & 1 \\ 1 & 1 & 1 & 1 & -1 & -1 & -1 & -1 \\ 1 & 1 & -1 & -1 & -1 & -1 & 1 & 1 \\ 1 & -1 & 1 & -1 & -1 & 1 & -1 & 1 \\ 1 & -1 & -1 & 1 & -1 & 1 & 1 & -1 \end{bmatrix} \begin{bmatrix} N_1 \\ N_2 \\ N_3 \\ N_4 \\ N_5 \\ N_6 \\ N_7 \\ N_8 \end{bmatrix} = \begin{bmatrix} 1 \\ \xi\eta \\ \xi\zeta \\ \eta\zeta \\ \xi\eta\zeta \\ \zeta \\ \eta \\ \xi \end{bmatrix}.$$

4.7 TWENTY-NODE RECTANGULAR HEXAHEDRAL ELEMENT

The twenty-node rectangular hexahedral element, shown in Fig. 4.7, is described by the group D_{2h} and it has four nodal sets

206   Formulation of shape functions in G-invariant subspaces   [Ch. 4

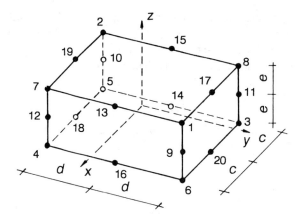

Fig. 4.7. The twenty-node rectangular hexahedral element with the nodal numbering of the group supermatrix procedure.

$$S_1(1, 2, 3, 4, 5, 6, 7, 8), \quad S_2(9, 10, 11, 12),$$
$$S_3(13, 14, 15, 16), \quad S_4(17, 18, 19, 20),$$

each set containing nodes that are permuted by action of symmetry operations of the group.

The unique nodal numbering for the eight-node rectangular hexahedral element, derived in section 3.8, is applied here for the set S_1 containing corner nodes. The nodes lying in midpoints of the edges of the hexahedron have the unique nodal numbering for the twelve-node truncated rectangular hexahedron, derived in section 3.8, encompassing the nodes contained in the nodal sets S_2, S_3, S_4.

The character table of the group D_{2h} (with Cartesian sets and products) and the idempotents of the centre of the group algebra are the same as for the eight-node rectangular hexahedral element in section 4.6. Applying the idempotents π_i to the nodal functions φ_j ($j = 1, 2, \ldots, 20$) derives the basis vectors $\bar{\varphi}_j$ of G-invariant subspaces U_i ($i = 1, 2, \ldots, 8$):

$$U_1: \quad \bar{\varphi}_1 = \tfrac{1}{8}(\varphi_1 + \varphi_2 + \varphi_3 + \varphi_4 + \varphi_5 + \varphi_6 + \varphi_7 + \varphi_8)$$
$$\bar{\varphi}_9 = \tfrac{1}{4}(\varphi_9 + \varphi_{10} + \varphi_{11} + \varphi_{12})$$
$$\bar{\varphi}_{13} = \tfrac{1}{4}(\varphi_{13} + \varphi_{14} + \varphi_{15} + \varphi_{16})$$
$$\bar{\varphi}_{17} = \tfrac{1}{4}(\varphi_{17} + \varphi_{18} + \varphi_{19} + \varphi_{20})$$
$$U_2: \quad \bar{\varphi}_2 = \tfrac{1}{8}(\varphi_1 + \varphi_2 - \varphi_3 - \varphi_4 + \varphi_5 + \varphi_6 - \varphi_7 - \varphi_8)$$
$$\bar{\varphi}_{10} = \tfrac{1}{4}(\varphi_9 + \varphi_{10} - \varphi_{11} - \varphi_{12})$$
$$U_3: \quad \bar{\varphi}_3 = \tfrac{1}{8}(\varphi_1 - \varphi_2 + \varphi_3 - \varphi_4 + \varphi_5 - \varphi_6 + \varphi_7 - \varphi_8)$$
$$\bar{\varphi}_{14} = \tfrac{1}{4}(\varphi_{13} + \varphi_{14} - \varphi_{15} - \varphi_{16})$$

U_4: $\quad \bar{\varphi}_4 = \frac{1}{8}(\varphi_1 - \varphi_2 - \varphi_3 + \varphi_4 + \varphi_5 - \varphi_6 - \varphi_7 + \varphi_8)$

$\quad \bar{\varphi}_{18} = \frac{1}{4}(\varphi_{17} + \varphi_{18} - \varphi_{19} - \varphi_{20})$

U_5: $\quad \bar{\varphi}_5 = \frac{1}{8}(\varphi_1 + \varphi_2 + \varphi_3 + \varphi_4 - \varphi_5 - \varphi_6 - \varphi_7 - \varphi_8)$

U_6: $\quad \bar{\varphi}_6 = \frac{1}{8}(\varphi_1 + \varphi_2 - \varphi_3 - \varphi_4 - \varphi_5 - \varphi_6 + \varphi_7 + \varphi_8)$

$\quad \bar{\varphi}_{15} = \frac{1}{4}(\varphi_{13} - \varphi_{14} + \varphi_{15} - \varphi_{16})$

$\quad \bar{\varphi}_{19} = \frac{1}{4}(\varphi_{17} - \varphi_{18} + \varphi_{19} - \varphi_{20})$

U_7: $\quad \bar{\varphi}_7 = \frac{1}{8}(\varphi_1 - \varphi_2 + \varphi_3 - \varphi_4 - \varphi_5 + \varphi_6 - \varphi_7 + \varphi_8)$

$\quad \bar{\varphi}_{11} = \frac{1}{4}(\varphi_9 - \varphi_{10} + \varphi_{11} - \varphi_{12})$

$\quad \bar{\varphi}_{20} = \frac{1}{4}(\varphi_{17} - \varphi_{18} - \varphi_{19} + \varphi_{20})$

U_8: $\quad \bar{\varphi}_8 = \frac{1}{8}(\varphi_1 - \varphi_2 - \varphi_3 + \varphi_4 - \varphi_5 + \varphi_6 + \varphi_7 - \varphi_8)$

$\quad \bar{\varphi}_{12} = \frac{1}{4}(\varphi_9 - \varphi_{10} - \varphi_{11} + \varphi_{12})$

$\quad \bar{\varphi}_{16} = \frac{1}{4}(\varphi_{13} - \varphi_{14} - \varphi_{15} + \varphi_{16}).$

Thus, the twenty-dimensional G-vector space is decomposed into eight G-invariant subspaces $U_1, U_2, U_3, U_4, U_5, U_6, U_7, U_8$ with 4, 2, 2, 2, 1, 3, 3, 3 dimensions respectively, spanned by the corresponding sets of basis vectors $\bar{\Phi}^{(1)}, \bar{\Phi}^{(2)}, \bar{\Phi}^{(3)}, \bar{\Phi}^{(4)}, \bar{\Phi}^{(5)}, \bar{\Phi}^{(6)}, \bar{\Phi}^{(7)}, \bar{\Phi}^{(8)}$:

$$\begin{bmatrix} \bar{\Phi}^{(1)} \\ \bar{\Phi}^{(2)} \\ \bar{\Phi}^{(3)} \\ \bar{\Phi}^{(4)} \\ \bar{\Phi}^{(5)} \\ \bar{\Phi}^{(6)} \\ \bar{\Phi}^{(7)} \\ \bar{\Phi}^{(8)} \end{bmatrix} = \begin{bmatrix} \bar{\varphi}_1 & \bar{\varphi}_9 & \bar{\varphi}_{13} & \bar{\varphi}_{17} \\ \bar{\varphi}_2 & \bar{\varphi}_{10} & & \\ \bar{\varphi}_3 & & \bar{\varphi}_{14} & \\ \bar{\varphi}_4 & & & \bar{\varphi}_{18} \\ \bar{\varphi}_5 & & & \\ \bar{\varphi}_6 & & \bar{\varphi}_{15} & \bar{\varphi}_{19} \\ \bar{\varphi}_7 & \bar{\varphi}_{11} & & \bar{\varphi}_{20} \\ \bar{\varphi}_8 & \bar{\varphi}_{12} & \bar{\varphi}_{16} & \end{bmatrix}.$$

The sets $\bar{\Phi}_1, \bar{\Phi}_2, \bar{\Phi}_3, \bar{\Phi}_4$ of basis vectors $\bar{\varphi}_j$ ($j = 1, 2, \ldots, 20$), ordered according to the numbering of the nodes in the nodal sets S_1, S_2, S_3, S_4, are given by

$$[\bar{\Phi}_1 \; \bar{\Phi}_2 \; \bar{\Phi}_3 \; \bar{\Phi}_4] = \begin{bmatrix} \bar{\varphi}_1 & & & \\ \bar{\varphi}_2 & & & \\ \bar{\varphi}_3 & \bar{\varphi}_9 & \bar{\varphi}_{13} & \bar{\varphi}_{17} \\ \bar{\varphi}_4 & \bar{\varphi}_{10} & \bar{\varphi}_{14} & \bar{\varphi}_{18} \\ \bar{\varphi}_5 & \bar{\varphi}_{11} & \bar{\varphi}_{15} & \bar{\varphi}_{19} \\ \bar{\varphi}_6 & \bar{\varphi}_{12} & \bar{\varphi}_{16} & \bar{\varphi}_{20} \\ \bar{\varphi}_7 & & & \\ \bar{\varphi}_8 & & & \end{bmatrix}.$$

The relation of the sets of basis vectors $\bar{\Phi}_l$ ($l = 1, 2, 3, 4$) is expressed by the following system of equations, with the supermatrix in diagonal form containing one transformation matrix of the group D_{2h} and three transformation matrices of the group C_{2v}:

$$\begin{bmatrix} \bar{\Phi}_1 \\ \bar{\Phi}_2 \\ \bar{\Phi}_3 \\ \bar{\Phi}_4 \end{bmatrix} = \begin{bmatrix} T_{D_{2h}} & & & \\ & T_{C_{2v}} & & \\ & & T_{C_{2v}} & \\ & & & T_{C_{2v}} \end{bmatrix} \begin{bmatrix} \Phi_1 \\ \Phi_2 \\ \Phi_3 \\ \Phi_4 \end{bmatrix} \quad \text{or} \quad \bar{\Phi} = \bar{T}\Phi,$$

with

$$[\Phi_1 \; \Phi_2 \; \Phi_3 \; \Phi_4] = \begin{bmatrix} \varphi_1 & & & \\ \varphi_2 & & & \\ \varphi_3 & \varphi_9 & \varphi_{13} & \varphi_{17} \\ \varphi_4 & \varphi_{10} & \varphi_{14} & \varphi_{18} \\ \varphi_5 & \varphi_{11} & \varphi_{15} & \varphi_{19} \\ \varphi_6 & \varphi_{12} & \varphi_{16} & \varphi_{20} \\ \varphi_7 & & & \\ \varphi_8 & & & \end{bmatrix}.$$

In expanded form this relation is (see p. 209).

Conversely, the relation of the set of nodal functions Φ_l to the sets of basis vectors $\bar{\Phi}_l$ is

$$\begin{bmatrix} \Phi_1 \\ \Phi_2 \\ \Phi_3 \\ \Phi_4 \end{bmatrix} = \begin{bmatrix} T_{D_{2h}}^{-1} & & & \\ & T_{C_{2v}}^{-1} & & \\ & & T_{C_{2v}}^{-1} & \\ & & & T_{C_{2v}}^{-1} \end{bmatrix} \begin{bmatrix} \bar{\Phi}_1 \\ \bar{\Phi}_2 \\ \bar{\Phi}_3 \\ \bar{\Phi}_4 \end{bmatrix} \quad \text{or} \quad \Phi = \bar{T}^{-1}\bar{\Phi},$$

with $T_{D_{2h}}^{-1} = 8 T_{D_{2h}}$ and $T_{C_{2v}}^{-1} = 4 T_{C_{2v}}$, or in expanded form (see p. 210).

Sec. 4.7] Twenty-node rectangular hexahedral element

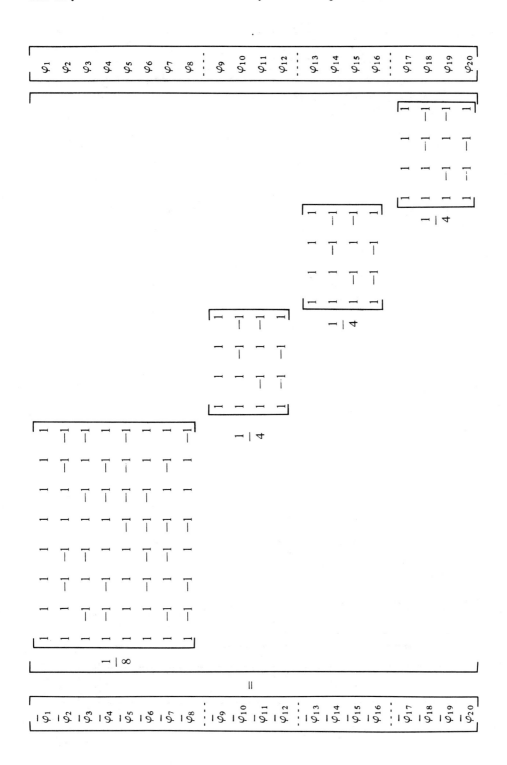

210 Formulation of shape functions in G-invariant subspaces [Ch. 4]

Twenty-node rectangular hexahedral element

The u-, v- or w-displacement function is usually assumed as

$$u(x, y, z) = \alpha_1 + \alpha_2 x + \alpha_3 y + \alpha_4 z + \alpha_5 x^2 + \alpha_6 y^2 + \alpha_7 z^2 + \alpha_8 xy + \alpha_9 yz$$
$$+ \alpha_{10} xz + \alpha_{11} x^2 y + \alpha_{12} x^2 z + \alpha_{13} y^2 x + \alpha_{14} y^2 z + \alpha_{15} z^2 x$$
$$+ \alpha_{16} z^2 y + \alpha_{17} xyz + \alpha_{18} x^2 yz + \alpha_{19} y^2 xz + \alpha_{20} z^2 xy.$$

For the group supermatrix procedure the terms in the above polynomial will be allocated by the following expression to their respective G-invariant subspaces, with the unique order of the terms that provides the block diagonal form of the matrix of the system of equations relating displacements $\bar{U}$ to coefficients A:

$$\begin{bmatrix} F_1 \\ F_2 \\ F_3 \\ F_4 \\ F_5 \\ F_6 \\ F_7 \\ F_8 \end{bmatrix} = \begin{bmatrix} 1 \\ xy \\ xz \\ yz \\ xyz \\ z \\ y \\ x \end{bmatrix} \begin{bmatrix} 1 & x^2 & y^2 & z^2 \end{bmatrix}$$

$$= \begin{bmatrix} 1 & x^2 & y^2 & z^2 \\ xy & x^3 y & y^3 x & z^2 xy \\ xz & x^3 z & y^2 xz & z^3 x \\ yz & x^2 yz & y^3 z & z^3 y \\ xyz & x^3 yz & y^3 xz & z^3 xy \\ z & x^2 z & y^2 z & z^3 \\ y & x^2 y & y^3 & z^2 y \\ x & x^3 & y^2 x & z^2 x \end{bmatrix} \quad \begin{array}{l} \text{Subspace} \\ U_1 \\ U_2 \\ U_3 \\ U_4 \\ U_5 \\ U_6 \\ U_7 \\ U_8 \end{array}$$

The column vector and the row vector are the Cartesian sets and products as they stand in addition to the character table of the group D_{2h} according to the pertinence of the terms to the symmetry types of group representations A_g, B_{1g}, B_{2g}, B_{3g}, A_u, B_{1u}, B_{2u}, B_{3u}.

By omitting the terms $x^3 y$, $y^3 x$, $x^3 z$, $z^3 x$, $y^3 z$, $z^3 y$, $x^3 yz$, $y^3 xz$, $z^3 xy$, z^3, y^3, x^3 that do not appear in the polynomial $u(x, y, z)$, one obtains

212   Formulation of shape functions in G-invariant subspaces   [Ch. 4

$$\begin{bmatrix} F_1' \\ F_2' \\ F_3' \\ F_4' \\ F_5' \\ F_6' \\ F_7' \\ F_8' \end{bmatrix} = \begin{bmatrix} 1 & x^2 & y^2 & z^2 \\ & xy & z^2xy \\ & xz & y^2xz \\ & yz & x^2yz \\ & & xyz \\ z & x^2z & y^2z \\ y & x^2y & z^2y \\ x & y^2x & z^2x \end{bmatrix} \begin{matrix} \text{Subspace} \\ U_1 \\ U_2 \\ U_3 \\ U_4 \\ U_5 \\ U_6 \\ U_7 \\ U_8 \end{matrix}$$

with the pertinence of the terms to the subspaces U_1, U_2, U_3, U_4, U_5, U_6, U_7, U_8 and the optimum order of the terms.

Consequently, the polynomial expression of the displacement function

$$u(x, y, z) = [1 \quad x^2 \quad y^2 \quad z^2 \mid xy \quad z^2xy \mid xz \quad y^2xz \mid yz \quad x^2yz \mid xyz$$
$$\mid z \quad x^2z \quad y^2z \mid y \quad x^2y \quad z^2y \mid x \quad y^2x \quad z^2x]$$
$$[\alpha_1 \quad \alpha_3 \quad \alpha_3 \quad \alpha_4 \mid \alpha_5 \quad \alpha_6 \mid \alpha_7 \quad \alpha_8 \mid \alpha_9 \quad \alpha_{10} \mid \alpha_{11}$$
$$\mid \alpha_{12} \quad \alpha_{13} \quad \alpha_{14} \mid \alpha_{15} \quad \alpha_{16} \quad \alpha_{17} \mid \alpha_{18} \quad \alpha_{19} \quad \alpha_{20}]^T$$
$$= [F_1' \mid F_2' \mid F_3' \mid F_4' \mid F_5' \mid F_6' \mid F_7' \mid F_8']$$
$$[A_1 \mid A_2 \mid A_3 \mid A_4 \mid A_5 \mid A_6 \mid A_7 \mid A_8]^T$$

is decomposed into polynomials pertaining to G-invariant subspaces U_1, U_2, U_3, U_4, U_5, U_6, U_7, U_8 with 4, 2, 2, 2, 1, 3, 3, 3 dimensions respectively, which coincides with the dimensions of the subspaces spanned by the basis vectors derived earlier.

Substitution of the nodal coordinates of the first nodes in the nodal sets $S_1(1, 2, 3, 4, 5, 6, 7, 8)$, $S_2(9, 10, 11, 12)$, $S_3(13, 14, 15, 16)$, $S_4(17, 18, 19, 20)$

	1	9	13	17
x	c	c	c	0
y	d	d	0	d
z	e	0	e	e

into the sections F_1', F_2', F_3', F_4', F_5', F_6', F_7', F_8' of $u(x, y, z)$ will provide relations of displacements $\bar{U}$ to coefficients A for each subspace separately:

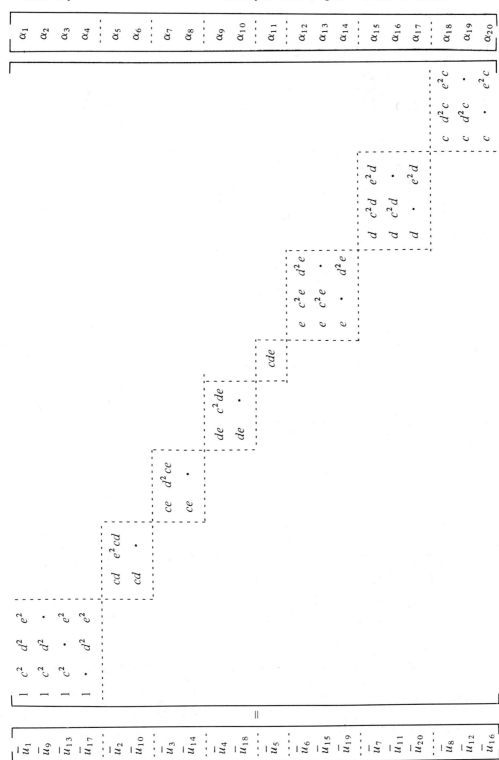

214 Formulation of shape functions in G-invariant subspaces [Ch. 4

or $\bar{U} = \bar{C}A$, where $\bar{U}$ is the set of displacements with the numbering of the basis vectors:

$$\begin{bmatrix} \bar{u}_1 \\ \bar{u}_9 \\ \bar{u}_{13} \\ \bar{u}_{17} \\ \hdashline \bar{u}_2 \\ \bar{u}_{10} \\ \hdashline \bar{u}_3 \\ \bar{u}_{14} \\ \hdashline \bar{u}_4 \\ \bar{u}_{18} \\ \hdashline \bar{u}_5 \\ \hdashline \bar{u}_6 \\ \bar{u}_{15} \\ \bar{u}_{19} \\ \hdashline \bar{u}_7 \\ \bar{u}_{11} \\ \bar{u}_{20} \\ \hdashline \bar{u}_8 \\ \bar{u}_{12} \\ \bar{u}_{16} \end{bmatrix} = \begin{bmatrix} \frac{1}{8}(u_1 + u_2 + u_3 + u_4 + u_5 + u_6 + u_7 + u_8) \\ \frac{1}{4}(u_9 + u_{10} + u_{11} + u_{12}) \\ \frac{1}{4}(u_{13} + u_{14} + u_{15} + u_{16}) \\ \frac{1}{4}(u_{17} + u_{18} + u_{19} + u_{20}) \\ \hdashline \frac{1}{8}(u_1 + u_2 - u_3 - u_4 + u_5 + u_6 - u_7 - u_8) \\ \frac{1}{4}(u_9 + u_{10} - u_{11} - u_{12}) \\ \hdashline \frac{1}{8}(u_1 - u_2 + u_3 - u_4 + u_5 - u_6 + u_7 - u_8) \\ \frac{1}{4}(u_{13} + u_{14} - u_{15} - u_{16}) \\ \hdashline \frac{1}{8}(u_1 - u_2 - u_3 + u_4 + u_5 - u_6 - u_7 + u_8) \\ \frac{1}{4}(u_{17} + u_{18} - u_{19} - u_{20}) \\ \hdashline \frac{1}{8}(u_1 + u_2 + u_3 + u_4 - u_5 - u_6 - u_7 - u_8) \\ \hdashline \frac{1}{8}(u_1 + u_2 - u_3 - u_4 - u_5 - u_6 + u_7 + u_8) \\ \frac{1}{4}(u_{13} - u_{14} + u_{15} - u_{16}) \\ \frac{1}{4}(u_{17} - u_{18} + u_{19} - u_{20}) \\ \hdashline \frac{1}{8}(u_1 - u_2 + u_3 - u_4 - u_5 + u_6 - u_7 + u_8) \\ \frac{1}{4}(u_9 - u_{10} + u_{11} - u_{12}) \\ \frac{1}{4}(u_{17} - u_{18} - u_{19} + u_{20}) \\ \hdashline \frac{1}{8}(u_1 - u_2 - u_3 + u_4 - u_5 + u_6 + u_7 - u_8) \\ \frac{1}{4}(u_9 - u_{10} - u_{11} + u_{12}) \\ \frac{1}{4}(u_{13} - u_{14} - u_{15} + u_{16}) \end{bmatrix}$$

Thus, the twenty-dimensional G-vector space of the problem is decomposed into eight G-invariant subspaces U_1, U_2, U_3, U_4, U_5, U_6, U_7, U_8 with 4, 2, 2, 2, 1, 3, 3, 3 dimensions respectively.

The coefficients $A_1, A_2, \ldots, A_8$ are found by inverting the respective matrices $\bar{C}_1, \bar{C}_2, \ldots, \bar{C}_8$ of the subspaces $U_1, U_2, \ldots, U_8$: (see pp. 216–217) or $A = \bar{C}^{-1}\bar{U}$.

The twenty-dimensional G-vector space of the displacement field Δ, decomposed into eight G-invariant subspaces $U_1, U_2, U_3, U_4, U_5, U_6, U_7, U_8$ with 4, 2, 2, 2, 1, 3, 3, 3 dimensions respectively, is given by

$$\Delta = T_D \bar{C}^{-1} \bar{U},$$

with

$$\begin{aligned} T_D &= \text{Diag} \, [1 \quad x^2 \quad y^2 \quad z^2 \mid xy \quad z^2 xy \mid xz \quad y^2 xz \mid yz \quad x^2 yz \\ &\qquad \mid xyz \mid z \quad x^2 z \quad y^2 z \mid y \quad x^2 y \quad z^2 y \mid x \quad y^2 x \quad z^2 x] \\ &= \text{Diag} \, [F'_1 \mid F'_2 \mid F'_3 \mid F'_4 \mid F'_5 \mid F'_6 \mid F'_7 \mid F'_8], \end{aligned}$$

providing (see pp. 218–219) and with substitutions $\xi = x/c$, $\eta = y/d$, $\zeta = z/e$ (see pp. 220–221) or

$$\begin{bmatrix} \Delta_1 \\ \Delta_2 \\ \Delta_3 \\ \Delta_4 \\ \Delta_5 \\ \Delta_6 \\ \Delta_7 \\ \Delta_8 \end{bmatrix} = \begin{bmatrix} \bar{N}_{\Delta_1} & & & & & & & \\ & \bar{N}_{\Delta_2} & & & & & & \\ & & \bar{N}_{\Delta_3} & & & & & \\ & & & \bar{N}_{\Delta_4} & & & & \\ & & & & \bar{N}_{\Delta_5} & & & \\ & & & & & \bar{N}_{\Delta_6} & & \\ & & & & & & \bar{N}_{\Delta_7} & \\ & & & & & & & \bar{N}_{\Delta_8} \end{bmatrix} \begin{bmatrix} \bar{U}^{(1)} \\ \bar{U}^{(2)} \\ \bar{U}^{(3)} \\ \bar{U}^{(4)} \\ \bar{U}^{(5)} \\ \bar{U}^{(6)} \\ \bar{U}^{(7)} \\ \bar{U}^{(8)} \end{bmatrix}$$

or $\Delta = \bar{N}_\Delta \bar{U}$.

Thus, the shape functions $\bar{N}^{(1)}, \bar{N}^{(2)}, \bar{N}^{(3)}, \bar{N}^{(4)}, \bar{N}^{(5)}, \bar{N}^{(6)}, \bar{N}^{(7)}, \bar{N}^{(8)}$ of G-invariant subspaces $U_1, U_2, U_3, U_4, U_5, U_6, U_7, U_8$ are obtained by $\bar{N} = S \bar{N}_\Delta$ with

$$S = [1 \quad 1 \quad 1 \quad 1 \mid 1 \quad 1 \mid 1 \quad 1 \mid 1 \quad 1 \mid 1 \mid 1 \quad 1 \quad 1 \\ \mid 1 \quad 1 \quad 1 \mid 1 \quad 1 \quad 1],$$

giving (see p. 222).

$$
\begin{bmatrix} \alpha_1 \\ \alpha_2 \\ \alpha_3 \\ \alpha_4 \\ \alpha_5 \\ \alpha_6 \\ \alpha_7 \\ \alpha_8 \\ \alpha_9 \\ \alpha_{10} \\ \alpha_{11} \\ \alpha_{12} \\ \alpha_{13} \\ \alpha_{14} \\ \alpha_{15} \\ \alpha_{16} \\ \alpha_{17} \\ \alpha_{18} \\ \alpha_{19} \\ \alpha_{20} \end{bmatrix}
=
\left[\begin{array}{cccc|cc|cc|cc|c}
-2 & 1 & 1 & 1 & & & & & & & \\
\dfrac{1}{c^2} & 0 & 0 & -\dfrac{1}{c^2} & & & & & & & \\
\dfrac{1}{d^2} & 0 & -\dfrac{1}{d^2} & 0 & & & & & & & \\
\dfrac{1}{e^2} & -\dfrac{1}{e^2} & 0 & 0 & & & & & & & \\
\hline
 & & & & 0 & \dfrac{1}{cd} & & & & & \\
 & & & & \dfrac{1}{e^2cd} & -\dfrac{1}{e^2cd} & & & & & \\
\hline
 & & & & & & 0 & \dfrac{1}{ce} & & & \\
 & & & & & & \dfrac{1}{d^2ce} & -\dfrac{1}{d^2ce} & & & \\
\hline
 & & & & & & & & 0 & \dfrac{1}{de} & \\
 & & & & & & & & \dfrac{1}{c^2de} & -\dfrac{1}{c^2de} & \\
\hline
 & & & & & & & & & & \\
\end{array}\right]
$$

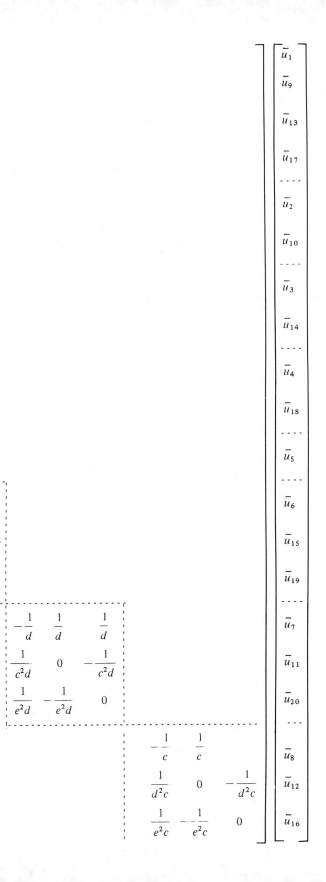

$$
\begin{bmatrix} \Delta_1 \\ \cdots \\ \Delta_2 \\ \cdots \\ \Delta_3 \\ \cdots \\ \Delta_4 \\ \cdots \\ \Delta_5 \\ \cdots \\ \Delta_6 \\ \cdots \\ \Delta_7 \\ \cdots \\ \Delta_8 \end{bmatrix} = \begin{bmatrix} -2 & 1 & 1 & 1 & & & & \\ \dfrac{x^2}{c^2} & 0 & 0 & -\dfrac{x^2}{c^2} & & & & \\ \dfrac{y^2}{d^2} & 0 & -\dfrac{y^2}{d^2} & 0 & & & & \\ \dfrac{z^2}{e^2} & -\dfrac{z^2}{e^2} & 0 & 0 & & & & \\ & & & & 0 & \dfrac{xy}{cd} & & \\ & & & & \dfrac{z^2 xy}{e^2 cd} & -\dfrac{z^2 xy}{e^2 cd} & & \\ & & & & & & 0 & \dfrac{xz}{ce} \\ & & & & & & \dfrac{y^2 xz}{d^2 ce} & -\dfrac{y^2 xz}{d^2 ce} \\ & & & & & & & & 0 & \dfrac{y}{c} \\ & & & & & & & & \dfrac{x^2 yz}{c^2 de} & \end{bmatrix}
$$

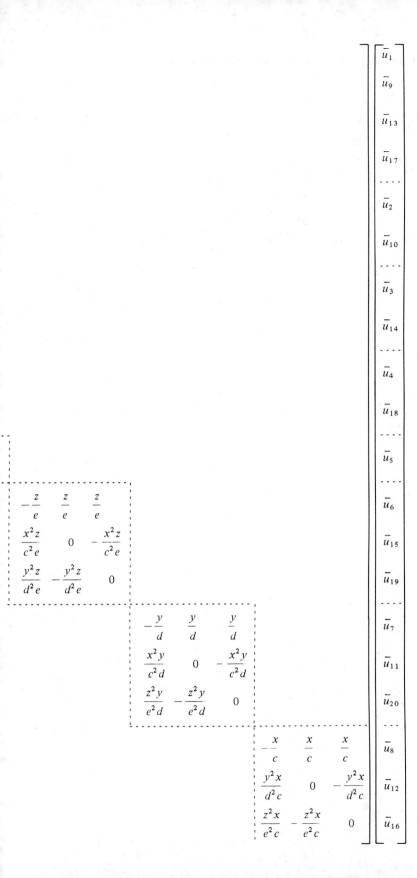

220 Formulation of shape functions in G-invariant subspaces [Ch. 4

$$
\begin{bmatrix} \Delta_1 \\ \cdots \\ \Delta_2 \\ \cdots \\ \Delta_3 \\ \cdots \\ \Delta_4 \\ \cdots \\ \Delta_5 \\ \cdots \\ \Delta_6 \\ \cdots \\ \Delta_7 \\ \cdots \\ \Delta_8 \end{bmatrix} = \begin{bmatrix} \begin{array}{cccc} -2 & 1 & 1 & 1 \\ \xi^2 & 0 & 0 & -\xi^2 \\ \eta^2 & 0 & -\eta^2 & 0 \\ \zeta & -\zeta^2 & 0 & 0 \end{array} & & & \\ & \begin{array}{cc} 0 & \xi\eta \\ \zeta^2\xi\eta & -\zeta^2\xi\eta \end{array} & & \\ & & \begin{array}{cc} 0 & \xi\zeta \\ \eta^2\xi\zeta & -\eta^2\xi\zeta \end{array} & \\ & & & \begin{array}{cc} 0 & \eta\zeta \\ \xi^2\eta\zeta & -\xi^2\eta\zeta \end{array} \end{bmatrix}
$$

Sec. 4.7] Twenty-node rectangular hexahedral element 221

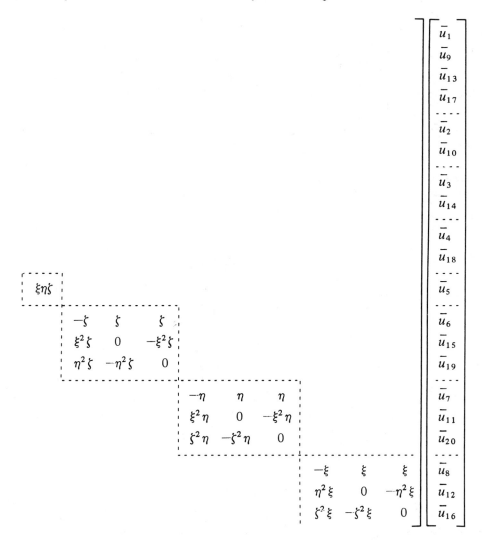

$$
\begin{bmatrix} \bar{N}_1 \\ \bar{N}_9 \\ \bar{N}_{13} \\ \bar{N}_{17} \\ \hdashline \bar{N}_2 \\ \bar{N}_{10} \\ \hdashline \bar{N}_3 \\ \bar{N}_{14} \\ \hdashline \bar{N}_4 \\ \bar{N}_{18} \\ \hdashline \bar{N}_5 \\ \hdashline \bar{N}_6 \\ \bar{N}_{15} \\ \bar{N}_{19} \\ \hdashline \bar{N}_7 \\ \bar{N}_{11} \\ \bar{N}_{20} \\ \hdashline \bar{N}_8 \\ \bar{N}_{12} \\ \bar{N}_{16} \end{bmatrix}
=
\begin{bmatrix}
-2+\xi^2+\eta^2+\zeta^2 \\ 1-\zeta^2 \\ 1-\eta^2 \\ 1-\xi^2 \\ \hdashline \zeta^2\xi\eta \\ \xi\eta-\zeta^2\xi\eta \\ \hdashline \eta^2\xi\zeta \\ \xi\zeta-\eta^2\xi\zeta \\ \hdashline \xi^2\eta\zeta \\ \eta\zeta-\xi^2\eta\zeta \\ \hdashline \xi\eta\zeta \\ \hdashline -\zeta+\xi^2\zeta+\eta^2\zeta \\ \zeta-\eta^2\zeta \\ \zeta-\xi^2\zeta \\ \hdashline -\eta+\xi^2\eta+\zeta^2\eta \\ \eta-\zeta^2\eta \\ \eta-\xi^2\eta \\ \hdashline -\xi+\eta^2\xi+\zeta^2\xi \\ \xi-\zeta^2\xi \\ \xi-\eta^2\xi
\end{bmatrix}
=
\begin{bmatrix}
N_1+N_2+N_3+N_4+N_5+N_6+N_7+N_8 \\
N_9+N_{10}+N_{11}+N_{12} \\
N_{13}+N_{14}+N_{15}+N_{16} \\
N_{17}+N_{18}+N_{19}+N_{20} \\ \hdashline
N_1+N_2-N_3-N_4+N_5+N_6-N_7-N_8 \\
N_9+N_{10}-N_{11}-N_{12} \\ \hdashline
N_1-N_2+N_3-N_4+N_5-N_6+N_7-N_8 \\
N_{13}+N_{14}-N_{15}-N_{16} \\ \hdashline
N_1-N_2-N_3+N_4+N_5-N_6-N_7+N_8 \\
N_{17}+N_{18}-N_{19}-N_{20} \\ \hdashline
N_1+N_2+N_3+N_4-N_5-N_6-N_7-N_8 \\ \hdashline
N_1+N_2-N_3-N_4-N_5-N_6+N_7+N_8 \\
N_{13}-N_{14}+N_{15}-N_{16} \\
N_{19}-N_{20} \\ \hdashline
N_1-N_2+N_3-N_4-N_5+N_6-N_7+N_8 \\
N_9-N_{10}+N_{11}-N_{12} \\
N_{17}-N_{18}+N_{19}-N_{20} \\ \hdashline
N_1-N_2-N_3+N_4-N_5+N_6+N_7-N_8 \\
N_9-N_{10}-N_{11}+N_{12} \\
N_{13}-N_{14}-N_{15}+N_{16}
\end{bmatrix}
$$

Sec. 4.7] Twenty-node rectangular hexahedral element

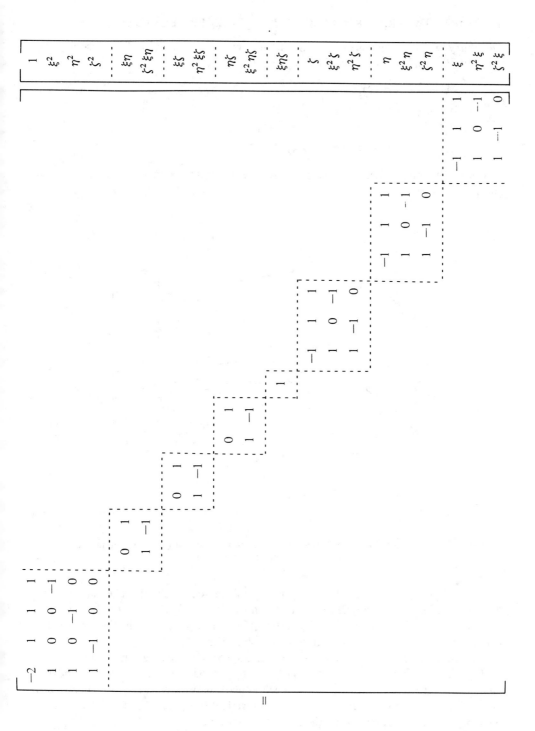

4.8 THIRTY-TWO-NODE RECTANGULAR HEXAHEDRAL ELEMENT

The thirty-two-node rectangular hexahedral element, shown in Fig. 4.8, is described by the group D_{2h} and it has four nodal sets:

$S_1(1, 2, 3, 4, 5, 6, 7, 8)$

$S_2(9, 10, 11, 12, 13, 14, 15, 16)$

$S_3(17, 18, 19, 20, 21, 22, 23, 24)$

$S_4(25, 26, 27, 28, 29, 30, 31, 32)$,

each set containing nodes that are permuted by action of symmetry operations of the group.

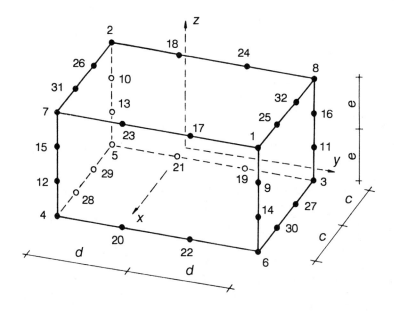

Fig. 4.8. The thirty-two-node rectangular hexahedral element with the nodal numbering of the group supermatrix procedure.

The unique nodal numbering for the eight-node rectangular hexahedral element, derived in section 3.8, is applied here for the nodal sets S_1, S_2, S_3, S_4. The first node of each set lies in the first octant, the second node of each set etc. in their respective octants. Nodes of each set are corner nodes of a rectangular hexahedron.

The coordinates of the nodes of the nodal sets S_1, S_2, S_3, S_4 are given in Table 4.4.

The character table of the group D_{2h} (with Cartesian sets and products) and the idempotents of the centre of the group algebra, given previously in section 4.6, are used here also for derivation of basis vectors $\overline{\Phi}$ of G-invariant subspaces U_i, which is performed by applying the idempotents π_i to the nodal functions φ_j:

Sec. 4.8]   Thirty-two-node rectangular hexahedral element   225

Table 4.4. Coordinates of nodes in the sets S_1, S_2, S_3, S_4

Set S_1	1	2	3	4	5	6	7	8
x	c	$-c$	$-c$	c	$-c$	c	c	$-c$
y	d	$-d$	d	$-d$	$-d$	d	$-d$	d
z	e	e	$-e$	$-e$	$-e$	$-e$	e	e
Set S_2	9	10	11	12	13	14	15	16
x	c	$-c$	$-c$	c	$-c$	c	c	$-c$
y	d	$-d$	d	$-d$	$-d$	d	$-d$	d
z	$e/3$	$e/3$	$-e/3$	$-e/3$	$-e/3$	$-e/3$	$e/3$	$e/3$
Set S_3	17	18	19	20	21	22	23	24
x	c	$-c$	$-c$	c	$-c$	c	c	$-c$
y	$d/3$	$-d/3$	$d/3$	$-d/3$	$-d/3$	$d/3$	$-d/3$	$d/3$
z	e	e	$-e$	$-e$	$-e$	$-e$	e	e
Set S_4	25	26	27	28	29	30	31	32
x	$c/3$	$-c/3$	$-c/3$	$c/3$	$-c/3$	$c/3$	$c/3$	$-c/3$
y	d	$-d$	d	$-d$	$-d$	d	$-d$	d
z	e	e	$-e$	$-e$	$-e$	$-e$	e	e

$$U_1: \quad \bar{\varphi}_1 = \tfrac{1}{8}(\varphi_1 + \varphi_2 + \varphi_3 + \varphi_4 + \varphi_5 + \varphi_6 + \varphi_7 + \varphi_8)$$
$$\bar{\varphi}_9 = \tfrac{1}{8}(\varphi_9 + \varphi_{10} + \varphi_{11} + \varphi_{12} + \varphi_{13} + \varphi_{14} + \varphi_{15} + \varphi_{16})$$
$$\bar{\varphi}_{17} = \tfrac{1}{8}(\varphi_{17} + \varphi_{18} + \varphi_{19} + \varphi_{20} + \varphi_{21} + \varphi_{22} + \varphi_{23} + \varphi_{24})$$
$$\bar{\varphi}_{25} = \tfrac{1}{8}(\varphi_{25} + \varphi_{26} + \varphi_{27} + \varphi_{28} + \varphi_{29} + \varphi_{30} + \varphi_{31} + \varphi_{32})$$

$$U_2: \quad \bar{\varphi}_2 = \tfrac{1}{8}(\varphi_1 + \varphi_2 - \varphi_3 - \varphi_4 + \varphi_5 + \varphi_6 - \varphi_7 - \varphi_8)$$
$$\bar{\varphi}_{10} = \tfrac{1}{8}(\varphi_9 + \varphi_{10} - \varphi_{11} - \varphi_{12} + \varphi_{13} + \varphi_{14} - \varphi_{15} - \varphi_{16})$$
$$\bar{\varphi}_{18} = \tfrac{1}{8}(\varphi_{17} + \varphi_{18} - \varphi_{19} - \varphi_{20} + \varphi_{21} + \varphi_{22} - \varphi_{23} - \varphi_{24})$$
$$\bar{\varphi}_{26} = \tfrac{1}{8}(\varphi_{25} + \varphi_{26} - \varphi_{27} - \varphi_{28} + \varphi_{29} + \varphi_{30} - \varphi_{31} - \varphi_{32})$$

U_3: $\bar{\varphi}_3 = \frac{1}{8}(\varphi_1 - \varphi_2 + \varphi_3 - \varphi_4 + \varphi_5 - \varphi_6 + \varphi_7 - \varphi_8)$
$\bar{\varphi}_{11} = \frac{1}{8}(\varphi_9 - \varphi_{10} + \varphi_{11} - \varphi_{12} + \varphi_{13} - \varphi_{14} + \varphi_{15} - \varphi_{16})$
$\bar{\varphi}_{19} = \frac{1}{8}(\varphi_{17} - \varphi_{18} + \varphi_{19} - \varphi_{20} + \varphi_{21} - \varphi_{22} + \varphi_{23} - \varphi_{24})$
$\bar{\varphi}_{27} = \frac{1}{8}(\varphi_{25} - \varphi_{26} + \varphi_{27} - \varphi_{28} + \varphi_{29} - \varphi_{30} + \varphi_{31} - \varphi_{32})$

U_4: $\bar{\varphi}_4 = \frac{1}{8}(\varphi_1 - \varphi_2 - \varphi_3 + \varphi_4 + \varphi_5 - \varphi_6 - \varphi_7 + \varphi_8)$
$\bar{\varphi}_{12} = \frac{1}{8}(\varphi_9 - \varphi_{10} - \varphi_{11} + \varphi_{12} + \varphi_{13} - \varphi_{14} - \varphi_{15} + \varphi_{16})$
$\bar{\varphi}_{20} = \frac{1}{8}(\varphi_{17} - \varphi_{18} - \varphi_{19} + \varphi_{20} + \varphi_{21} - \varphi_{22} - \varphi_{23} + \varphi_{24})$
$\bar{\varphi}_{28} = \frac{1}{8}(\varphi_{25} - \varphi_{26} - \varphi_{27} + \varphi_{28} + \varphi_{29} - \varphi_{30} - \varphi_{31} + \varphi_{32})$

U_5: $\bar{\varphi}_5 = \frac{1}{8}(\varphi_1 + \varphi_2 + \varphi_3 + \varphi_4 - \varphi_5 - \varphi_6 - \varphi_7 - \varphi_8)$
$\bar{\varphi}_{13} = \frac{1}{8}(\varphi_9 + \varphi_{10} + \varphi_{11} + \varphi_{12} - \varphi_{13} - \varphi_{14} - \varphi_{15} - \varphi_{16})$
$\bar{\varphi}_{21} = \frac{1}{8}(\varphi_{17} + \varphi_{18} + \varphi_{19} + \varphi_{20} - \varphi_{21} - \varphi_{22} - \varphi_{23} - \varphi_{24})$
$\bar{\varphi}_{29} = \frac{1}{8}(\varphi_{25} + \varphi_{26} + \varphi_{27} + \varphi_{28} - \varphi_{29} - \varphi_{30} - \varphi_{31} - \varphi_{32})$

U_6: $\bar{\varphi}_6 = \frac{1}{8}(\varphi_1 + \varphi_2 - \varphi_3 - \varphi_4 - \varphi_5 - \varphi_6 + \varphi_7 + \varphi_8)$
$\bar{\varphi}_{14} = \frac{1}{8}(\varphi_9 + \varphi_{10} - \varphi_{11} - \varphi_{12} - \varphi_{13} - \varphi_{14} + \varphi_{15} + \varphi_{16})$
$\bar{\varphi}_{22} = \frac{1}{8}(\varphi_{17} + \varphi_{18} - \varphi_{19} - \varphi_{20} - \varphi_{21} - \varphi_{22} + \varphi_{23} + \varphi_{24})$
$\bar{\varphi}_{30} = \frac{1}{8}(\varphi_{25} + \varphi_{26} - \varphi_{27} - \varphi_{28} - \varphi_{29} - \varphi_{30} + \varphi_{31} + \varphi_{32})$

U_7: $\bar{\varphi}_7 = \frac{1}{8}(\varphi_1 - \varphi_2 + \varphi_3 - \varphi_4 - \varphi_5 + \varphi_6 - \varphi_7 + \varphi_8)$
$\bar{\varphi}_{15} = \frac{1}{8}(\varphi_9 - \varphi_{10} + \varphi_{11} - \varphi_{12} - \varphi_{13} + \varphi_{14} - \varphi_{15} + \varphi_{16})$
$\bar{\varphi}_{23} = \frac{1}{8}(\varphi_{17} - \varphi_{18} + \varphi_{19} - \varphi_{20} - \varphi_{21} + \varphi_{22} - \varphi_{23} + \varphi_{24})$
$\bar{\varphi}_{31} = \frac{1}{8}(\varphi_{25} - \varphi_{26} + \varphi_{27} - \varphi_{28} - \varphi_{29} + \varphi_{30} - \varphi_{31} + \varphi_{32})$

U_8: $\bar{\varphi}_8 = \frac{1}{8}(\varphi_1 - \varphi_2 - \varphi_3 + \varphi_4 - \varphi_5 + \varphi_6 + \varphi_7 - \varphi_8)$
$\bar{\varphi}_{16} = \frac{1}{8}(\varphi_9 - \varphi_{10} - \varphi_{11} + \varphi_{12} - \varphi_{13} + \varphi_{14} + \varphi_{15} - \varphi_{16})$
$\bar{\varphi}_{24} = \frac{1}{8}(\varphi_{17} - \varphi_{18} - \varphi_{19} + \varphi_{20} - \varphi_{21} + \varphi_{22} + \varphi_{23} - \varphi_{24})$
$\bar{\varphi}_{32} = \frac{1}{8}(\varphi_{25} - \varphi_{26} - \varphi_{27} + \varphi_{28} - \varphi_{29} + \varphi_{30} + \varphi_{31} - \varphi_{32})$.

Thus the thirty-two-dimensional G-vector space is decomposed into eight four-dimensional G-invariant subspaces U_1, U_2, U_3, U_4, U_5, U_6, U_7, U_8 spanned by the respective sets of basis vectors $\bar{\Phi}^{(1)}$, $\bar{\Phi}^{(2)}$, $\bar{\Phi}^{(3)}$, $\bar{\Phi}^{(4)}$, $\bar{\Phi}^{(5)}$, $\bar{\Phi}^{(6)}$, $\bar{\Phi}^{(7)}$, $\bar{\Phi}^{(8)}$:

Thirty-two-node rectangular hexahedral element

$$\begin{bmatrix} \bar{\Phi}^{(1)} \\ \bar{\Phi}^{(2)} \\ \bar{\Phi}^{(3)} \\ \bar{\Phi}^{(4)} \\ \bar{\Phi}^{(5)} \\ \bar{\Phi}^{(6)} \\ \bar{\Phi}^{(7)} \\ \bar{\Phi}^{(8)} \end{bmatrix} = \begin{bmatrix} \bar{\varphi}_1 & \bar{\varphi}_9 & \bar{\varphi}_{17} & \bar{\varphi}_{25} \\ \bar{\varphi}_2 & \bar{\varphi}_{10} & \bar{\varphi}_{18} & \bar{\varphi}_{26} \\ \bar{\varphi}_3 & \bar{\varphi}_{11} & \bar{\varphi}_{19} & \bar{\varphi}_{27} \\ \bar{\varphi}_4 & \bar{\varphi}_{12} & \bar{\varphi}_{20} & \bar{\varphi}_{28} \\ \bar{\varphi}_5 & \bar{\varphi}_{13} & \bar{\varphi}_{21} & \bar{\varphi}_{29} \\ \bar{\varphi}_6 & \bar{\varphi}_{14} & \bar{\varphi}_{22} & \bar{\varphi}_{30} \\ \bar{\varphi}_7 & \bar{\varphi}_{15} & \bar{\varphi}_{23} & \bar{\varphi}_{31} \\ \bar{\varphi}_8 & \bar{\varphi}_{16} & \bar{\varphi}_{24} & \bar{\varphi}_{32} \end{bmatrix}.$$

The sets $\bar{\Phi}_1, \bar{\Phi}_2, \bar{\Phi}_3, \bar{\Phi}_4$ of basis vectors $\bar{\varphi}_j$ ($j = 1, 2, \ldots, 32$), ordered according to the numbering of the nodes in the nodal sets S_1, S_2, S_3, S_4, are given by

$$[\bar{\Phi}_1 \ \bar{\Phi}_2 \ \bar{\Phi}_3 \ \bar{\Phi}_4] = \begin{bmatrix} \bar{\varphi}_1 & \bar{\varphi}_9 & \bar{\varphi}_{17} & \bar{\varphi}_{25} \\ \bar{\varphi}_2 & \bar{\varphi}_{10} & \bar{\varphi}_{18} & \bar{\varphi}_{26} \\ \bar{\varphi}_3 & \bar{\varphi}_{11} & \bar{\varphi}_{19} & \bar{\varphi}_{27} \\ \bar{\varphi}_4 & \bar{\varphi}_{12} & \bar{\varphi}_{20} & \bar{\varphi}_{28} \\ \bar{\varphi}_5 & \bar{\varphi}_{13} & \bar{\varphi}_{21} & \bar{\varphi}_{29} \\ \bar{\varphi}_6 & \bar{\varphi}_{14} & \bar{\varphi}_{22} & \bar{\varphi}_{30} \\ \bar{\varphi}_7 & \bar{\varphi}_{15} & \bar{\varphi}_{23} & \bar{\varphi}_{31} \\ \bar{\varphi}_8 & \bar{\varphi}_{16} & \bar{\varphi}_{24} & \bar{\varphi}_{32} \end{bmatrix}.$$

The relation of the sets of basis vectors $\bar{\Phi}_l$ to the sets of nodal functions Φ_l ($l = 1, 2, 3, 4$) is expressed by the following system of equations with the supermatrix in diagonal form containing four transformation matrices of the group D_{2h}:

$$\begin{bmatrix} \bar{\Phi}_1 \\ \bar{\Phi}_2 \\ \bar{\Phi}_3 \\ \bar{\Phi}_4 \end{bmatrix} = \begin{bmatrix} T & & & \\ & T & & \\ & & T & \\ & & & T \end{bmatrix} \begin{bmatrix} \Phi_1 \\ \Phi_2 \\ \Phi_3 \\ \Phi_4 \end{bmatrix} \quad \text{or} \quad \bar{\Phi} = \bar{T}\Phi,$$

with

$$T = \frac{1}{8} \begin{bmatrix} 1 & 1 & 1 & 1 & 1 & 1 & 1 & 1 \\ 1 & 1 & -1 & -1 & 1 & 1 & -1 & -1 \\ 1 & -1 & 1 & -1 & 1 & -1 & 1 & -1 \\ 1 & -1 & -1 & 1 & 1 & -1 & -1 & 1 \\ 1 & 1 & 1 & 1 & -1 & -1 & -1 & -1 \\ 1 & 1 & -1 & -1 & -1 & -1 & 1 & 1 \\ 1 & -1 & 1 & -1 & -1 & 1 & -1 & 1 \\ 1 & -1 & -1 & 1 & -1 & 1 & 1 & -1 \end{bmatrix}$$

and

$$[\Phi_1 \quad \Phi_2 \quad \Phi_3 \quad \Phi_4] = \begin{bmatrix} \varphi_1 & \varphi_9 & \varphi_{17} & \varphi_{25} \\ \varphi_2 & \varphi_{10} & \varphi_{18} & \varphi_{26} \\ \varphi_3 & \varphi_{11} & \varphi_{19} & \varphi_{27} \\ \varphi_4 & \varphi_{12} & \varphi_{20} & \varphi_{28} \\ \varphi_5 & \varphi_{13} & \varphi_{21} & \varphi_{29} \\ \varphi_6 & \varphi_{14} & \varphi_{22} & \varphi_{30} \\ \varphi_7 & \varphi_{15} & \varphi_{23} & \varphi_{31} \\ \varphi_8 & \varphi_{16} & \varphi_{24} & \varphi_{32} \end{bmatrix}.$$

Conversely, the relation of the sets of nodal functions Φ_I to the sets of basis vectors $\overline{\Phi}_I$ is

$$\begin{bmatrix} \Phi_1 \\ \Phi_2 \\ \Phi_3 \\ \Phi_4 \end{bmatrix} = 8 \begin{bmatrix} T & & & \\ & T & & \\ & & T & \\ & & & T \end{bmatrix} \begin{bmatrix} \overline{\Phi}_1 \\ \overline{\Phi}_2 \\ \overline{\Phi}_3 \\ \overline{\Phi}_4 \end{bmatrix} \quad \text{or} \quad \Phi = 8\overline{T}\overline{\Phi},$$

since $T^{-1} = 8T$.

The u-, v- or w-displacement function is usually assumed to be

$$\begin{aligned} u(x, y, z) &= \alpha_1 + \alpha_2 x + \alpha_3 y + \alpha_4 z + \alpha_5 x^2 + \alpha_6 y^2 + \alpha_7 z^2 + \alpha_8 xy + \alpha_9 yz \\ &+ \alpha_{10} xz + \alpha_{11} x^3 + \alpha_{12} y^3 + \alpha_{13} z^3 + \alpha_{14} x^2 y + \alpha_{15} x^2 z + \alpha_{16} y^2 x \\ &+ \alpha_{17} y^2 z + \alpha_{18} z^2 x + \alpha_{19} z^2 y + \alpha_{20} xyz + \alpha_{21} x^3 y + \alpha_{22} x^3 z \\ &+ \alpha_{23} y^3 z + \alpha_{24} y^3 x + \alpha_{25} z^3 x + \alpha_{26} z^3 y + \alpha_{27} x^2 yz + \alpha_{28} y^2 xz \\ &+ \alpha_{29} z^2 xy + \alpha_{30} x^3 yz + \alpha_{31} y^3 xz + \alpha_{32} z^3 xy. \end{aligned}$$

For the group supermatrix procedure the terms in the above polynomial will be allocated by the following expression to their respective G-invariant subspaces, with the unique order of the terms that gives the block diagonal form of the matrix of equations relating displacements $\bar{U}$ to coefficients A:

$$\begin{bmatrix} F_1 \\ F_2 \\ F_3 \\ F_4 \\ F_5 \\ F_6 \\ F_7 \\ F_8 \end{bmatrix} = \begin{bmatrix} 1 \\ xy \\ xz \\ yz \\ xyz \\ z \\ y \\ x \end{bmatrix} [1 \quad x^2 \quad y^2 \quad z^2]$$

$$= \begin{bmatrix} 1 & x^2 & y^2 & z^2 \\ xy & x^3y & y^3x & z^2xy \\ xz & x^3z & y^2xz & z^3x \\ yz & x^2yz & y^3z & z^3y \\ xyz & x^3yz & y^3xz & z^3xy \\ z & x^2z & y^2z & z^3 \\ y & x^2y & y^3 & z^2y \\ x & x^3 & y^2x & z^2x \end{bmatrix} \begin{matrix} \text{Subspace} \\ U_1 \\ U_2 \\ U_3 \\ U_4 \\ U_5 \\ U_6 \\ U_7 \\ U_8 \end{matrix}$$

The column and the row vector are Cartesian sets and products as they stand in addition to the character table of the group D_{2h} according to the pertinence of the terms to the symmetry types of group representations A_g, B_{1g}, B_{2g}, B_{3g}, A_u, B_{1u}, B_{2u}, B_{3u}.

Consequently, the polynomial expression of the displacement function

$$u(x, y, z) = [1 \quad x^2 \quad y^2 \quad z^2 \mid xy \quad x^3y \quad y^3x \quad z^2xy \mid xz \quad x^3z \quad y^2xz \quad z^3x$$
$$\mid yz \quad x^2yz \quad y^3z \quad z^3y \mid xyz \quad x^3yz \quad y^3xz \quad z^3xy$$
$$\mid z \quad x^2z \quad y^2z \quad z^3 \mid y \quad x^2y \quad y^3 \quad z^2y \mid x \quad x^3 \quad y^2x \quad z^2x]$$
$$[\alpha_1 \ldots \alpha_4 \mid \alpha_5 \ldots \alpha_8 \mid \alpha_9 \ldots \alpha_{12} \mid \alpha_{13} \ldots \alpha_{16}$$
$$\mid \alpha_{17} \ldots \alpha_{20} \mid \alpha_{21} \ldots \alpha_{24} \mid \alpha_{25} \ldots \alpha_{28} \mid \alpha_{29} \ldots \alpha_{32}]^T$$
$$= [F_1 \mid F_2 \mid F_3 \mid F_4 \mid F_5 \mid F_6 \mid F_7 \mid F_8]$$
$$[A_1 \mid A_2 \mid A_3 \mid A_4 \mid A_5 \mid A_6 \mid A_7 \mid A_8]^T$$

is decomposed into eight four-dimensional subspaces U_1, U_2, U_3, U_4, U_5, U_6, U_7, U_8, which coincides with dimensions of the subspaces spanned by basis vectors derived earlier.

Substitution of the nodal coordinates of the first nodes in the nodal sets S_1, S_2, S_3, S_4

Node	1	9	17	25
x	c	c	c	$c/3$
y	d	d	$d/3$	d
z	e	$e/3$	e	e

into the sections F_1, F_2, F_3, F_4, F_5, F_6, F_7, F_8 of $u(x, y, z)$, pertaining to subspaces U_1, U_2, U_3, U_4, U_5, U_6, U_7, U_8, will give the relations of displacements $\bar{U}$ to coefficients A for each subspace separately:

$$\begin{bmatrix} \bar{U}^{(1)} \\ \bar{U}^{(2)} \\ \bar{U}^{(3)} \\ \bar{U}^{(4)} \\ \bar{U}^{(5)} \\ \bar{U}^{(6)} \\ \bar{U}^{(7)} \\ \bar{U}^{(8)} \end{bmatrix} = \begin{bmatrix} \bar{C}_1 & & & & & & & \\ & \bar{C}_2 & & & & & & \\ & & \bar{C}_3 & & & & & \\ & & & \bar{C}_4 & & & & \\ & & & & \bar{C}_5 & & & \\ & & & & & \bar{C}_6 & & \\ & & & & & & \bar{C}_7 & \\ & & & & & & & \bar{C}_8 \end{bmatrix} \begin{bmatrix} A_1 \\ A_2 \\ A_3 \\ A_4 \\ A_5 \\ A_6 \\ A_7 \\ A_8 \end{bmatrix}$$

or (see pp. 232–233),

with

$$\bar{C}_1 = \begin{bmatrix} 1 & c^2 & d^2 & e^2 \\ 1 & c^2 & d^2 & \dfrac{e^2}{9} \\ 1 & c^2 & \dfrac{d^2}{9} & e^2 \\ 1 & \dfrac{c^2}{9} & d^2 & e^2 \end{bmatrix} \qquad \bar{C}_2 = \begin{bmatrix} cd & c^3d & d^3c & e^2cd \\ cd & c^3d & d^3c & \dfrac{e^2cd}{9} \\ \dfrac{cd}{3} & \dfrac{c^3d}{3} & \dfrac{d^3c}{27} & \dfrac{e^2cd}{3} \\ \dfrac{cd}{3} & \dfrac{c^3d}{27} & \dfrac{d^3c}{3} & \dfrac{e^2cd}{3} \end{bmatrix}$$

$$\bar{C}_3 = \begin{bmatrix} ce & c^3e & d^2ce & e^3c \\ \dfrac{ce}{3} & \dfrac{c^3e}{3} & \dfrac{d^2ce}{3} & \dfrac{e^3c}{27} \\ ce & c^3e & \dfrac{d^2ce}{9} & e^3c \\ \dfrac{ce}{3} & \dfrac{c^3e}{27} & \dfrac{d^2ce}{3} & \dfrac{e^3c}{3} \end{bmatrix} \qquad \bar{C}_4 = \begin{bmatrix} de & c^2de & d^3e & e^3d \\ \dfrac{de}{3} & \dfrac{c^2de}{3} & \dfrac{d^3e}{3} & \dfrac{e^3d}{27} \\ \dfrac{de}{3} & \dfrac{c^2de}{3} & \dfrac{d^3e}{27} & \dfrac{e^3d}{3} \\ de & \dfrac{c^2de}{9} & d^3e & e^3d \end{bmatrix}$$

$$\bar{C}_5 = \begin{bmatrix} cde & c^3de & d^3ce & e^3cd \\ \dfrac{cde}{3} & \dfrac{c^3de}{3} & \dfrac{d^3ce}{3} & \dfrac{e^3cd}{27} \\ \dfrac{cde}{3} & \dfrac{c^3de}{3} & \dfrac{d^3ce}{27} & \dfrac{e^3cd}{3} \\ \dfrac{cde}{3} & \dfrac{c^3de}{27} & \dfrac{d^3ce}{3} & \dfrac{e^3cd}{3} \end{bmatrix} \qquad \bar{C}_6 = \begin{bmatrix} e & c^2e & d^2e & e^3 \\ \dfrac{e}{3} & \dfrac{c^2e}{3} & \dfrac{d^2e}{3} & \dfrac{e^3}{27} \\ e & c^2e & \dfrac{d^2e}{9} & e^3 \\ e & \dfrac{c^2e}{9} & d^2e & e^3 \end{bmatrix}$$

$$\bar{C}_7 = \begin{bmatrix} d & c^2d & d^3 & e^2d \\ d & c^2d & d^3 & \dfrac{e^2d}{9} \\ \dfrac{d}{3} & \dfrac{c^2d}{3} & \dfrac{d^3}{27} & \dfrac{e^2d}{3} \\ d & \dfrac{c^2d}{9} & d^3 & e^2d \end{bmatrix} \qquad \bar{C}_8 = \begin{bmatrix} c & c^3 & d^2c & e^2c \\ c & c^3 & d^2c & \dfrac{e^2c}{9} \\ c & c^3 & \dfrac{d^2c}{9} & e^2c \\ \dfrac{c}{3} & \dfrac{c^3}{27} & \dfrac{d^2c}{3} & \dfrac{e^2c}{3} \end{bmatrix}$$

where $\bar{U}^{(1)}$, $\bar{U}^{(2)}$, $\bar{U}^{(3)}$, $\bar{U}^{(4)}$, $\bar{U}^{(5)}$, $\bar{U}^{(6)}$, $\bar{U}^{(7)}$, $\bar{U}^{(8)}$ are the sets of displacements with the numbering according to the numbering of basis vectors derived earlier. The thirty-two-dimensional G-vector space of the problem is decomposed now into eight four-dimensional G-invariant subspaces U_1, U_2, U_3, U_4, U_5, U_6, U_7, U_8.

The coefficients A are determined by inverting the matrices $\bar{C}_1, \bar{C}_2, \ldots, \bar{C}_8$ of subspaces $U_1, U_2, \ldots, U_8$: (see p. 234)

$$\begin{bmatrix} \bar{u}_1 \\ \bar{u}_9 \\ \bar{u}_{17} \\ \bar{u}_{25} \\ \hdashline \bar{u}_2 \\ \bar{u}_{10} \\ \bar{u}_{18} \\ \bar{u}_{26} \\ \hdashline \bar{u}_3 \\ \bar{u}_{11} \\ \bar{u}_{19} \\ \bar{u}_{27} \\ \hdashline \bar{u}_4 \\ \bar{u}_{12} \\ \bar{u}_{20} \\ \bar{u}_{28} \\ \hdashline \bar{u}_5 \\ \bar{u}_{13} \\ \bar{u}_{21} \\ \bar{u}_{29} \\ \hdashline \bar{u}_6 \\ \bar{u}_{14} \\ \bar{u}_{22} \\ \bar{u}_{30} \\ \hdashline \bar{u}_7 \\ \bar{u}_{15} \\ \bar{u}_{23} \\ \bar{u}_{31} \\ \hdashline \bar{u}_8 \\ \bar{u}_{16} \\ \bar{u}_{24} \\ \bar{u}_{32} \end{bmatrix} = \frac{1}{8} \begin{bmatrix} u_1 + u_2 + u_3 + u_4 + u_5 + u_6 + u_7 + u_8 \\ u_9 + u_{10} + u_{11} + u_{12} + u_{13} + u_{14} + u_{15} + u_{16} \\ u_{17} + u_{18} + u_{19} + u_{20} + u_{21} + u_{22} + u_{23} + u_{24} \\ u_{25} + u_{26} + u_{27} + u_{28} + u_{29} + u_{30} + u_{31} + u_{32} \\ \hdashline u_1 + u_2 - u_3 - u_4 + u_5 + u_6 - u_7 - u_8 \\ u_9 + u_{10} - u_{11} - u_{12} + u_{13} + u_{14} - u_{15} - u_{16} \\ u_{17} + u_{18} - u_{19} - u_{20} + u_{21} + u_{22} - u_{23} - u_{24} \\ u_{25} + u_{26} - u_{27} - u_{28} + u_{29} + u_{30} - u_{31} - u_{32} \\ \hdashline u_1 - u_2 + u_3 - u_4 + u_5 - u_6 + u_7 - u_8 \\ u_9 - u_{10} + u_{11} - u_{12} + u_{13} - u_{14} + u_{15} - u_{16} \\ u_{17} - u_{18} + u_{19} - u_{20} + u_{21} - u_{22} + u_{23} - u_{24} \\ u_{25} - u_{26} + u_{27} - u_{28} + u_{29} - u_{30} + u_{31} - u_{32} \\ \hdashline u_1 - u_2 - u_3 + u_4 + u_5 - u_6 - u_7 + u_8 \\ u_9 - u_{10} - u_{11} + u_{12} + u_{13} - u_{14} - u_{15} + u_{16} \\ u_{17} - u_{18} - u_{19} + u_{20} + u_{21} - u_{22} - u_{23} + u_{24} \\ u_{25} - u_{26} - u_{27} + u_{28} + u_{29} - u_{30} - u_{31} + u_{32} \\ \hdashline u_1 + u_2 + u_3 + u_4 - u_5 - u_6 - u_7 - u_8 \\ u_9 + u_{10} + u_{11} + u_{12} - u_{13} - u_{14} - u_{15} - u_{16} \\ u_{17} + u_{18} + u_{19} + u_{20} - u_{21} - u_{22} - u_{23} - u_{24} \\ u_{25} + u_{26} + u_{27} + u_{28} - u_{29} - u_{30} - u_{31} - u_{32} \\ \hdashline u_1 + u_2 - u_3 - u_4 - u_5 - u_6 + u_7 + u_8 \\ u_9 + u_{10} - u_{11} - u_{12} - u_{13} - u_{14} + u_{15} + u_{16} \\ u_{17} + u_{18} - u_{19} - u_{20} - u_{21} - u_{22} + u_{23} + u_{24} \\ u_{25} + u_{26} - u_{27} - u_{28} - u_{29} - u_{30} + u_{31} + u_{32} \\ \hdashline u_1 - u_2 + u_3 - u_4 - u_5 + u_6 - u_7 + u_8 \\ u_9 - u_{10} + u_{11} - u_{12} - u_{13} + u_{14} - u_{15} + u_{16} \\ u_{17} - u_{18} + u_{19} - u_{20} - u_{21} + u_{22} - u_{23} + u_{24} \\ u_{25} - u_{26} + u_{27} - u_{28} - u_{29} + u_{30} - u_{31} + u_{32} \\ \hdashline u_1 - u_2 - u_3 + u_4 - u_5 + u_6 + u_7 - u_8 \\ u_9 - u_{10} - u_{11} + u_{12} - u_{13} + u_{14} + u_{15} - u_{16} \\ u_{17} - u_{18} - u_{19} + u_{20} - u_{21} + u_{22} + u_{23} - u_{24} \\ u_{25} - u_{26} - u_{27} + u_{28} - u_{29} + u_{30} + u_{31} - u_{32} \end{bmatrix}$$

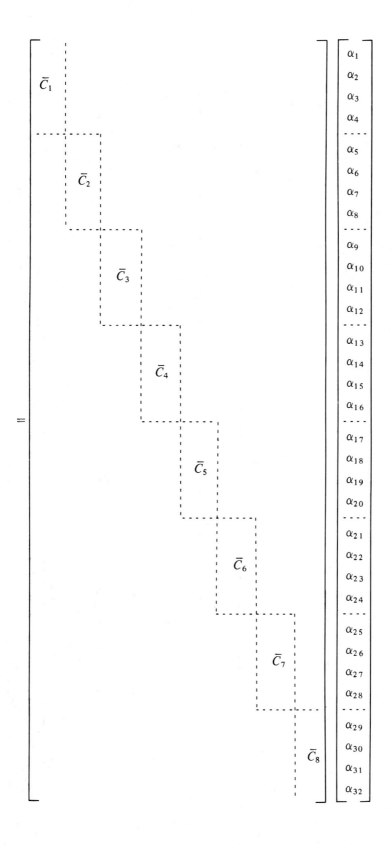

$$
\begin{bmatrix} \alpha_1 \\ \alpha_2 \\ \alpha_3 \\ \alpha_4 \\ \hdashline \alpha_5 \\ \alpha_6 \\ \alpha_7 \\ \alpha_8 \\ \hdashline \alpha_9 \\ \alpha_{10} \\ \alpha_{11} \\ \alpha_{12} \\ \hdashline \alpha_{13} \\ \alpha_{14} \\ \alpha_{15} \\ \alpha_{16} \\ \hdashline \alpha_{17} \\ \alpha_{18} \\ \alpha_{19} \\ \alpha_{20} \\ \hdashline \alpha_{21} \\ \alpha_{22} \\ \alpha_{23} \\ \alpha_{24} \\ \hdashline \alpha_{25} \\ \alpha_{26} \\ \alpha_{27} \\ \alpha_{28} \\ \hdashline \alpha_{29} \\ \alpha_{30} \\ \alpha_{31} \\ \alpha_{32} \end{bmatrix} = \mathrm{Diag}\,\frac{1}{8}\,
\begin{bmatrix} 1 \\ 1/c^2 \\ 1/d^2 \\ 1/e^2 \\ \hdashline 1/cd \\ 1/c^3 d \\ 1/d^3 c \\ 1/e^2 cd \\ \hdashline 1/ce \\ 1/c^3 e \\ 1/d^2 ce \\ 1/e^3 c \\ \hdashline 1/de \\ 1/c^2 de \\ 1/d^3 e \\ 1/e^3 d \\ \hdashline 1/cde \\ 1/c^3 de \\ 1/d^3 ce \\ 1/e^3 cd \\ \hdashline 1/e \\ 1/c^2 e \\ 1/d^2 e \\ 1/e^3 \\ \hdashline 1/d \\ 1/c^2 d \\ 1/d^3 \\ 1/e^2 d \\ \hdashline 1/c \\ 1/c^3 \\ 1/d^2 c \\ 1/e^2 c \end{bmatrix}\,\mathrm{Diag}\,
\begin{bmatrix} -19 & 9 & 9 & 9 \\ 9 & 0 & 0 & -9 \\ 9 & 0 & -9 & 0 \\ 9 & -9 & 0 & 0 \\ \hdashline -19 & 9 & 27 & 27 \\ 9 & 0 & 0 & -27 \\ 9 & 0 & -27 & 0 \\ 9 & -9 & 0 & 0 \\ \hdashline -19 & 27 & 9 & 27 \\ 9 & 0 & 0 & -27 \\ 9 & 0 & -9 & 0 \\ 9 & -27 & 0 & 0 \\ \hdashline -19 & 27 & 27 & 9 \\ 9 & 0 & 0 & -9 \\ 9 & 0 & -27 & 0 \\ 9 & -27 & 0 & 0 \\ \hdashline -19 & 27 & 27 & 27 \\ 9 & 0 & 0 & -27 \\ 9 & 0 & -27 & 0 \\ 9 & -27 & 0 & 0 \\ \hdashline -19 & 27 & 9 & 9 \\ 9 & 0 & 0 & -9 \\ 9 & 0 & -9 & 0 \\ 9 & -27 & 0 & 0 \\ \hdashline -19 & 9 & 27 & 9 \\ 9 & 0 & 0 & -9 \\ 9 & 0 & -27 & 0 \\ 9 & -9 & 0 & 0 \\ \hdashline -19 & 9 & 9 & 27 \\ 9 & 0 & 0 & -27 \\ 9 & 0 & -9 & 0 \\ 9 & -9 & 0 & 0 \end{bmatrix}
\begin{bmatrix} \bar{u}_1 \\ \bar{u}_9 \\ \bar{u}_{17} \\ \bar{u}_{25} \\ \hdashline \bar{u}_2 \\ \bar{u}_{10} \\ \bar{u}_{18} \\ \bar{u}_{26} \\ \hdashline \bar{u}_3 \\ \bar{u}_{11} \\ \bar{u}_{19} \\ \bar{u}_{27} \\ \hdashline \bar{u}_4 \\ \bar{u}_{12} \\ \bar{u}_{20} \\ \bar{u}_{28} \\ \hdashline \bar{u}_5 \\ \bar{u}_{13} \\ \bar{u}_{21} \\ \bar{u}_{29} \\ \hdashline \bar{u}_6 \\ \bar{u}_{14} \\ \bar{u}_{22} \\ \bar{u}_{30} \\ \hdashline \bar{u}_7 \\ \bar{u}_{15} \\ \bar{u}_{23} \\ \bar{u}_{31} \\ \hdashline \bar{u}_8 \\ \bar{u}_{16} \\ \bar{u}_{24} \\ \bar{u}_{32} \end{bmatrix}
$$

Sec. 4.8] Thirty-two-node rectangular hexahedral element

The thirty-two-dimensional G-vector space of the displacement field Δ, decomposed into eight four-dimensional G-invariant subspaces $U_1, U_2, U_3, U_4, U_5, U_6, U_7, U_8$, is given by

$$\Delta = T_D \bar{C}^{-1} \bar{U},$$

with

$$T_D = \text{Diag}\,[F_1 \mid F_2 \mid F_3 \mid F_4 \mid F_5 \mid F_6 \mid F_7 \mid F_8]$$

and $\xi = x/c$, $\eta = y/d$, $\zeta = z/e$, giving

$$\Delta_1 = \text{Diag}\,\left[1 \quad \frac{x^2}{c^2} \quad \frac{y^2}{d^2} \quad \frac{z^2}{e^2}\right] \frac{1}{8} \begin{bmatrix} -19 & 9 & 9 & 9 \\ 9 & 0 & 0 & -9 \\ 9 & 0 & -9 & 0 \\ 9 & -9 & 0 & 0 \end{bmatrix} \begin{bmatrix} \bar{u}_1 \\ \bar{u}_9 \\ \bar{u}_{17} \\ \bar{u}_{25} \end{bmatrix}$$

$$= \frac{1}{8} \begin{bmatrix} -19 & 9 & 9 & 9 \\ 9\xi^2 & 0 & 0 & -9\xi^2 \\ 9\eta^2 & 0 & -9\eta^2 & 0 \\ 9\zeta^2 & -9\zeta^2 & 0 & 0 \end{bmatrix} \begin{bmatrix} \bar{u}_1 \\ \bar{u}_9 \\ \bar{u}_{17} \\ \bar{u}_{25} \end{bmatrix}$$

$$\Delta_2 = \text{Diag}\,\left[\frac{xy}{cd} \quad \frac{x^3 y}{c^3 d} \quad \frac{y^3 x}{d^3 c} \quad \frac{z^2 xy}{e^2 cd}\right] \frac{1}{8} \begin{bmatrix} -19 & 9 & 27 & 27 \\ 9 & 0 & 0 & -27 \\ 9 & 0 & -27 & 0 \\ 9 & -9 & 0 & 0 \end{bmatrix} \begin{bmatrix} \bar{u}_2 \\ \bar{u}_{10} \\ \bar{u}_{18} \\ \bar{u}_{26} \end{bmatrix}$$

$$= \frac{1}{8} \begin{bmatrix} -19\xi\eta & 9\xi\eta & 27\xi\eta & 27\xi\eta \\ 9\xi^3\eta & 0 & 0 & -27\xi^3\eta \\ 9\eta^3\xi & 0 & -27\eta^3\xi & 0 \\ 9\zeta^2\xi\eta & -9\zeta^2\xi\eta & 0 & 0 \end{bmatrix} \begin{bmatrix} \bar{u}_2 \\ \bar{u}_{10} \\ \bar{u}_{18} \\ \bar{u}_{26} \end{bmatrix}$$

$$\Delta_3 = \text{Diag}\,\left[\frac{xz}{ce} \quad \frac{x^3 z}{c^3 e} \quad \frac{y^2 xz}{d^2 ce} \quad \frac{z^3 x}{e^3 c}\right] \frac{1}{8} \begin{bmatrix} -19 & 27 & 9 & 27 \\ 9 & 0 & 0 & -27 \\ 9 & 0 & -9 & 0 \\ 9 & -27 & 0 & 0 \end{bmatrix} \begin{bmatrix} \bar{u}_3 \\ \bar{u}_{11} \\ \bar{u}_{19} \\ \bar{u}_{27} \end{bmatrix}$$

$$= \frac{1}{8} \begin{bmatrix} -19\xi\zeta & 27\xi\zeta & 9\xi\zeta & 27\xi\zeta \\ 9\xi^3\zeta & 0 & 0 & -27\xi^3\zeta \\ 9\eta^2\xi\zeta & 0 & -9\eta^2\xi\zeta & 0 \\ 9\zeta^3\xi & -27\zeta^3\xi & 0 & 0 \end{bmatrix} \begin{bmatrix} \bar{u}_3 \\ \bar{u}_{11} \\ \bar{u}_{19} \\ \bar{u}_{27} \end{bmatrix}$$

$$\Delta_4 = \text{Diag} \begin{bmatrix} \dfrac{yz}{de} & \dfrac{x^2 yz}{c^2 de} & \dfrac{y^3 z}{d^3 e} & \dfrac{z^3 y}{e^3 d} \end{bmatrix} \dfrac{1}{8} \begin{bmatrix} -19 & 27 & 27 & 9 \\ 9 & 0 & 0 & -9 \\ 9 & 0 & -27 & 0 \\ 9 & -27 & 0 & 0 \end{bmatrix} \begin{bmatrix} \bar{u}_4 \\ \bar{u}_{12} \\ \bar{u}_{20} \\ \bar{u}_{28} \end{bmatrix}$$

$$= \frac{1}{8} \begin{bmatrix} -19\eta\zeta & 27\eta\zeta & 27\eta\zeta & 9\eta\zeta \\ 9\xi^2\eta\zeta & 0 & 0 & -9\xi^2\eta\zeta \\ 9\eta^3\zeta & 0 & -27\eta^3\zeta & 0 \\ 9\zeta^3\eta & -27\zeta^3\eta & 0 & 0 \end{bmatrix} \begin{bmatrix} \bar{u}_4 \\ \bar{u}_{12} \\ \bar{u}_{20} \\ \bar{u}_{28} \end{bmatrix}$$

$$\Delta_5 = \text{Diag} \begin{bmatrix} \dfrac{xyz}{cde} & \dfrac{x^3 yz}{c^3 de} & \dfrac{y^3 xz}{d^3 ce} & \dfrac{z^3 xy}{e^3 cd} \end{bmatrix} \dfrac{1}{8} \begin{bmatrix} -19 & 27 & 27 & 27 \\ 9 & 0 & 0 & -27 \\ 9 & 0 & -27 & 0 \\ 9 & -27 & 0 & 0 \end{bmatrix} \begin{bmatrix} \bar{u}_5 \\ \bar{u}_{13} \\ \bar{u}_{21} \\ \bar{u}_{29} \end{bmatrix}$$

$$= \frac{1}{8} \begin{bmatrix} -19\xi\eta\zeta & 27\xi\eta\zeta & 27\xi\eta\zeta & 27\xi\eta\zeta \\ 9\xi^3\eta\zeta & 0 & 0 & -27\xi^3\eta\zeta \\ 9\eta^3\xi\zeta & 0 & -27\eta^3\xi\zeta & 0 \\ 9\zeta^3\xi\eta & -27\zeta^3\xi\eta & 0 & 0 \end{bmatrix} \begin{bmatrix} \bar{u}_5 \\ \bar{u}_{13} \\ \bar{u}_{21} \\ \bar{u}_{29} \end{bmatrix}$$

$$\Delta_6 = \text{Diag} \begin{bmatrix} \dfrac{z}{e} & \dfrac{x^2 z}{c^2 e} & \dfrac{y^2 z}{d^2 e} & \dfrac{z^3}{e^3} \end{bmatrix} \dfrac{1}{8} \begin{bmatrix} -19 & 27 & 9 & 9 \\ 9 & 0 & 0 & -9 \\ 9 & 0 & -9 & 0 \\ 9 & -27 & 0 & 0 \end{bmatrix} \begin{bmatrix} \bar{u}_6 \\ \bar{u}_{14} \\ \bar{u}_{22} \\ \bar{u}_{30} \end{bmatrix}$$

$$= \frac{1}{8} \begin{bmatrix} -19\zeta & 27\zeta & 9\zeta & 9\zeta \\ 9\xi^2\zeta & 0 & 0 & -9\xi^2\zeta \\ 9\eta^2\zeta & 0 & -9\eta^2\zeta & 0 \\ 9\zeta^3 & -27\zeta^3 & 0 & 0 \end{bmatrix} \begin{bmatrix} \bar{u}_6 \\ \bar{u}_{14} \\ \bar{u}_{22} \\ \bar{u}_{30} \end{bmatrix}$$

Sec. 4.8]   Thirty-two-node rectangular hexahedral element   237

$$\Delta_7 = \text{Diag}\left[\frac{y}{d} \quad \frac{x^2 y}{c^2 d} \quad \frac{y^3}{d^3} \quad \frac{z^2 y}{e^2 d}\right] \frac{1}{8} \begin{bmatrix} -19 & 9 & 27 & 9 \\ 9 & 0 & 0 & -9 \\ 9 & 0 & -27 & 0 \\ 9 & -9 & 0 & 0 \end{bmatrix} \begin{bmatrix} \bar{u}_7 \\ \bar{u}_{15} \\ \bar{u}_{23} \\ \bar{u}_{31} \end{bmatrix}$$

$$= \frac{1}{8} \begin{bmatrix} -19\eta & 9\eta & 27\eta & 9\eta \\ 9\xi^2\eta & 0 & 0 & -9\xi^2\eta \\ 9\eta^3 & 0 & -27\eta^3 & 0 \\ 9\zeta^2\eta & -9\zeta^2\eta & 0 & 0 \end{bmatrix} \begin{bmatrix} \bar{u}_7 \\ \bar{u}_{15} \\ \bar{u}_{23} \\ \bar{u}_{31} \end{bmatrix}$$

$$\Delta_8 = \text{Diag}\left[\frac{x}{c} \quad \frac{x^3}{c^3} \quad \frac{y^2 x}{d^2 c} \quad \frac{z^2 x}{e^2 c}\right] \frac{1}{8} \begin{bmatrix} -19 & 9 & 9 & 27 \\ 9 & 0 & 0 & -27 \\ 9 & 0 & -9 & 0 \\ 9 & -9 & 0 & 0 \end{bmatrix} \begin{bmatrix} \bar{u}_8 \\ \bar{u}_{16} \\ \bar{u}_{24} \\ \bar{u}_{32} \end{bmatrix}$$

$$= \frac{1}{8} \begin{bmatrix} -19\xi & 9\xi & 9\xi & 27\xi \\ 9\xi^3 & 0 & 0 & -27\xi^3 \\ 9\eta^2\xi & 0 & -9\eta^2\xi & 0 \\ 9\zeta^2\xi & -9\zeta^2\xi & 0 & 0 \end{bmatrix} \begin{bmatrix} \bar{u}_8 \\ \bar{u}_{16} \\ \bar{u}_{24} \\ \bar{u}_{32} \end{bmatrix}$$

or $\Delta = \bar{N}_\Delta \bar{U}$, and expressed in supermatrix form this is

$$\begin{bmatrix} \Delta_1 \\ \Delta_2 \\ \Delta_3 \\ \Delta_4 \\ \Delta_5 \\ \Delta_6 \\ \Delta_7 \\ \Delta_8 \end{bmatrix} = \begin{bmatrix} \bar{N}_{\Delta_1} & & & & & & & \\ & \bar{N}_{\Delta_2} & & & & & & \\ & & \bar{N}_{\Delta_3} & & & & & \\ & & & \bar{N}_{\Delta_4} & & & & \\ & & & & \bar{N}_{\Delta_5} & & & \\ & & & & & \bar{N}_{\Delta_6} & & \\ & & & & & & \bar{N}_{\Delta_7} & \\ & & & & & & & \bar{N}_{\Delta_8} \end{bmatrix} \begin{bmatrix} \bar{U}^{(1)} \\ \bar{U}^{(2)} \\ \bar{U}^{(3)} \\ \bar{U}^{(4)} \\ \bar{U}^{(5)} \\ \bar{U}^{(6)} \\ \bar{U}^{(7)} \\ \bar{U}^{(8)} \end{bmatrix}.$$

The shape functions $\bar{N}$ of G-invariant subspaces $U_1, U_2, U_3, U_4, U_4, U_5, U_6, U_7, U_8$ are obtained by

$$\bar{N} = S\bar{N}_\Delta,$$

with

238   Formulation of shape functions in G-invariant subspaces   [Ch. 4

$$S = [1\ 1\ 1\ 1\ \vdots\ 1\ 1\ 1\ 1\ \vdots\ 1\ 1\ 1\ 1\ \vdots\ 1\ 1\ 1\ 1\ \vdots\ 1\ 1\ 1\ 1$$
$$\vdots\ 1\ 1\ 1\ 1\ \vdots\ 1\ 1\ 1\ 1\ \vdots\ 1\ 1\ 1\ 1]$$

Subspace U_1
$$\begin{bmatrix} \bar{N}_1 \\ \bar{N}_9 \\ \bar{N}_{17} \\ \bar{N}_{25} \end{bmatrix} = \frac{1}{8} \begin{bmatrix} -19 + 9\xi^2 + 9\eta^2 + 9\zeta^2 \\ 9 - 9\zeta^2 \\ 9 - 9\eta^2 \\ 9 - 9\xi^2 \end{bmatrix}$$

$$= \frac{1}{8} \begin{bmatrix} -19 & 9 & 9 & 9 \\ 9 & 0 & 0 & -9 \\ 9 & 0 & -9 & 0 \\ 9 & -9 & 0 & 0 \end{bmatrix} \begin{bmatrix} 1 \\ \xi^2 \\ \eta^2 \\ \zeta^2 \end{bmatrix}$$

Subspace U_2
$$\begin{bmatrix} \bar{N}_2 \\ \bar{N}_{10} \\ \bar{N}_{18} \\ \bar{N}_{26} \end{bmatrix} = \frac{1}{8} \begin{bmatrix} -19\xi\eta + 9\xi^3\eta + 9\eta^3\xi + 9\zeta^2\xi\eta \\ 9\xi\eta - 9\zeta^2\xi\eta \\ 27\xi\eta - 27\eta^3\xi \\ 27\xi\eta - 27\xi^3\eta \end{bmatrix}$$

$$= \frac{1}{8} \begin{bmatrix} -19 & 9 & 9 & 9 \\ 9 & 0 & 0 & -9 \\ 27 & 0 & -27 & 0 \\ 27 & -27 & 0 & 0 \end{bmatrix} \begin{bmatrix} \xi\eta \\ \xi^3\eta \\ \eta^3\xi \\ \zeta^2\xi\eta \end{bmatrix}$$

Subspace U_3
$$\begin{bmatrix} \bar{N}_3 \\ \bar{N}_{11} \\ \bar{N}_{19} \\ \bar{N}_{27} \end{bmatrix} = \frac{1}{8} \begin{bmatrix} -19\xi\zeta + 9\xi^3\zeta + 9\eta^2\xi\zeta + 9\zeta^3\xi \\ 27\xi\zeta - 27\zeta^3\xi \\ 9\xi\zeta - 9\eta^2\xi\zeta \\ 27\xi\zeta - 27\xi^3\zeta \end{bmatrix}$$

$$= \frac{1}{8} \begin{bmatrix} -19 & 9 & 9 & 9 \\ 27 & 0 & 0 & -27 \\ 9 & 0 & -9 & 0 \\ 27 & -27 & 0 & 0 \end{bmatrix} \begin{bmatrix} \xi\zeta \\ \xi^3\zeta \\ \eta^2\xi\zeta \\ \zeta^3\xi \end{bmatrix}$$

Subspace U_4
$$\begin{bmatrix} \bar{N}_4 \\ \bar{N}_{12} \\ \bar{N}_{20} \\ \bar{N}_{28} \end{bmatrix} = \frac{1}{8} \begin{bmatrix} -19\eta\zeta + 9\xi^2\eta\zeta + 9\eta^3\zeta + 9\zeta^3\eta \\ 27\eta\zeta - 27\xi^3\eta \\ 27\eta\zeta - 27\zeta^2\eta \\ 9\eta\zeta - 9\xi^2\eta\zeta \end{bmatrix}$$

$$= \frac{1}{8} \begin{bmatrix} -19 & 9 & 9 & 9 \\ 27 & 0 & 0 & -27 \\ 27 & 0 & -27 & 0 \\ 9 & -9 & 0 & 0 \end{bmatrix} \begin{bmatrix} \eta\zeta \\ \xi^2\eta\zeta \\ \eta^3\zeta \\ \xi^3\eta \end{bmatrix}$$

Subspace U_5
$$\begin{bmatrix} \bar{N}_5 \\ \bar{N}_{13} \\ \bar{N}_{21} \\ \bar{N}_{29} \end{bmatrix} = \frac{1}{8} \begin{bmatrix} -19\xi\eta\zeta + 9\xi^3\eta\zeta + 9\eta^3\xi\zeta + 9\zeta^3\xi\eta \\ 27\xi\eta\zeta - 27\zeta^3\xi\eta \\ 27\xi\eta\zeta - 27\eta^3\xi\zeta \\ 27\xi\eta\zeta - 27\xi^3\eta\zeta \end{bmatrix}$$

$$= \frac{1}{8} \begin{bmatrix} -19 & 9 & 9 & 9 \\ 27 & 0 & 0 & -27 \\ 27 & 0 & -27 & 0 \\ 27 & -27 & 0 & 0 \end{bmatrix} \begin{bmatrix} \xi\eta\zeta \\ \xi^3\eta\zeta \\ \eta^3\xi\zeta \\ \zeta^3\xi\eta \end{bmatrix}$$

Subspace U_6
$$\begin{bmatrix} \bar{N}_6 \\ \bar{N}_{14} \\ \bar{N}_{22} \\ \bar{N}_{30} \end{bmatrix} = \frac{1}{8} \begin{bmatrix} -19\zeta + 9\xi^2\zeta + 9\eta^2\zeta + 9\zeta^3 \\ 27\zeta - 27\zeta^3 \\ 9\zeta - 9\eta^2\zeta \\ 9\zeta - 9\xi^2\zeta \end{bmatrix}$$

$$= \frac{1}{8} \begin{bmatrix} -19 & 9 & 9 & 9 \\ 27 & 0 & 0 & -27 \\ 9 & 0 & -9 & 0 \\ 9 & -9 & 0 & 0 \end{bmatrix} \begin{bmatrix} \zeta \\ \xi^2\zeta \\ \eta^2\zeta \\ \zeta^3 \end{bmatrix}$$

Subspace U_7

$$\begin{bmatrix} \bar{N}_7 \\ \bar{N}_{15} \\ \bar{N}_{23} \\ \bar{N}_{31} \end{bmatrix} = \frac{1}{8} \begin{bmatrix} -19\eta + 9\xi^2\eta + 9\eta^3 + 9\zeta^2\eta \\ 9\eta - 9\zeta^2\eta \\ 27\eta - 27\eta^3 \\ 9\eta - 9\xi^2\eta \end{bmatrix}$$

$$= \frac{1}{8} \begin{bmatrix} -19 & 9 & 9 & 9 \\ 9 & 0 & 0 & -9 \\ 27 & 0 & -27 & 0 \\ 9 & -9 & 0 & 0 \end{bmatrix} \begin{bmatrix} \eta \\ \xi^2\eta \\ \eta^3 \\ \zeta^2\eta \end{bmatrix}$$

Subspace U_8

$$\begin{bmatrix} \bar{N}_8 \\ \bar{N}_{16} \\ \bar{N}_{24} \\ \bar{N}_{32} \end{bmatrix} = \frac{1}{8} \begin{bmatrix} -19\xi + 9\xi^3 + 9\eta^2\xi + 9\zeta^2\xi \\ 9\xi - 9\zeta^2\xi \\ 9\xi - 9\eta^2\xi \\ 27\xi - 27\xi^3 \end{bmatrix}$$

$$= \frac{1}{8} \begin{bmatrix} -19 & 9 & 9 & 9 \\ 9 & 0 & 0 & -9 \\ 9 & 0 & -9 & 0 \\ 27 & -27 & 0 & 0 \end{bmatrix} \begin{bmatrix} \xi \\ \xi^3 \\ \eta^2\xi \\ \zeta^2\xi \end{bmatrix}.$$

These shape functions $\bar{N}$ can be written in supermatrix form:

$$\begin{bmatrix} \bar{N}^{(1)} \\ \bar{N}^{(2)} \\ \bar{N}^{(3)} \\ \bar{N}^{(4)} \\ \bar{N}^{(5)} \\ \bar{N}^{(6)} \\ \bar{N}^{(7)} \\ \bar{N}^{(8)} \end{bmatrix} = \begin{bmatrix} A & & & & & & & \\ & B & & & & & & \\ & & C & & & & & \\ & & & D & & & & \\ & & & & F & & & \\ & & & & & G & & \\ & & & & & & H & \\ & & & & & & & I \end{bmatrix} \begin{bmatrix} F_1 \\ F_2 \\ F_3 \\ F_4 \\ F_5 \\ F_6 \\ F_7 \\ F_8 \end{bmatrix}$$

and transformed by using the expression that relates the nodal functions Φ to the basis vectors $\bar{\Phi}$ (derived earlier):

$$\begin{bmatrix} E & E & E & E & E & E & E & E \\ E & E & -E & -E & E & E & -E & -E \\ E & -E & E & -E & E & -E & E & -E \\ E & -E & -E & E & E & -E & -E & E \\ E & E & E & E & -E & -E & -E & -E \\ E & E & -E & -E & -E & -E & E & E \\ E & -E & E & -E & -E & E & -E & E \\ E & -E & -E & E & -E & E & E & -E \end{bmatrix} \begin{bmatrix} \bar{N}^{(1)} \\ \bar{N}^{(2)} \\ \bar{N}^{(3)} \\ \bar{N}^{(4)} \\ \bar{N}^{(5)} \\ \bar{N}^{(6)} \\ \bar{N}^{(7)} \\ \bar{N}^{(8)} \end{bmatrix}$$

$$= \begin{bmatrix} E & E & E & E & E & E & E & E \\ E & E & -E & -E & E & E & -E & -E \\ E & -E & E & -E & E & -E & E & -E \\ E & -E & -E & E & E & -E & -E & E \\ E & E & E & E & -E & -E & -E & -E \\ E & E & -E & -E & -E & -E & E & E \\ E & -E & E & -E & -E & E & -E & E \\ E & -E & -E & E & -E & E & E & -E \end{bmatrix} \text{Diag} \begin{bmatrix} A \\ B \\ C \\ D \\ F \\ G \\ H \\ I \end{bmatrix} \begin{bmatrix} F_1 \\ F_2 \\ F_3 \\ F_4 \\ F_5 \\ F_6 \\ F_7 \\ F_8 \end{bmatrix}$$

into the nodal shape functions N:

$$\begin{bmatrix} N^{(1)} \\ N^{(2)} \\ N^{(3)} \\ N^{(4)} \\ N^{(5)} \\ N^{(6)} \\ N^{(7)} \\ N^{(8)} \end{bmatrix} = \begin{bmatrix} A & B & C & D & F & G & H & I \\ A & B & -C & -D & F & G & -H & -I \\ A & -B & C & -D & F & -G & H & -I \\ A & -B & -C & D & F & -G & -H & I \\ A & B & C & D & -F & -G & -H & -I \\ A & B & -C & -D & -F & -G & H & I \\ A & -B & C & -D & -F & G & -H & I \\ A & -B & -C & D & -F & G & H & -I \end{bmatrix} \begin{bmatrix} F_1 \\ F_2 \\ F_3 \\ F_4 \\ F_5 \\ F_6 \\ F_7 \\ F_8 \end{bmatrix}.$$

4.9 SIXTY-FOUR-NODE RECTANGULAR HEXAHEDRAL ELEMENT

The sixty-four-node rectangular hexahedral element, shown in Fig. 4.9, is described by the group D_{2h} and it has eight nodal sets

$S_1(1, 2, 3, 4, 5, 6, 7, 8)$

$S_2(9, 10, 11, 12, 13, 14, 15, 16)$

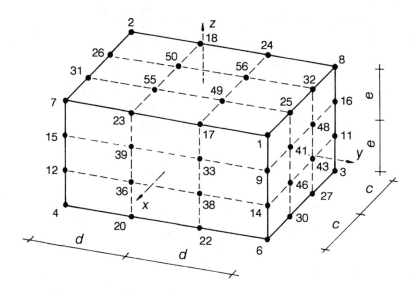

Fig. 4.9. The sixty-four-node rectangular hexahedral element with the nodal numbering of the group supermatrix procedure.

$S_3(17, 18, 19, 20, 21, 22, 23, 24)$

$S_4(25, 26, 27, 28, 29, 30, 31, 32)$

$S_5(33, 34, 35, 36, 37, 38, 39, 40)$

$S_6(41, 42, 43, 44, 45, 46, 47, 48)$

$S_7(49, 50, 51, 52, 53, 54, 55, 56)$

$S_8(57, 58, 59, 60, 61, 62, 63, 64)$.

The unique nodal numbering for the eight-node rectangular hexahedral element, derived in section 3.8, is applied here for the above nodal sets. Each set contains nodes that are permuted by action of symmetry operations of the group. The first node of each set lies in the first octant, the second nodes etc. in their respective octants. Nodes of each set are corner nodes of a rectangular hexahedron.

The coordinates of the nodes are given in Table 4.5.

The character table of the group D_{2h} (with Cartesian sets and products) and the idempotents of the centre of the group algebra, given previously in section 4.6, are used here also for derivation of basis vectors $\bar{\varphi}_j$ ($j = 1, 2, \ldots, 64$) of G-invariant subspaces U_i, which is performed by application of the idempotents π_i to the nodal functions $\varphi_1, \varphi_2, \ldots, \varphi_{64}$:

Sixty-four-node rectangular hexahedral element

Table 4.5. Nodal coordinates of the 64-node rectangular hexahedral element with the unique group supermatrix nodal numbering

Set S_1	1	2	3	4	5	6	7	8
x	c	$-c$	$-c$	c	$-c$	c	c	$-c$
y	d	$-d$	d	$-d$	$-d$	d	$-d$	d
z	e	e	$-e$	$-e$	$-e$	$-e$	e	e
Set S_2	9	10	11	12	13	14	15	16
x	c	$-c$	$-c$	c	$-c$	c	c	$-c$
y	d	$-d$	d	$-d$	$-d$	d	$-d$	d
z	$e/3$	$e/3$	$-e/3$	$-e/3$	$-e/3$	$-e/3$	$e/3$	$e/3$
Set S_3	17	18	19	20	21	22	23	24
x	c	$-c$	$-c$	c	$-c$	c	c	$-c$
y	$d/3$	$-d/3$	$d/3$	$-d/3$	$-d/3$	$d/3$	$-d/3$	$d/3$
z	e	e	$-e$	$-e$	$-e$	$-e$	e	e
Set S_4	25	26	27	28	29	30	31	32
x	$c/3$	$-c/3$	$-c/3$	$c/3$	$-c/3$	$c/3$	$c/3$	$-c/3$
y	d	$-d$	d	$-d$	$-d$	d	$-d$	d
z	e	e	$-e$	$-e$	$-e$	$-e$	e	e
Set S_5	33	34	35	36	37	38	39	40
x	c	$-c$	$-c$	c	$-c$	c	c	$-c$
y	$d/33$	$-d/3$	$d/3$	$-d/3$	$-d/3$	$d/3$	$-d/3$	$d/3$
z	$e/3$	$e/3$	$-e/3$	$-e/3$	$-e/3$	$-e/3$	$e/3$	$e/3$
Set S_6	41	42	43	44	45	46	47	48
x	$c/3$	$-c/3$	$-c/3$	$c/3$	$-c/3$	$c/3$	$c/3$	$-c/3$
y	d	$-d$	d	$-d$	$-d$	d	$-d$	d
z	$e/3$	$e/3$	$-e/3$	$-e/3$	$-e/3$	$-e/3$	$e/3$	$e/3$

Table 4.5 (continued)

Set S_7	49	50	51	52	53	54	55	56
x	$c/3$	$-c/3$	$-c/3$	$c/3$	$-c/3$	$c/3$	$c/3$	$-c/3$
y	$d/3$	$-d/3$	$d/3$	$-d/3$	$-d/3$	$d/3$	$-d/3$	$d/3$
z	e	e	$-e$	$-e$	$-e$	$-e$	e	e
Set S_8	57	58	59	60	61	62	63	64
x	$c/3$	$-c/3$	$-c/3$	$c/3$	$-c/3$	$c/3$	$c/3$	$-c/3$
y	$d/3$	$-d/3$	$d/3$	$-d/3$	$-d/3$	$d/3$	$-d/3$	$d/3$
z	$e/3$	$e/3$	$-e/3$	$-e/3$	$-e/3$	$-e/3$	$e/3$	$e/3$

$$U_1: \bar{\varphi}_1 = \tfrac{1}{8}(\varphi_1 + \varphi_2 + \varphi_3 + \varphi_4 + \varphi_5 + \varphi_6 + \varphi_7 + \varphi_8)$$

$$\bar{\varphi}_9 = \tfrac{1}{8}(\varphi_9 + \varphi_{10} + \varphi_{11} + \varphi_{12} + \varphi_{13} + \varphi_{14} + \varphi_{15} + \varphi_{16})$$

$$\bar{\varphi}_{17} = \tfrac{1}{8}(\varphi_{17} + \varphi_{18} + \varphi_{19} + \varphi_{20} + \varphi_{21} + \varphi_{22} + \varphi_{23} + \varphi_{24})$$

$$\bar{\varphi}_{25} = \tfrac{1}{8}(\varphi_{25} + \varphi_{26} + \varphi_{27} + \varphi_{28} + \varphi_{29} + \varphi_{30} + \varphi_{31} + \varphi_{32})$$

$$\bar{\varphi}_{33} = \tfrac{1}{8}(\varphi_{33} + \varphi_{34} + \varphi_{35} + \varphi_{36} + \varphi_{37} + \varphi_{38} + \varphi_{39} + \varphi_{40})$$

$$\bar{\varphi}_{41} = \tfrac{1}{8}(\varphi_{41} + \varphi_{42} + \varphi_{43} + \varphi_{44} + \varphi_{45} + \varphi_{46} + \varphi_{47} + \varphi_{48})$$

$$\bar{\varphi}_{49} = \tfrac{1}{8}(\varphi_{49} + \varphi_{50} + \varphi_{51} + \varphi_{52} + \varphi_{53} + \varphi_{54} + \varphi_{55} + \varphi_{56})$$

$$\bar{\varphi}_{57} = \tfrac{1}{8}(\varphi_{57} + \varphi_{58} + \varphi_{59} + \varphi_{60} + \varphi_{61} + \varphi_{62} + \varphi_{63} + \varphi_{64})$$

$$U_2: \bar{\varphi}_2 = \tfrac{1}{8}(\varphi_1 + \varphi_2 - \varphi_3 - \varphi_4 + \varphi_5 + \varphi_6 - \varphi_7 - \varphi_8)$$

$$\bar{\varphi}_{10} = \tfrac{1}{8}(\varphi_9 + \varphi_{10} - \varphi_{11} - \varphi_{12} + \varphi_{13} + \varphi_{14} - \varphi_{15} - \varphi_{16})$$

$$\bar{\varphi}_{18} = \tfrac{1}{8}(\varphi_{17} + \varphi_{18} - \varphi_{19} - \varphi_{20} + \varphi_{21} + \varphi_{22} - \varphi_{23} - \varphi_{24})$$

$$\bar{\varphi}_{26} = \tfrac{1}{8}(\varphi_{25} + \varphi_{26} - \varphi_{27} - \varphi_{28} + \varphi_{29} + \varphi_{30} - \varphi_{31} - \varphi_{32})$$

$$\bar{\varphi}_{34} = \tfrac{1}{8}(\varphi_{33} + \varphi_{34} - \varphi_{35} - \varphi_{36} + \varphi_{37} + \varphi_{38} - \varphi_{39} - \varphi_{40})$$

$$\bar{\varphi}_{42} = \tfrac{1}{8}(\varphi_{41} + \varphi_{42} - \varphi_{43} - \varphi_{44} + \varphi_{45} + \varphi_{46} - \varphi_{47} - \varphi_{48})$$

$$\bar{\varphi}_{50} = \tfrac{1}{8}(\varphi_{49} + \varphi_{50} - \varphi_{51} - \varphi_{52} + \varphi_{53} + \varphi_{54} - \varphi_{55} - \varphi_{56})$$

$$\bar{\varphi}_{58} = \tfrac{1}{8}(\varphi_{57} + \varphi_{58} - \varphi_{59} - \varphi_{60} + \varphi_{61} + \varphi_{62} - \varphi_{63} - \varphi_{64}),$$

and similarly for the subspaces U_3, U_4, U_5, U_6, U_7, U_8.

Thus, the sixty-four-dimensional G-vector space is decomposed into eight eight-dimensional G-invariant subspaces U_1, U_2, U_3, U_4, U_5, U_6, U_7, U_8 spanned by the respective sets of basis vectors $\bar{\Phi}^{(1)}$, $\bar{\Phi}^{(2)}$, $\bar{\Phi}^{(3)}$, $\bar{\Phi}^{(4)}$, $\bar{\Phi}^{(5)}$, $\bar{\Phi}^{(6)}$, $\bar{\Phi}^{(7)}$, $\bar{\Phi}^{(8)}$:

Sec. 4.9] Sixty-four-node rectangular hexahedral element 245

$$
\begin{bmatrix} \bar{\Phi}^{(1)} \\ \bar{\Phi}^{(2)} \\ \bar{\Phi}^{(3)} \\ \bar{\Phi}^{(4)} \\ \bar{\Phi}^{(5)} \\ \bar{\Phi}^{(6)} \\ \bar{\Phi}^{(7)} \\ \bar{\Phi}^{(8)} \end{bmatrix}^{T} = \begin{bmatrix} \bar{\varphi}_1 & \bar{\varphi}_2 & \bar{\varphi}_3 & \bar{\varphi}_4 & \bar{\varphi}_5 & \bar{\varphi}_6 & \bar{\varphi}_7 & \bar{\varphi}_8 \\ \bar{\varphi}_9 & \bar{\varphi}_{10} & \bar{\varphi}_{11} & \bar{\varphi}_{12} & \bar{\varphi}_{13} & \bar{\varphi}_{14} & \bar{\varphi}_{15} & \bar{\varphi}_{16} \\ \bar{\varphi}_{17} & \bar{\varphi}_{18} & \bar{\varphi}_{19} & \bar{\varphi}_{20} & \bar{\varphi}_{21} & \bar{\varphi}_{22} & \bar{\varphi}_{23} & \bar{\varphi}_{24} \\ \bar{\varphi}_{25} & \bar{\varphi}_{26} & \bar{\varphi}_{27} & \bar{\varphi}_{28} & \bar{\varphi}_{29} & \bar{\varphi}_{30} & \bar{\varphi}_{31} & \bar{\varphi}_{32} \\ \bar{\varphi}_{33} & \bar{\varphi}_{34} & \bar{\varphi}_{35} & \bar{\varphi}_{36} & \bar{\varphi}_{37} & \bar{\varphi}_{38} & \bar{\varphi}_{39} & \bar{\varphi}_{40} \\ \bar{\varphi}_{41} & \bar{\varphi}_{42} & \bar{\varphi}_{43} & \bar{\varphi}_{44} & \bar{\varphi}_{45} & \bar{\varphi}_{46} & \bar{\varphi}_{47} & \bar{\varphi}_{48} \\ \bar{\varphi}_{49} & \bar{\varphi}_{50} & \bar{\varphi}_{51} & \bar{\varphi}_{52} & \bar{\varphi}_{53} & \bar{\varphi}_{54} & \bar{\varphi}_{55} & \bar{\varphi}_{56} \\ \bar{\varphi}_{57} & \bar{\varphi}_{58} & \bar{\varphi}_{59} & \bar{\varphi}_{60} & \bar{\varphi}_{61} & \bar{\varphi}_{62} & \bar{\varphi}_{63} & \bar{\varphi}_{64} \end{bmatrix}.
$$

The sets $\bar{\Phi}_1, \bar{\Phi}_2, \bar{\Phi}_3, \bar{\Phi}_4, \bar{\Phi}_5, \bar{\Phi}_6, \bar{\Phi}_7, \bar{\Phi}_8$ of basis vectors $\bar{\varphi}_j$ ($j = 1, 2, \ldots, 64$), ordered according to the numbering of the nodes in the nodal sets $S_1, S_2, S_3, S_4, S_5, S_6, S_7, S_8$, are given by

$$
\begin{bmatrix} \bar{\Phi}_1 \\ \bar{\Phi}_2 \\ \bar{\Phi}_3 \\ \bar{\Phi}_4 \\ \bar{\Phi}_5 \\ \bar{\Phi}_6 \\ \bar{\Phi}_7 \\ \bar{\Phi}_8 \end{bmatrix}^{T} = \begin{bmatrix} \bar{\varphi}_1 & \bar{\varphi}_9 & \bar{\varphi}_{17} & \bar{\varphi}_{25} & \bar{\varphi}_{33} & \bar{\varphi}_{41} & \bar{\varphi}_{49} & \bar{\varphi}_{57} \\ \bar{\varphi}_2 & \bar{\varphi}_{10} & \bar{\varphi}_{18} & \bar{\varphi}_{26} & \bar{\varphi}_{34} & \bar{\varphi}_{42} & \bar{\varphi}_{50} & \bar{\varphi}_{58} \\ \bar{\varphi}_3 & \bar{\varphi}_{11} & \bar{\varphi}_{19} & \bar{\varphi}_{27} & \bar{\varphi}_{35} & \bar{\varphi}_{43} & \bar{\varphi}_{51} & \bar{\varphi}_{59} \\ \bar{\varphi}_4 & \bar{\varphi}_{12} & \bar{\varphi}_{20} & \bar{\varphi}_{28} & \bar{\varphi}_{36} & \bar{\varphi}_{44} & \bar{\varphi}_{52} & \bar{\varphi}_{60} \\ \bar{\varphi}_5 & \bar{\varphi}_{13} & \bar{\varphi}_{21} & \bar{\varphi}_{29} & \bar{\varphi}_{37} & \bar{\varphi}_{45} & \bar{\varphi}_{53} & \bar{\varphi}_{61} \\ \bar{\varphi}_6 & \bar{\varphi}_{14} & \bar{\varphi}_{22} & \bar{\varphi}_{30} & \bar{\varphi}_{38} & \bar{\varphi}_{46} & \bar{\varphi}_{54} & \bar{\varphi}_{62} \\ \bar{\varphi}_7 & \bar{\varphi}_{15} & \bar{\varphi}_{23} & \bar{\varphi}_{31} & \bar{\varphi}_{39} & \bar{\varphi}_{47} & \bar{\varphi}_{55} & \bar{\varphi}_{63} \\ \bar{\varphi}_8 & \bar{\varphi}_{16} & \bar{\varphi}_{24} & \bar{\varphi}_{32} & \bar{\varphi}_{40} & \bar{\varphi}_{48} & \bar{\varphi}_{56} & \bar{\varphi}_{64} \end{bmatrix}.
$$

The relation of the sets of basis vectors $\bar{\Phi}_l$ to the sets of nodal functions Φ_l ($l = 1, 2, \ldots, 8$) is expressed by the following system of equations, with the supermatrix in diagonal form containing eight transformation matrices T of the group D_{2h}

$$
\begin{bmatrix} \bar{\Phi}_1 \\ \bar{\Phi}_2 \\ \bar{\Phi}_3 \\ \bar{\Phi}_4 \\ \bar{\Phi}_5 \\ \bar{\Phi}_6 \\ \bar{\Phi}_7 \\ \bar{\Phi}_8 \end{bmatrix} = \begin{bmatrix} T & & & & & & & \\ & T & & & & & & \\ & & T & & & & & \\ & & & T & & & & \\ & & & & T & & & \\ & & & & & T & & \\ & & & & & & T & \\ & & & & & & & T \end{bmatrix} \begin{bmatrix} \Phi_1 \\ \Phi_2 \\ \Phi_3 \\ \Phi_4 \\ \Phi_5 \\ \Phi_6 \\ \Phi_7 \\ \Phi_8 \end{bmatrix} \quad \text{or} \quad \bar{\Phi} = \bar{T}\Phi,
$$

with

$$T = \frac{1}{8}\begin{bmatrix} 1 & 1 & 1 & 1 & 1 & 1 & 1 & 1 \\ 1 & 1 & -1 & -1 & 1 & 1 & -1 & -1 \\ 1 & -1 & 1 & -1 & 1 & -1 & 1 & -1 \\ 1 & -1 & -1 & 1 & 1 & -1 & -1 & 1 \\ 1 & 1 & 1 & 1 & -1 & -1 & -1 & -1 \\ 1 & 1 & -1 & -1 & -1 & -1 & 1 & 1 \\ 1 & -1 & 1 & -1 & -1 & 1 & -1 & 1 \\ 1 & -1 & -1 & 1 & -1 & 1 & 1 & -1 \end{bmatrix}$$

and

$$\begin{bmatrix} \bar{\Phi}_1 \\ \bar{\Phi}_2 \\ \bar{\Phi}_3 \\ \bar{\Phi}_4 \\ \bar{\Phi}_5 \\ \bar{\Phi}_6 \\ \bar{\Phi}_7 \\ \bar{\Phi}_8 \end{bmatrix}^T = \begin{bmatrix} \varphi_1 & \varphi_9 & \varphi_{17} & \varphi_{25} & \varphi_{33} & \varphi_{41} & \varphi_{49} & \varphi_{57} \\ \varphi_2 & \varphi_{10} & \varphi_{18} & \varphi_{26} & \varphi_{34} & \varphi_{42} & \varphi_{50} & \varphi_{58} \\ \varphi_3 & \varphi_{11} & \varphi_{19} & \varphi_{27} & \varphi_{35} & \varphi_{43} & \varphi_{51} & \varphi_{59} \\ \varphi_4 & \varphi_{12} & \varphi_{20} & \varphi_{28} & \varphi_{36} & \varphi_{44} & \varphi_{52} & \varphi_{60} \\ \varphi_5 & \varphi_{13} & \varphi_{21} & \varphi_{29} & \varphi_{37} & \varphi_{45} & \varphi_{53} & \varphi_{61} \\ \varphi_6 & \varphi_{14} & \varphi_{22} & \varphi_{30} & \varphi_{38} & \varphi_{46} & \varphi_{54} & \varphi_{62} \\ \varphi_7 & \varphi_{15} & \varphi_{23} & \varphi_{31} & \varphi_{39} & \varphi_{47} & \varphi_{55} & \varphi_{63} \\ \varphi_8 & \varphi_{16} & \varphi_{24} & \varphi_{32} & \varphi_{40} & \varphi_{48} & \varphi_{56} & \varphi_{64} \end{bmatrix}$$

Conversely, the relation of the sets of nodal functions Φ_l to the sets of basis vectors $\bar{\Phi}_l$ is

$$\begin{bmatrix} \Phi_1 \\ \Phi_2 \\ \Phi_3 \\ \Phi_4 \\ \Phi_5 \\ \Phi_6 \\ \Phi_7 \\ \Phi_8 \end{bmatrix} = 8 \begin{bmatrix} T & & & & & & & \\ & T & & & & & & \\ & & T & & & & & \\ & & & T & & & & \\ & & & & T & & & \\ & & & & & T & & \\ & & & & & & T & \\ & & & & & & & T \end{bmatrix} \begin{bmatrix} \bar{\Phi}_1 \\ \bar{\Phi}_2 \\ \bar{\Phi}_3 \\ \bar{\Phi}_4 \\ \bar{\Phi}_5 \\ \bar{\Phi}_6 \\ \bar{\Phi}_7 \\ \bar{\Phi}_8 \end{bmatrix} \quad \text{or} \quad \Phi = 8\bar{T}\bar{\Phi},$$

since $T^{-1} = 8T$.

The 64 terms for the u-, v- or w-displacement function, as given in Yang (1986), result from the product

$$(a_1 + a_2\xi + a_3\xi^2 + a_4\xi^3)(b_1 + b_2\eta + b_3\eta^2 + b_4\eta^3)(c_1 + c_2\zeta + c_3\zeta^2 + c_4\zeta^3).$$

Sec. 4.9] Sixty-four-node rectangular hexahedral element 247

For the group supermatrix procedure the order of the 64 terms in the polynomial of the displacement function will be determined according to the pertinence of the terms to the symmetry types of the group D_{2h}:

$$\begin{bmatrix} F_1 \\ F_2 \\ F_3 \\ F_4 \\ F_5 \\ F_6 \\ F_7 \\ F_8 \end{bmatrix} = \begin{bmatrix} 1 \\ xy \\ xz \\ yz \\ xyz \\ z \\ y \\ x \end{bmatrix} \begin{bmatrix} 1 & x^2 & y^2 & z^2 & x^2y^2z^2 & x^2y^2 & x^2z^2 & y^2z^2 \end{bmatrix}$$

$$= \begin{bmatrix} 1 & x^2 & y^2 & z^2 & x^2y^2z^2 & x^2y^2 & x^2z^2 & y^2z^2 \\ xy & x^3y & y^3x & z^2xy & x^3y^3z^2 & x^3y^3 & x^3yz^2 & xy^3z^2 \\ xz & x^3z & y^2xz & z^3x & x^3y^2z^3 & x^3y^2z & x^3z^3 & xy^2z^3 \\ yz & x^2yz & y^3z & z^3y & x^2y^3z^3 & x^2y^3z & x^2yz^3 & y^3z^3 \\ xyz & x^3yz & y^3xz & z^3xy & x^3y^3z^3 & x^3y^3z & x^3yz^3 & xy^3z^3 \\ z & x^2z & y^2z & z^3 & x^2y^2z^3 & x^2y^2z & x^2z^3 & y^2z^3 \\ y & x^2y & y^3 & z^2y & x^2y^3z^2 & x^2y^3 & x^2yz^2 & y^3z^2 \\ x & x^3 & y^2x & z^2x & x^3y^2z^2 & x^3y^2 & x^3z^2 & xy^2z^2 \end{bmatrix}.$$

The column vector and the row vector are Cartesian sets and products as they stand in addition to the character table of the group D_{2h} according to the pertinence of the terms to the symmetry types of group representations A_g, B_{1g}, B_{2g}, B_{3g}, A_u, B_{1u}, B_{2u}, B_{3u}.

Consequently, the polynomial expression of the displacement function

$$u(x,y,z) = [1 \quad x^2 \quad y^2 \quad z^2 \quad x^2y^2z^2 \quad x^2y^2 \quad x^2z^2 \quad y^2z^2$$
$$xy \quad x^3y \quad y^3x \quad z^2xy \quad x^3y^3z^2 \quad x^3y^3 \quad x^3yz^2 \quad xy^3z^2$$
$$xz \quad x^3z \quad y^2xz \quad z^3x \quad x^3y^2z^3 \quad x^3y^2z \quad x^3z^3 \quad xy^2z^3$$
$$yz \quad x^2yz \quad y^3z \quad z^3y \quad x^2y^3z^3 \quad x^2y^3z \quad x^2yz^3 \quad y^3z^3$$
$$xyz \quad x^3yz \quad y^3xz \quad z^3xy \quad x^3y^3z^3 \quad x^3y^3z \quad x^3yz^3 \quad xy^3z^3$$
$$z \quad x^2z \quad y^2z \quad z^3 \quad x^2y^2z^3 \quad x^2y^2z \quad x^2z^3 \quad y^2z^3$$
$$y \quad x^2y \quad y^3 \quad z^2y \quad x^2y^3z^2 \quad x^2y^3 \quad x^2yz^2 \quad y^3z^2$$
$$x \quad x^3 \quad y^2x \quad z^2x \quad x^3y^2z^2 \quad x^3y^2 \quad x^3z^2 \quad xy^2z^2]\cdot$$

248 Formulation of shape functions in G-invariant subspaces [Ch. 4

$$\cdot [\alpha_1 \ldots \alpha_8 \mid \alpha_9 \ldots \alpha_{16} \mid \alpha_{17} \ldots \alpha_{24} \mid \alpha_{25} \ldots \alpha_{32}$$
$$\mid \alpha_{33} \ldots \alpha_{40} \mid \alpha_{41} \ldots \alpha_{48} \mid \alpha_{49} \ldots \alpha_{56} \mid \alpha_{57} \ldots \alpha_{64}]^T$$
$$= [F_1 \mid F_2 \mid F_3 \mid F_4 \mid F_5 \mid F_6 \mid F_7 \mid F_8]$$
$$[A_1 \mid A_2 \mid A_3 \mid A_4 \mid A_5 \mid A_6 \mid A_7 \mid A_8]^T$$

is decomposed into eight eight-dimensional G-invariant subspaces U_1, U_2, U_3, U_4, U_5, U_6, U_7, U_8.

Substitution of the nodal coordinates of the first nodes in the nodal sets S_1, S_2, S_3, S_4, S_5, S_6, S_7, S_8

Node	1	9	17	25	33	41	49	57
x	c	c	c	$c/3$	c	$c/3$	$c/3$	$c/3$
y	d	d	$d/3$	d	$d/3$	d	$d/3$	$d/3$
z	e	$e/3$	e	e	$e/3$	$e/3$	e	$e/3$

into particular sections of $u(x, y, z)$ pertaining to subspaces U_i will provide relations of displacements $\bar{U}$ to coefficients A for each subspace U_1, U_2, U_3, U_4, U_5, U_6, U_7, U_8 separately:

$$\begin{bmatrix} \bar{U}^{(1)} \\ \bar{U}^{(2)} \\ \bar{U}^{(3)} \\ \bar{U}^{(4)} \\ \bar{U}^{(5)} \\ \bar{U}^{(6)} \\ \bar{U}^{(7)} \\ \bar{U}^{(8)} \end{bmatrix} = \begin{bmatrix} \bar{C}_1 & & & & & & & \\ & \bar{C}_2 & & & & & & \\ & & \bar{C}_3 & & & & & \\ & & & \bar{C}_4 & & & & \\ & & & & \bar{C}_5 & & & \\ & & & & & \bar{C}_6 & & \\ & & & & & & \bar{C}_7 & \\ & & & & & & & \bar{C}_8 \end{bmatrix} \begin{bmatrix} A_1 \\ A_2 \\ A_3 \\ A_4 \\ A_5 \\ A_6 \\ A_7 \\ A_8 \end{bmatrix}.$$

In the subspace U_1, by using the first row of the transformation matrix T, one obtains

Sec. 4.9] Sixty-four-node rectangular hexahedral element

$$\bar{U}^{(1)} = \begin{bmatrix} \bar{u}_1 \\ \bar{u}_9 \\ \bar{u}_{17} \\ \bar{u}_{25} \\ \bar{u}_{33} \\ \bar{u}_{41} \\ \bar{u}_{49} \\ \bar{u}_{57} \end{bmatrix} = \frac{1}{8} \begin{bmatrix} 1 \\ 1 \\ 1 \\ 1 \\ 1 \\ 1 \\ 1 \\ 1 \end{bmatrix}^T \begin{bmatrix} u_1 & u_9 & u_{17} & u_{25} & u_{33} & u_{41} & u_{49} & u_{57} \\ u_2 & u_{10} & u_{18} & u_{26} & u_{34} & u_{42} & u_{50} & u_{58} \\ u_3 & u_{11} & u_{19} & u_{27} & u_{35} & u_{43} & u_{51} & u_{59} \\ u_4 & u_{12} & u_{20} & u_{28} & u_{36} & u_{44} & u_{52} & u_{60} \\ u_5 & u_{13} & u_{21} & u_{29} & u_{37} & u_{45} & u_{53} & u_{61} \\ u_6 & u_{14} & u_{22} & u_{30} & u_{38} & u_{46} & u_{54} & u_{62} \\ u_7 & u_{15} & u_{23} & u_{31} & u_{39} & u_{47} & u_{55} & u_{63} \\ u_8 & u_{16} & u_{24} & u_{32} & u_{40} & u_{48} & u_{56} & u_{64} \end{bmatrix}$$

$$= \frac{1}{8} \begin{bmatrix} u_1 + u_2 + u_3 + u_4 + u_5 + u_6 + u_7 + u_8 \\ u_9 + u_{10} + u_{11} + u_{12} + u_{13} + u_{14} + u_{15} + u_{16} \\ u_{17} + u_{18} + u_{19} + u_{20} + u_{21} + u_{22} + u_{23} + u_{24} \\ u_{25} + u_{26} + u_{27} + u_{28} + u_{29} + u_{30} + u_{31} + u_{32} \\ u_{33} + u_{34} + u_{35} + u_{36} + u_{37} + u_{38} + u_{39} + u_{40} \\ u_{41} + u_{42} + u_{43} + u_{44} + u_{45} + u_{46} + u_{47} + u_{48} \\ u_{49} + u_{50} + u_{51} + u_{52} + u_{53} + u_{54} + u_{55} + u_{56} \\ u_{57} + u_{58} + u_{59} + u_{60} + u_{61} + u_{62} + u_{63} + u_{64} \end{bmatrix}.$$

Substitution of the nodal coordinates of the first nodes in the nodal sets $S_1, S_2, S_3, S_4, S_5, S_6, S_7, S_8$, i.e. the nodes $1, 9, 17, 25, 33, 41, 49, 56$ into

$$u^{(1)}(x, y, z) = [1 \quad x^2 \quad y^2 \quad z^2 \quad x^2y^2z^2 \quad x^2y^2 \quad x^2z^2 \quad y^2z^2] \cdot$$
$$\cdot [\alpha_1 \quad \alpha_2 \quad \alpha_3 \quad \alpha_4 \quad \alpha_5 \quad \alpha_6 \quad \alpha_7 \quad \alpha_8]^T$$

gives (see p. 250) or $\bar{U}^{(1)} = \bar{C}_1 A_1$.

The relation of A_1 to $\bar{U}_1$ is obtained by $A_1 = \bar{C}_1^{-1} \bar{U}_1$, or (see p. 250).

The eight-dimensional subspace U_1 of the displacement field Δ_1 is given by

$$\Delta_1 = T_{D_1} \bar{C}_1^{-1} \bar{U}_1,$$

with

$$T_{D_1} = \text{Diag}\, [F_1] = [1 \quad x^2 \quad y^2 \quad z^2 \quad x^2y^2z^2 \quad x^2y^2 \quad x^2z^2 \quad y^2z^2]$$

and $\xi = x/c$, $\eta = y/d$, $\zeta = z/e$, providing (see pp. 251–252)

$$
\begin{bmatrix} \bar{u}_1 \\ \bar{u}_9 \\ \bar{u}_{19} \\ \bar{u}_{25} \\ \bar{u}_{33} \\ \bar{u}_{41} \\ \bar{u}_{49} \\ \bar{u}_{57} \end{bmatrix} = \begin{bmatrix} 1 & c^2 & d^2 & e^2 & c^2d^2e^2 & c^2d^2 & c^2e^2 & d^2e^2 \\ 1 & c^2 & d^2 & \dfrac{e^2}{9} & \dfrac{c^2d^2e^2}{9} & c^2d^2 & \dfrac{c^2e^2}{9} & \dfrac{d^2e^2}{9} \\ 1 & c^2 & \dfrac{d^2}{9} & e^2 & \dfrac{c^2d^2e^2}{9} & \dfrac{c^2d^2}{9} & c^2e^2 & \dfrac{d^2e^2}{9} \\ 1 & \dfrac{c^2}{9} & d^2 & e^2 & \dfrac{c^2d^2e^2}{9} & \dfrac{c^2d^2}{9} & \dfrac{c^2e^2}{9} & d^2e^2 \\ 1 & c^2 & \dfrac{d^2}{9} & \dfrac{e^2}{9} & \dfrac{c^2d^2e^2}{81} & \dfrac{c^2d^2}{9} & \dfrac{c^2e^2}{9} & \dfrac{d^2e^2}{81} \\ 1 & \dfrac{c^2}{9} & d^2 & \dfrac{e^2}{9} & \dfrac{c^2d^2e^2}{81} & \dfrac{c^2d^2}{9} & \dfrac{c^2e^2}{81} & \dfrac{d^2e^2}{9} \\ 1 & \dfrac{c^2}{9} & \dfrac{d^2}{9} & e^2 & \dfrac{c^2d^2e^2}{81} & \dfrac{c^2d^2}{81} & \dfrac{c^2e^2}{9} & \dfrac{d^2e^2}{9} \\ 1 & \dfrac{c^2}{9} & \dfrac{d^2}{9} & \dfrac{e^2}{9} & \dfrac{c^2d^2e^2}{729} & \dfrac{c^2d^2}{81} & \dfrac{c^2e^2}{81} & \dfrac{d^2e^2}{81} \end{bmatrix} \begin{bmatrix} \alpha_1 \\ \alpha_2 \\ \alpha_3 \\ \alpha_4 \\ \alpha_5 \\ \alpha_6 \\ \alpha_7 \\ \alpha_8 \end{bmatrix}
$$

$$
\begin{bmatrix} \alpha_1 \\ \alpha_2 \\ \alpha_3 \\ \alpha_4 \\ \alpha_5 \\ \alpha_6 \\ \alpha_7 \\ \alpha_8 \end{bmatrix} = \text{Diag} \begin{bmatrix} 1 \\ 1/c^2 \\ 1/d^2 \\ 1/e^2 \\ 1/c^2d^2e^2 \\ 1/c^2d^2 \\ 1/c^2e^2 \\ 1/d^2e^2 \end{bmatrix} \dfrac{1}{512} \begin{bmatrix} -1 & 9 & 9 & 9 & -81 & -81 & -81 & 729 \\ 9 & -81 & -81 & -9 & 729 & 81 & 81 & -729 \\ 9 & -81 & -9 & -81 & 81 & 729 & 81 & -729 \\ 9 & -9 & -81 & -81 & 81 & 81 & 729 & -729 \\ 729 & -729 & -729 & -729 & 729 & -729 & -729 & 729 \\ -81 & 729 & 81 & 81 & -729 & 729 & -81 & 729 \\ -81 & 81 & 729 & 81 & -729 & -81 & -729 & 729 \\ -81 & 81 & 81 & 729 & -81 & -729 & -729 & 729 \end{bmatrix} \begin{bmatrix} \bar{u}_1 \\ \bar{u}_9 \\ \bar{u}_{17} \\ \bar{u}_{25} \\ \bar{u}_{33} \\ \bar{u}_{41} \\ \bar{u}_{49} \\ \bar{u}_{57} \end{bmatrix}
$$

Sec. 4.9] Sixty-four-node rectangular hexahedral element

$$\Delta_1 = \text{Diag} \begin{bmatrix} 1 \\ \dfrac{x^2}{c^2} \\ \dfrac{y^2}{d^2} \\ \dfrac{z^2}{e^2} \\ \dfrac{x^2 y^2 z^2}{c^2 d^2 e^2} \\ \dfrac{x^2 y^2}{c^2 d^2} \\ \dfrac{x^2 z^2}{c^2 e^2} \\ \dfrac{y^2 z^2}{d^2 e^2} \end{bmatrix} \dfrac{1}{512} \begin{bmatrix} -1 & 9 & 9 & 9 & -81 & -81 & -81 & 729 \\ 9 & -81 & -81 & -9 & 729 & 81 & 81 & -729 \\ 9 & -81 & -9 & -81 & 81 & 729 & 81 & -729 \\ 9 & -9 & -81 & -81 & 81 & 81 & 729 & -729 \\ 729 & -729 & -729 & -729 & 729 & 729 & 729 & -729 \\ -81 & 729 & 81 & 81 & -729 & -729 & -81 & 729 \\ -81 & 81 & 729 & 81 & -729 & -81 & -729 & 729 \\ -81 & 81 & 81 & 729 & -81 & -729 & -729 & 729 \end{bmatrix} \begin{bmatrix} \bar{u}_1 \\ \bar{u}_9 \\ \bar{u}_{17} \\ \bar{u}_{25} \\ \bar{u}_{33} \\ \bar{u}_{41} \\ \bar{u}_{49} \\ \bar{u}_{57} \end{bmatrix}$$

$$= \frac{1}{512} \begin{bmatrix} -1 & 9 & 9 & 9 & -81 \\ 9\xi^2 & -81\xi^2 & -81\xi^2 & -9\xi^2 & 729\xi^2 \\ 9\eta^2 & -81\eta^2 & -9\eta^2 & -81\eta^2 & 81\eta^2 \\ 9\zeta^2 & -9\zeta^2 & -81\zeta^2 & -81\zeta^2 & 81\zeta^2 \\ 729\xi^2\eta^2\zeta^2 & -729\xi^2\eta^2\zeta^2 & -729\xi^2\eta^2\zeta^2 & -729\xi^2\eta^2\zeta^2 & 729\xi^2\eta^2\zeta^2 \\ -81\xi^2\eta^2 & 729\xi^2\eta^2 & 81\xi^2\eta^2 & 81\xi^2\eta^2 & -729\xi^2\eta^2 \\ -81\xi^2\zeta^2 & 81\xi^2\zeta^2 & 729\xi^2\zeta^2 & 81\xi^2\zeta^2 & -729\xi^2\zeta^2 \\ -81\eta^2\zeta^2 & 81\eta^2\zeta^2 & 81\eta^2\zeta^2 & 729\eta^2\zeta^2 & -81\eta^2\zeta^2 \end{bmatrix}$$

$$\begin{bmatrix} -81 & -81 & 729 \\ 81\xi^2 & 81\xi^2 & -729\xi^2 \\ 729\eta^2 & 81\eta^2 & -729\eta^2 \\ 81\zeta^2 & 729\zeta^2 & -729\zeta^2 \\ 729\xi^2\eta^2\zeta^2 & 729\xi^2\eta^2\zeta^2 & -729\xi^2\eta^2\zeta^2 \\ -729\xi^2\eta^2 & -81\xi^2\eta^2 & 729\xi^2\eta^2 \\ -81\xi^2\zeta^2 & -729\xi^2\zeta^2 & 729\xi^2\zeta^2 \\ -729\eta^2\zeta^2 & -729\eta^2\zeta^2 & 729\eta^2\zeta^2 \end{bmatrix} \begin{bmatrix} \bar{u}_1 \\ \bar{u}_9 \\ \bar{u}_{17} \\ \bar{u}_{25} \\ \bar{u}_{33} \\ \bar{u}_{41} \\ \bar{u}_{49} \\ \bar{u}_{57} \end{bmatrix}$$

or $\Delta_1 = \bar{N}_{\Delta_1}\bar{U}_1$.

The shape functions $\bar{N}^{(1)}$ of the G-invariant subspace U_1 are obtained by

$$\bar{N}^{(1)} = S_1 \bar{N}_{\Delta_1},$$

with

$$S_1 = [1 \quad 1 \quad 1 \quad 1 \quad 1 \quad 1 \quad 1 \quad 1],$$

$$\begin{bmatrix} \bar{N}_1 \\ \bar{N}_9 \\ \bar{N}_{17} \\ \bar{N}_{25} \\ \bar{N}_{33} \\ \bar{N}_{41} \\ \bar{N}_{49} \\ \bar{N}_{57} \end{bmatrix} = \frac{1}{512} \begin{bmatrix} -1+ 9\xi^2+ 9\eta^2+ 9\zeta^2+729\xi^2\eta^2\zeta^2- 81\xi^2\eta^2- 81\xi^2\zeta^2- 81\eta^2\zeta^2 \\ 9- 81\xi^2- 81\eta^2- 9\zeta^2-729\xi^2\eta^2\zeta^2+729\xi^2\eta^2+ 81\xi^2\zeta^2+ 81\eta^2\zeta^2 \\ 9- 81\xi^2- 9\eta^2- 81\zeta^2-729\xi^2\eta^2\zeta^2+ 81\xi^2\eta^2+729\xi^2\zeta^2+ 81\eta^2\zeta^2 \\ 9- 9\xi^2- 81\eta^2- 81\zeta^2-729\xi^2\eta^2\zeta^2+ 81\xi^2\eta^2+ 81\xi^2\zeta^2+729\eta^2\zeta^2 \\ -81+729\xi^2+ 81\eta^2+ 81\zeta^2+729\xi^2\eta^2\zeta^2-729\xi^2\eta^2-729\xi^2\zeta^2- 81\eta^2\zeta^2 \\ -81+ 81\xi^2+729\eta^2+ 81\zeta^2+729\xi^2\eta^2\zeta^2-729\xi^2\eta^2- 81\xi^2\zeta^2-729\eta^2\zeta^2 \\ -81+ 81\xi^2+ 81\eta^2+729\zeta^2+729\xi^2\eta^2\zeta^2- 81\xi^2\eta^2-729\xi^2\zeta^2-729\eta^2\zeta^2 \\ 729-729\xi^2-729\eta^2-729\zeta^2-729\xi^2\eta^2\zeta^2+729\xi^2\eta^2+729\xi^2\zeta^2+729\eta^2\zeta^2 \end{bmatrix}$$

$$= \frac{1}{512} \begin{bmatrix} -1 & 9 & 9 & 9 & 729 & -81 & -81 & -81 \\ 9 & -81 & -81 & -9 & -729 & 729 & 81 & 81 \\ 9 & -81 & -9 & -81 & -729 & 81 & 729 & 81 \\ 9 & -9 & -81 & -81 & -729 & 81 & 81 & 729 \\ -81 & 729 & 81 & 81 & 729 & -729 & -729 & -81 \\ -81 & 81 & 729 & 81 & 729 & -729 & -81 & -729 \\ -81 & 81 & 81 & 729 & 729 & -81 & -729 & -729 \\ 729 & -729 & -729 & -729 & -729 & 729 & 729 & 729 \end{bmatrix} \begin{bmatrix} 1 \\ \xi^2 \\ \eta^2 \\ \zeta^2 \\ \xi^2\eta^2\zeta^2 \\ \xi^2\eta^2 \\ \xi^2\zeta^2 \\ \eta^2\zeta^2 \end{bmatrix}.$$

Shape functions $\bar{N}^{(2)}, \bar{N}^{(3)}, \bar{N}^{(4)}, \bar{N}^{(5)}, \bar{N}^{(6)}, \bar{N}^{(7)}, \bar{N}^{(8)}$ of the subspaces $U_2, U_3, U_4, U_5, U_6, U_7, U_8$ are derived in the same ways as $\bar{N}^{(1)}$.

5

Stiffness equations in G-invariant subspaces

5.1 GROUP SUPERMATRIX PROCEDURE FOR DERIVATION OF STIFFNESS EQUATIONS IN G-INVARIANT SUBSPACES

When symmetry properties of a finite element with its nodal pattern can be described by a group G, it is possible to derive the element stiffness equations with the matrix in block diagonal form which formulates stiffness equations in G-invariant subspaces.

The group supermatrix procedure for derivation of stiffness equations in G-invariant subspaces is systematized into 11 consecutive steps. Steps 1 to 8 correspond to the procedure for derivation of shape functions in G-invariant subspaces, given in section 4.1, with additional formulations concerning nodal degrees of freedom. Steps 9 to 11 contain derivation of stiffness matrices in G-invariant subspaces and stiffness equations with stiffness matrices in normal form, as well as transformations into stiffness equations in conventional form and vice versa.

(1) *Introduction of the unique group supermatrix nodal numberings*

The unique nodal numberings for elements with symmetry properties described by various groups are derived in Chapter 3. When a group G describes the symmetry properties of an element and its nodal pattern, the nodes can be grouped into one or more nonoverlapping sets $S_1, S_2, \ldots, S_l$, which figure in formulations of optimum nodal numberings.

Since the coordinates of all nodes in a nodal set can be generated by action of symmetry operations of the group G on a single node, in all analyses in the group supermatrix procedures only the first nodes in the nodal sets $S_1, S_2, \ldots, S_l$ will be used.

(2) *Derivation of basis vectors of G-invariant subspaces*

The set of positive directions of generalized nodal displacements and forces of the element must suit the first symmetry type of the group, so that the nodal configuration is transformed into itself by every symmetry operation of the group.

When in a nodal pattern n_n is the total number of nodes that, by action of elements of G, are permuted within a single nodal set, and when l is the number of degrees of freedom in all nodes, the total number of degrees of freedom and the dimension of the G-vector space is $n = n_n l$. Then each specific generalized displacement forms its sets of nodal functions, thus formulating the sets $S_1, S_2, \ldots, S_l$ that correspond to component displacement functions. The sequence of these sets and their functions are arranged in correspondence to the numbering system given in section 4.1, step (2), as well as in section 3.2, 3.5, 3.8 for the groups C_2, C_{2v}, D_{2h} respectively.

The character table of the group G, with Cartesian sets and products which are used in step 4 for derivation of displacement field functions in G-invariant subspaces, gives the characters of irreducible group representations for determination of the idempotents of the centre of the group algebra, as explained in section 4.1, step (2).

By application of the idempotents π_i ($i = 1, 2, \ldots, k$), where k is the number of irreducible group representations, to the nodal functions φ_j ($j = 1, 2, \ldots, n$), the basis vectors $\bar{\varphi}_j$ of G-invariant subspaces are derived.

(3) *Formulation of relations of basis vectors and nodal functions as systems of equations with the supermatrix in diagonal form*
These relations are explained in section 4.1, step (3).

(4) *Derivation of the function of the displacement field decomposed into G-invariant subspaces*
Application of the displacement field polynomial with its terms arranged according to their pertinence to the symmetry types of the group, with the objective of producing displacement field functions in G-invariant subspaces, is derived in Chapter 4 for elements having nodes with one degree of freedom, as given in section 4.1, step (4). When nodes possess l degrees of freedom, this procedure provides an $ln_i \times ln_i$ matrix for the subspace U_i, where n_i is the dimension of the subspace when the nodes possess one degree of freedom.

(5) *Derivation of relations of generalized displacements $\bar{\Phi}$ to coefficients A in G-invariant subspaces*
Substitution of the nodal coordinates of the first nodes in the nodal sets into the sections of the polynomial displacement function that pertain to the subspaces $U_1, U_2, \ldots, U_k$ provides relations of generalized displacements $\bar{\Phi}$ to the coefficients A for each subspace separately, i.e. the system of equations with the supermatrix in diagonal form (see section 4.1, step (5)).

(6) *Determination of the coefficients A in G-invariant subspaces*
The coefficients A_i are obtained by inverting the matrices $\bar{C}_i$ of the subspaces U_i, as given in section 4.1, step (6).

(7) *Formulation of displacement fields in G-invariant subspaces*
The n-dimensional G-vector space of the displacement field Δ is decomposed into k G-invariant subspaces $U_1, U_2, \ldots, U_k$ with $n_1, n_2, \ldots, n_k$ dimensions respectively, as shown in section 4.1, step (7).

(8) *Derivation of element shape functions $\bar{N}$ in G-invariant subspaces*
The shape functions $\bar{N}^{(i)}$ are obtained for each subspace U_i separately, as shown in section 4.1, step (8).

(9) *Derivation of stiffness matrices $\bar{K}_i$ in G-invariant subspaces*
By using the shape functions $\bar{N}^{(i)}$ in G-invariant subspaces U_i, the stiffness matrices

$\bar{K}_i$ pertaining to the subspaces U_i are obtained, so that the element stiffness group supermatrix in diagonal form is

$$\bar{K} = \begin{bmatrix} \bar{K}_1 & & & \\ & \bar{K}_2 & & \\ & & \ddots & \\ & & & \bar{K}_k \end{bmatrix} = \text{Diag}\,[\bar{K}_1 \quad \bar{K}_2 \quad \ldots \quad \bar{K}_k].$$

(10) *Transformation of the stiffness equations with the stiffness group supermatrix in diagonal form into the stiffness equation with the stiffness group supermatrix in normal form*

The stiffness equation with the stiffness group supermatrix in diagonal form

$$\begin{bmatrix} \bar{K}_1 & & & \\ & \bar{K}_2 & & \\ & & \ddots & \\ & & & \bar{K}_k \end{bmatrix} \begin{bmatrix} \bar{\Phi}^{(1)} \\ \bar{\Phi}^{(2)} \\ \vdots \\ \bar{\Phi}^{(k)} \end{bmatrix} = \begin{bmatrix} \bar{P}^{(1)} \\ \bar{P}^{(2)} \\ \vdots \\ \bar{P}^{(k)} \end{bmatrix} \quad \text{or} \quad \bar{K}\bar{\Phi} = \bar{P},$$

is transformed into the stiffness equation $K\Phi = P$, with the stiffness group supermatrix in normal form, by the group supermatrix transformation

$$T^{-1}\bar{K}TT^{-1}\bar{\Phi} = T^{-1}\bar{P},$$

which is derived in section 3.1.

(11) *Transformation of the stiffness equation with the stiffness group supermatrix in normal form into the conventional stiffness equation*

This transformation is performed by changing the nodal numbering, introduced by the group supermatrix analysis, into the conventional numbering, and by changing the set of positive directions of generalized nodal displacements that suit the first symmetry type of the group into the conventional set of positive directions of generalized nodal displacements.

By using the respective opposite transformations in the procedures of steps (11) and (10) in reverse order, the conventional stiffness equation is transformed into the stiffness equation with the stiffness group supermatrix in normal form. Thereafter, this equation is transformed into the stiffness equation with the stiffness group supermatrix in diagonal form (as derived in section 3.1).

In the following sections the group supermatrix procedure is applied for derivation of stiffness equations in G-invariant subspaces for one-dimensional elements, the beam element, the rectangular element for planar analysis and the rectangular element for plate flexure, and comparisons are made of derivations by the conventional method with those by the group supermatrix procedure.

The group supermatrix procedure for derivation of stiffness equations in G-invariant subspaces is developed on the basis of finite elements introduced by

— Argyris, J. H. and Kelsey, S. (1960): the four-node rectangular element for planar analysis,
— Adini, A. and Clough, R. W. (1961): 12-d.-o.-f. rectangular element for plate flexure,
— Bogner, F. K. et al. (1966): the four-node 16-d.-o.-f. rectangular element for plate flexure.

In addition, some explanations and data from Dawe, D. J. (1984), Rockey, et al. (1975) and Yang (1986) were used.

5.2 STIFFNESS EQUATIONS IN G–INVARIANT SUBSPACES FOR ONE-DIMENSIONAL ELEMENTS

The two-node bar element, as given in Fig. 5.1, with the nodal numbering established in section 3.2 (the node 1 lying on the positive branch of the x axis), is described by the group C_2. The nodes of its nodal set $S(1, 2)$ are permuted by action of symmetry operations E (identity) and C_2 (rotation through $180°$ about the z axis).

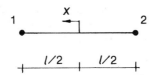

Fig. 5.1. The two-node bar element with the nodal numbering of the group supermatrix procedure.

The character table of the group C_2 (with Cartesian sets and products) is

C_2	E	C_2		
A	1	1	z	x^2, y^2, z^2, xy
B	1	-1	x, y	yz, xz

The idempotents of the centre of the group algebra are obtained by

$$\pi_i = \frac{h_i}{h} \sum_\sigma \chi_i(\sigma^{-1})\sigma$$

with h_i, the dimension of the ith character, given by $h_i = \chi_i(E)$, h the order, χ_i the ith character, σ the element, σ^{-1} its inverse and $i = 1, 2$. Thus

$$\pi_1 = \tfrac{1}{2}(E + C_2), \quad \pi_2 = \tfrac{1}{2}(E - C_2), \quad \text{or} \quad \begin{bmatrix} \pi_1 \\ \pi_2 \end{bmatrix} = \tfrac{1}{2} \begin{bmatrix} 1 & 1 \\ 1 & -1 \end{bmatrix} \begin{bmatrix} E \\ C_2 \end{bmatrix}$$

or $\Pi = T\Sigma$, with $T^{-1} = 2T$.

By application of the idempotents π_i ($i = 1, 2$) to the nodal functions φ_j ($j = 1, 2$), the basis vectors $\bar{\varphi}_j$ of G-invariant subspaces U_i are derived:

$$U_1: \quad \bar{\varphi}_1 = \pi_1 \varphi_1 = \pi_1 \varphi_2 = \tfrac{1}{2}(\varphi_1 + \varphi_2)$$
$$U_2: \quad \bar{\varphi}_2 = \pi_2 \varphi_1 = -\pi_2 \varphi_2 = \tfrac{1}{2}(\varphi_1 - \varphi_2)$$

or

$$\begin{bmatrix} \bar{\varphi}_1 \\ \bar{\varphi}_2 \end{bmatrix} = \tfrac{1}{2} \begin{bmatrix} 1 & 1 \\ 1 & -1 \end{bmatrix} \begin{bmatrix} \varphi_1 \\ \varphi_2 \end{bmatrix} \quad \text{or} \quad \bar{\Phi} = T\Phi.$$

Conversely, the relation of the nodal functions φ_j to the basis vectors $\bar{\varphi}_j$ is

$$\begin{bmatrix} \varphi_1 \\ \varphi_2 \end{bmatrix} = \begin{bmatrix} 1 & 1 \\ 1 & -1 \end{bmatrix} \begin{bmatrix} \bar{\varphi}_1 \\ \bar{\varphi}_2 \end{bmatrix} \quad \text{or} \quad \Phi = T^{-1}\bar{\Phi} = 2T\bar{\Phi}.$$

The linear u-displacement equation is usually assumed as

$$u(x) = \alpha_1 + \alpha_2 x.$$

For the group supermatrix procedure the order of the terms in this polynomial must be pertinent to the order of symmetry types of irreducible group representations A, B of the group C_2, as given in addition to its character table. Therefore, in the polynomial

$$u(x) = \begin{bmatrix} 1 & x \end{bmatrix} \begin{bmatrix} \alpha_1 & \alpha_2 \end{bmatrix}^T$$

the terms $1, x$ belong to the subspaces U_1, U_2 respectively.

In the subspace U_1

$$\bar{u}_1 = \tfrac{1}{2}(u_1 + u_2) = \alpha_1 \cdot 1 = \alpha_1 \quad \text{and} \quad \Delta_1 = \bar{N}_1 \bar{u}_1 = 1 \cdot \bar{u}_1, \text{ i.e. } \bar{N}_1 = 1,$$

$$B_1 = \frac{d\bar{N}_1}{dx} = \frac{d}{dx} 1 = 0$$

and the stiffness matrix in the subspace U_1 is $\bar{K}_1 = 0$.

In the subspace U_2 substitution of the nodal coordinate $x = l/2$ of the node 1 in the second section of $u(x)$ containing x gives

$$\bar{u}_2 = \tfrac{1}{2}(u_1 - u_2) = \frac{l}{2}\alpha_2, \quad \alpha_2 = \frac{2}{l}\bar{u}_2$$

and

$$\Delta_2 = \frac{2}{l} x \bar{u}_2 = \bar{N}_2 \bar{u}_2$$

$$B_2 = \frac{d\bar{N}_2}{dx} = \frac{d}{dx}\frac{2}{l}x = \frac{2}{l}.$$

With the element cross-section area A and Young's modulus E' the stiffness matrix of the element in the subspace U_2 is

Sec. 5.2] Stiffness equations in G-invariant subspaces for 1-D elements 259

$$\bar{K}_2 = AE' \int_0^{1/2} \frac{2}{l} \frac{2}{l} dx = 2 \frac{AE'}{l}.$$

Thus the stiffness equation with the matrix in block diagonal form is

$$\frac{AE'}{l} \begin{bmatrix} 0 & 0 \\ 0 & 2 \end{bmatrix} \begin{bmatrix} \bar{u}_1 \\ \bar{u}_2 \end{bmatrix} = \begin{bmatrix} \bar{P}_1 \\ \bar{P}_2 \end{bmatrix} \quad \text{or} \quad \bar{K}\bar{U} = \bar{P},$$

with

$$\bar{P} = \begin{bmatrix} \bar{P}_1 \\ \bar{P}_2 \end{bmatrix} = \frac{1}{2} \begin{bmatrix} P_1 + P_2 \\ P_1 - P_2 \end{bmatrix}.$$

By applying the group supermatrix transformation

$$T^{-1}\bar{K}TT^{-1}\bar{U} = T^{-1}\bar{P},$$

derived in section 3.1, one obtains

$$\begin{bmatrix} 1 & 1 \\ 1 & -1 \end{bmatrix} \frac{AE'}{l} \begin{bmatrix} 0 & 0 \\ 0 & 2 \end{bmatrix} \frac{1}{2} \begin{bmatrix} 1 & 1 \\ 1 & -1 \end{bmatrix} \begin{bmatrix} 1 & 1 \\ 1 & -1 \end{bmatrix} \begin{bmatrix} \bar{u}_1 \\ \bar{u}_2 \end{bmatrix} = \begin{bmatrix} 1 & 1 \\ 1 & -1 \end{bmatrix} \begin{bmatrix} \bar{P}_1 \\ \bar{P}_2 \end{bmatrix}$$

producing the stiffness equation with the stiffness group matrix in normal form, which, in this case, coincides with the conventional stiffness equation

$$\frac{AE'}{l} \begin{bmatrix} 1 & -1 \\ -1 & 1 \end{bmatrix} \begin{bmatrix} u_1 \\ u_2 \end{bmatrix} = \begin{bmatrix} P_1 \\ P_2 \end{bmatrix} \quad \text{or} \quad KU = P.$$

Conversely, this stiffness equation is transformed into the stiffness equation with the stiffness matrix in block diagonal form by the group supermatrix transformation

$$TKT^{-1}TU = TP$$

and

$$\frac{1}{2}\begin{bmatrix} 1 & 1 \\ 1 & -1 \end{bmatrix} \frac{AE'}{l} \begin{bmatrix} 1 & -1 \\ -1 & 1 \end{bmatrix} \begin{bmatrix} 1 & 1 \\ 1 & -1 \end{bmatrix} \frac{1}{2}\begin{bmatrix} 1 & 1 \\ 1 & -1 \end{bmatrix} \begin{bmatrix} u_1 \\ u_2 \end{bmatrix} = \frac{1}{2}\begin{bmatrix} 1 & 1 \\ 1 & -1 \end{bmatrix} \begin{bmatrix} P_1 \\ P_2 \end{bmatrix}$$

giving

$$\frac{AE'}{l} \begin{bmatrix} 0 & 0 \\ 0 & 2 \end{bmatrix} \begin{bmatrix} \bar{u}_1 \\ \bar{u}_2 \end{bmatrix} = \begin{bmatrix} \bar{P}_1 \\ \bar{P}_2 \end{bmatrix}.$$

The four-node bar element, as shown in Fig. 5.2 with the nodal numbering derived in 3.2, is described by the group C_2 and it has two nodal sets $S_1(1, 2)$ and $S_2(3, 4)$, each set containing nodes that are permuted by action of symmetry operations of the group.

As in the case of the two-node bar element, the idempotents of the centre of group algebra are

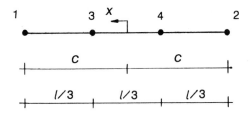

Fig. 5.2. The four-node bar element with the nodal numbering of the group supermatrix procedure.

$$\begin{bmatrix} \pi_1 \\ \pi_2 \end{bmatrix} = \frac{1}{2} \begin{bmatrix} 1 & 1 \\ 1 & -1 \end{bmatrix} \begin{bmatrix} E \\ C_2 \end{bmatrix} \quad \text{or} \quad \Pi = T\Sigma, \quad \text{with} \quad T^{-1} = 2T.$$

By application of the idempotents π_i ($i = 1, 2$) to the nodal functions φ_j ($j = 1, 2, 3, 4$) the basis vectors $\bar{\varphi}_j$ of G-invariant subspaces U_i are derived:

$$\begin{bmatrix} \bar{\Phi}^{(1)} \\ \bar{\Phi}^{(2)} \end{bmatrix} = \begin{bmatrix} \bar{\varphi}_1 & \bar{\varphi}_3 \\ \bar{\varphi}_2 & \bar{\varphi}_4 \end{bmatrix} = \frac{1}{2} \begin{bmatrix} 1 & 1 \\ 1 & -1 \end{bmatrix} \begin{bmatrix} \varphi_1 & \varphi_3 \\ \varphi_2 & \varphi_4 \end{bmatrix} = \frac{1}{2} \begin{bmatrix} \varphi_1 + \varphi_2 & \varphi_3 + \varphi_4 \\ \varphi_1 - \varphi_2 & \varphi_3 - \varphi_4 \end{bmatrix}$$

or $\bar{\Phi} = T\Phi$, with $T^{-1} = 2T$.

Conversely, the relation of the nodal functions $\Phi^{(i)}$ to the basis vectors $\bar{\Phi}^{(i)}$ is

$$\begin{bmatrix} \Phi^{(1)} \\ \Phi^{(2)} \end{bmatrix} = \begin{bmatrix} \varphi_1 & \varphi_3 \\ \varphi_2 & \varphi_4 \end{bmatrix} = \begin{bmatrix} 1 & 1 \\ 1 & -1 \end{bmatrix} \begin{bmatrix} \bar{\varphi}_1 & \bar{\varphi}_3 \\ \bar{\varphi}_2 & \bar{\varphi}_4 \end{bmatrix} = \begin{bmatrix} \bar{\varphi}_1 + \bar{\varphi}_2 & \bar{\varphi}_3 + \bar{\varphi}_4 \\ \bar{\varphi}_1 - \bar{\varphi}_2 & \bar{\varphi}_3 - \bar{\varphi}_4 \end{bmatrix}$$

or $\Phi = T^{-1}\bar{\Phi} = 2T\bar{\Phi}$.

The sets $\bar{\Phi}_l$ of basis vectors $\bar{\varphi}_j$ ($l = 1, 2; j = 1, 2, 3, 4$), ordered according to the numbering of the nodes in the nodal sets S_1, S_2, are given by

$$[\bar{\Phi}_1 \quad \bar{\Phi}_2] = \begin{bmatrix} \bar{\varphi}_1 & \bar{\varphi}_3 \\ \bar{\varphi}_2 & \bar{\varphi}_4 \end{bmatrix}.$$

The relation of the sets of basis vectors $\bar{\Phi}_l$ to the sets of the nodal functions Φ_l is expressed by the following system of equations, with the supermatrix in diagonal form containing two transformation matrices T of the group C_2:

$$\begin{bmatrix} \bar{\Phi}_1 \\ \bar{\Phi}_2 \end{bmatrix} = \begin{bmatrix} T & \\ & T \end{bmatrix} \begin{bmatrix} \Phi_1 \\ \Phi_2 \end{bmatrix} \quad \text{or} \quad \bar{\Phi}_l = \bar{T}\Phi_l,$$

with

$$T = \frac{1}{2} \begin{bmatrix} 1 & 1 \\ 1 & -1 \end{bmatrix} \quad \text{and} \quad [\Phi_1 \quad \Phi_2] = \begin{bmatrix} \varphi_1 & \varphi_3 \\ \varphi_2 & \varphi_4 \end{bmatrix}.$$

Conversely, the relation of the sets of the nodal functions Φ_l to the sets of basis vectors $\bar{\Phi}_l$ is

Sec. 5.2] Stiffness equations in G-invariant subspaces for 1-D elements

$$\begin{bmatrix} \Phi_1 \\ \Phi_2 \end{bmatrix} = 2 \begin{bmatrix} T & \\ & T \end{bmatrix} \begin{bmatrix} \overline{\Phi}_1 \\ \overline{\Phi}_2 \end{bmatrix} \quad \text{or} \quad \Phi_l = 2\overline{T}\overline{\Phi}_l,$$

since $T^{-1} = 2T$.

The cubic displacement function of the bar element is usually assumed to be

$$u(x) = \alpha_1 + \alpha_2 x + \alpha_3 x^2 + \alpha_4 x^3.$$

For the group supermatrix procedure by the following expression the terms of the above polynomial will be allocated to their respective G-invariant subspaces, with the unique order of the terms that gives the block diagonal form of the matrix of the system of equations relating displacements $\overline{U}$ to coefficients A:

$$\begin{bmatrix} F_1 \\ F_2 \end{bmatrix} = \begin{bmatrix} 1 \\ x \end{bmatrix} \begin{bmatrix} 1 & x^2 \end{bmatrix} = \begin{bmatrix} 1 & x^2 \\ x & x^3 \end{bmatrix} \quad \begin{matrix} \text{Subspace} \\ U_1 \\ U_2 \end{matrix}$$

The column vector and the row vector are taken in accordance to the Cartesian sets as they stand in addition to the character table of the group C_2 with the pertinence of the terms to the symmetry types of group representations A, B.

Thus, the displacement function

$$u(x) = \begin{bmatrix} 1 & x^2 & \vdots & x & x^3 \end{bmatrix} \begin{bmatrix} \alpha_1 & \alpha_2 & \vdots & \alpha_3 & \alpha_4 \end{bmatrix}^{\text{T}} = [F_1 \vdots F_2][A_1 \vdots A_2]^{\text{T}}$$

is decomposed into two two-dimensional G-invariant subspaces U_1, U_2.

Substitution of the nodal coordinates of the first nodes in the nodal sets $S_1(1, 2)$, $S_2(3, 4)$

Node	1	3
x	c	$c/3$

into the sections F_1, F_2 of $u(x)$, pertaining to the subspaces U_1, U_2, will give the relations of the displacements $\overline{U}$ to the coefficients A for each subspace separately:

$$\begin{bmatrix} \overline{u}_1 \\ \overline{u}_3 \\ \hdashline \overline{u}_2 \\ \overline{u}_4 \end{bmatrix} = \frac{1}{2} \begin{bmatrix} u_1 + u_2 \\ u_3 + u_4 \\ \hdashline u_1 - u_2 \\ u_3 - u_4 \end{bmatrix} = \begin{bmatrix} 1 & c^2 & & \\ 1 & \dfrac{c^2}{9} & & \\ \hdashline & & c & c^3 \\ & & \dfrac{c}{3} & \dfrac{c^3}{27} \end{bmatrix} \begin{bmatrix} \alpha_1 \\ \alpha_2 \\ \hdashline \alpha_3 \\ \alpha_4 \end{bmatrix}$$

or

$$\begin{bmatrix} \bar{U}^{(1)} \\ \bar{U}^{(2)} \end{bmatrix} = \begin{bmatrix} \bar{C}_1 & \\ & \bar{C}_2 \end{bmatrix} \begin{bmatrix} A_1 \\ A_2 \end{bmatrix}$$

or $\bar{U} = \bar{C}A$, where $\bar{U}$ is the set of displacements $\bar{U}^{(1)}, \bar{U}^{(2)}$, with the numbering according to that of the sets of basis vectors $\bar{\Phi}^{(1)}, \bar{\Phi}^{(2)}$ derived earlier. The four-dimensional G-vector space of the problem is decomposed now into two two-dimensional G-invariant subspaces U_1, U_2.

The coefficients A are determined by inverting the matrices $\bar{C}_1, \bar{C}_2$ of the subspaces U_1, U_2:

$$\begin{bmatrix} \alpha_1 \\ \alpha_2 \\ \cdots \\ \alpha_3 \\ \alpha_4 \end{bmatrix} = \text{Diag} \begin{bmatrix} 1 \\ \dfrac{1}{c^2} \\ \cdots \\ \dfrac{1}{c} \\ \dfrac{1}{c^3} \end{bmatrix} \frac{1}{8} \left[\begin{array}{cc|cc} -1 & 9 & & \\ 9 & -9 & & \\ \hline & & -1 & 27 \\ & & 9 & -27 \end{array} \right] \begin{bmatrix} \bar{u}_1 \\ \bar{u}_3 \\ \cdots \\ \bar{u}_2 \\ \bar{u}_4 \end{bmatrix}.$$

The four-dimensional G-vector space of the displacement field Δ, decomposed into two two-dimensional G-invariant subspaces U_1, U_2, is

$$\Delta = T_D \bar{C}^{-1} \bar{U},$$

with

$$T_D = \text{Diag} \begin{bmatrix} 1 & x^2 & | & x & x^3 \end{bmatrix} = \text{Diag} \begin{bmatrix} F_1 & | & F_2 \end{bmatrix}$$

and $\xi = x/c$, giving

$$\begin{bmatrix} \Delta_1 \\ \Delta_2 \end{bmatrix} = \text{Diag} \begin{bmatrix} 1 \\ \dfrac{x^2}{c^2} \\ \cdots \\ \dfrac{x}{c} \\ \dfrac{x^3}{c^3} \end{bmatrix} \frac{1}{8} \left[\begin{array}{cc|cc} -1 & 9 & & \\ 9 & -9 & & \\ \hline & & -1 & 27 \\ & & 9 & -27 \end{array} \right] \begin{bmatrix} \bar{u}_1 \\ \bar{u}_3 \\ \cdots \\ \bar{u}_2 \\ \bar{u}_4 \end{bmatrix}.$$

Sec. 5.2]   Stiffness equations in G-invariant subspaces for 1-D elements   263

$$= \frac{1}{8} \left[\begin{array}{cc|cc} -1 & 9 & & \\ 9\xi^2 & -9\xi^2 & & \\ \hline & & -\xi & 27\xi \\ & & 9\xi^3 & -27\xi^3 \end{array} \right] \left[\begin{array}{c} \bar{u}_1 \\ \bar{u}_3 \\ \hline \bar{u}_2 \\ \bar{u}_4 \end{array} \right]$$

or $\Delta = \bar{N}_\Delta \bar{U}$.

The shape functions $\bar{N}$ of G-invariant subspaces are obtained by

$$\bar{N} = S \bar{N}_\Delta,$$

with

$$S = [1 \quad 1 \;\vdots\; 1 \quad 1]$$

$$\left[\begin{array}{c} \bar{N}^{(1)} \\ \cdots \\ \bar{N}^{(2)} \end{array} \right] = \left[\begin{array}{c} \bar{N}_1 \\ \bar{N}_3 \\ \cdots \\ \bar{N}_2 \\ \bar{N}_4 \end{array} \right] = \left[\begin{array}{c} N_1 + N_2 \\ N_3 + N_4 \\ \cdots \\ N_1 - N_2 \\ N_3 - N_4 \end{array} \right]$$

$$= \left[\begin{array}{c} 1 \\ 1 \\ \cdots \\ 1 \\ 1 \end{array} \right]^T \frac{1}{8} \left[\begin{array}{cc|cc} -1 & 9 & & \\ 9\xi^2 & -9\xi^2 & & \\ \hline & & -\xi & 27\xi \\ & & 9\xi^3 & -27\xi^3 \end{array} \right] = \frac{1}{8} \left[\begin{array}{c} -1 + 9\xi^2 \\ 9 - 9\xi^2 \\ \cdots \\ -\xi + 9\xi^3 \\ 27\xi - 27\xi^3 \end{array} \right],$$

or

$$\left[\begin{array}{c} \bar{N}^{(1)} \\ \cdots \\ \bar{N}^{(2)} \end{array} \right] = \left[\begin{array}{c} \bar{N}_1 \\ \bar{N}_3 \\ \cdots \\ \bar{N}_2 \\ \bar{N}_4 \end{array} \right] = \frac{1}{8} \left[\begin{array}{cc|cc} -1 & 9 & & \\ 9 & -9 & & \\ \hline & & -1 & 9 \\ & & 27 & -27 \end{array} \right] \left[\begin{array}{c} 1 \\ \xi^2 \\ \xi \\ \xi^3 \end{array} \right].$$

In the subspace U_1, with the bar element constant cross-section area A and Young's modulus E' (since $\xi = x/c = 2(x/l)$)

$$\bar{B}_1 = \frac{d\bar{N}^{(1)}}{dx} = \frac{d}{dx} \frac{1}{8} \left[-1 + 9 \times 2^2 \frac{x^2}{l^2} \quad 9 - 9 \times 2^2 \frac{x^2}{l^2} \right] = \frac{9}{l^2} [x \quad -x],$$

and the stiffness matrix $\bar{K}_1$ of the bar element in the subspace U_1 is

$$\bar{K}_1 = AE' \int_0^{l/2} \bar{B}_1^T \bar{B}_1 \, dx = AE' \int_0^{l/2} \frac{9}{l^2} \begin{bmatrix} x \\ -x \end{bmatrix} \frac{9}{l^2} [x \; -x] \, dx$$

$$= AE' \frac{81}{l^4} \int_0^{l/2} x^2 \begin{bmatrix} 1 & -1 \\ -1 & 1 \end{bmatrix} dx = \frac{AE'}{40l} \begin{bmatrix} 135 & -135 \\ -135 & 135 \end{bmatrix}.$$

In the subspace U_2

$$\bar{B}_2 = \frac{d\bar{N}^{(2)}}{dx} = \frac{d}{dx} \frac{1}{8} \left[-2\frac{x}{l} + 9 \times 2^3 \frac{x^3}{l^3} \quad 27 \times 2\frac{x}{l} - 27 \times 2^3 \frac{x^3}{l^3} \right]$$

$$= \frac{1}{l} \left[-\frac{1}{4} + 27\frac{x^2}{l^2} \quad \frac{27}{4} - 81\frac{x^2}{l^2} \right]$$

and the stiffness matrix $\bar{K}_2$ of the bar element in the subspace U_2 is

$$\bar{K}_2 = AE' \int_0^{l/2} \bar{B}_2^T \bar{B}_2 \, dx = \frac{AE'}{l^2} \int_0^{l/2} \begin{bmatrix} \left(-\frac{1}{4} + \frac{27}{4}\frac{x^2}{l^2}\right)^2 & \left(-\frac{1}{4} + 27\frac{x^2}{l^2}\right)\left(\frac{27}{4} - 81\frac{x^2}{l^2}\right) \\ \left(-\frac{1}{4} + 27\frac{x^2}{l^2}\right)\left(\frac{27}{4} - 81\frac{x^2}{l^2}\right) & \left(\frac{27}{4} - 81\frac{x^2}{l^2}\right)^2 \end{bmatrix} dx$$

$$= \frac{AE'}{40l} \begin{bmatrix} 161 & -243 \\ -243 & 729 \end{bmatrix}.$$

Thus, the stiffness equation with the stiffness matrix in block diagonal form, i.e. with stiffness matrices in G-invariant subspaces U_1, U_2, is obtained:

$$\frac{AE'}{40} \begin{bmatrix} 135 & -135 & & \\ -135 & 135 & & \\ & & 161 & -243 \\ & & -243 & 729 \end{bmatrix} \begin{bmatrix} \bar{u}_1 \\ \bar{u}_3 \\ \bar{u}_2 \\ \bar{u}_4 \end{bmatrix} = \begin{bmatrix} \bar{P}_1 \\ \bar{P}_3 \\ \bar{P}_2 \\ \bar{P}_4 \end{bmatrix} \quad \text{or} \quad \bar{K}\bar{U} = \bar{P},$$

which, as in section 3.2, can be written in the form

Sec. 5.2] Stiffness equations in G-invariant subspaces for 1-D elements

$$\begin{bmatrix} A+B & \\ & A-B \end{bmatrix} \begin{bmatrix} \bar{U}^{(1)} \\ \bar{U}^{(2)} \end{bmatrix} = \begin{bmatrix} \bar{P}^{(1)} \\ \bar{P}^{(2)} \end{bmatrix}.$$

The stiffness equation with the stiffness group supermatrix in normal form is obtained by applying the group supermatrix transformation (derived in section 3.1)

$$T^{-1}\bar{K}TT^{-1}\bar{\Phi} = T^{-1}\bar{P},$$

giving

$$\begin{bmatrix} E & E \\ E & -E \end{bmatrix} \begin{bmatrix} A+B & \\ & A-B \end{bmatrix} \frac{1}{2} \begin{bmatrix} E & E \\ E & -E \end{bmatrix} \begin{bmatrix} E & E \\ E & -E \end{bmatrix} \frac{1}{2} \begin{bmatrix} U^{(1)} + U^{(2)} \\ U^{(1)} - U^{(2)} \end{bmatrix}$$

$$= \begin{bmatrix} E & E \\ E & -E \end{bmatrix} \frac{1}{2} \begin{bmatrix} P^{(1)} + P^{(2)} \\ P^{(1)} - P^{(2)} \end{bmatrix}$$

$$\begin{bmatrix} A & B \\ B & A \end{bmatrix} \begin{bmatrix} U^{(1)} \\ U^{(2)} \end{bmatrix} = \begin{bmatrix} P^{(1)} \\ P^{(2)} \end{bmatrix}$$

and

$$A = \tfrac{1}{2}[(A+B) + (A-B)], \quad B = \tfrac{1}{2}[(A+B) - (A-B)]$$

$$A = \frac{1}{2} \frac{AE'}{40} \left\{ \begin{bmatrix} 135 & -135 \\ -135 & 135 \end{bmatrix} + \begin{bmatrix} 161 & -243 \\ -243 & 729 \end{bmatrix} \right\} = \frac{AE'}{40} \begin{bmatrix} 148 & -189 \\ -189 & 432 \end{bmatrix}$$

$$B = \frac{1}{2} \frac{AE'}{40} \left\{ \begin{bmatrix} 135 & -135 \\ -135 & 135 \end{bmatrix} - \begin{bmatrix} 161 & -243 \\ -243 & 729 \end{bmatrix} \right\} = \frac{AE'}{40} \begin{bmatrix} -13 & 54 \\ 54 & -297 \end{bmatrix}$$

and the stiffness matrix in shape of a supermatrix in normal form is

$$K = \begin{bmatrix} A & B \\ B & A \end{bmatrix} = \frac{AE'}{40} \left[\begin{array}{cc|cc} 148 & -189 & -13 & 54 \\ -189 & 432 & 54 & -297 \\ \hline -13 & 54 & 148 & -189 \\ 54 & -297 & -189 & 432 \end{array} \right].$$

This stiffness matrix corresponds to the nodal numbering in Fig. 5.2 with the sequence of the nodes 1, 3, 4, 2 (from left to right). When the usual sequence 1, 2, 3, 4 is adopted, the stiffness matrix K is transformed into the conventional stiffness matrix of the four-node bar element

$$K_c = \frac{AE'}{40} \begin{bmatrix} 148 & -189 & 54 & -13 \\ -189 & 432 & -297 & 54 \\ 54 & -297 & 432 & -189 \\ -13 & 54 & -189 & 148 \end{bmatrix}.$$

Conversely, by changing the nodal numbering 1, 2, 3, 4 to 1, 3, 4, 2, the conventional stiffness matrix K_c is transformed into the stiffness group supermatrix in normal form, which becomes the stiffness supermatrix in diagonal form by the group supermatrix transformation TKT^{-1}

$$\bar{K} = \frac{1}{2} \begin{bmatrix} E & E \\ E & -E \end{bmatrix} \begin{bmatrix} A & B \\ B & A \end{bmatrix} \begin{bmatrix} E & E \\ E & -E \end{bmatrix} = \begin{bmatrix} A+B & \\ & A-B \end{bmatrix}$$

$$= \frac{AE'}{40} \begin{bmatrix} 135 & -135 & & \\ -135 & 135 & & \\ & & 161 & -243 \\ & & -243 & 729 \end{bmatrix}.$$

5.3 STIFFNESS EQUATIONS IN G-INVARIANT SUBSPACES FOR THE BEAM ELEMENT

The two-node beam element, as given in Fig. 5.3, with the nodal numbering established in section 3.2 (the node 1 lying on the positive branch of the x axis), is described by the group C_2, with positive directions of w_1, w_2, θ_{y1}, θ_{y2}, P_{z1}, P_{z2}, M_{y1}, M_{y2} that suit the symmetry type of the first irreducible representation of the group. As in the case of the

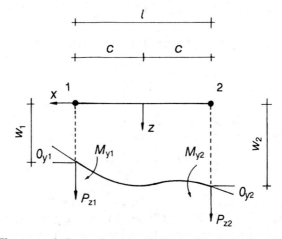

Fig. 5.3. The two-node beam element with the nodal numbering of the group supermatrix procedure.

Sec. 5.3] Stiffness equations in G-invariant subspaces for the beam element

two-node bar element in section 5.2, the two-node beam element has one nodal set $S(1, 2)$, its nodes permuted by symmetry operations of the group.

By application of the idempotents of the centre of the group algebra

$$\begin{bmatrix} \pi_1 \\ \pi_2 \end{bmatrix} = \frac{1}{2} \begin{bmatrix} 1 & 1 \\ 1 & -1 \end{bmatrix} \begin{bmatrix} E \\ C_2 \end{bmatrix} \quad \text{or} \quad \Pi = T\Sigma, \quad \text{with} \quad T^{-1} = 2T,$$

to the nodal functions $\Phi^{(i)}$ the basis vectors $\bar{\Phi}^{(i)}$ of G-invariant subspaces are derived:

$$\begin{bmatrix} \bar{\Phi}^{(1)} \\ \bar{\Phi}^{(2)} \end{bmatrix} = \begin{bmatrix} \bar{\varphi}_1 & \bar{\varphi}_3 \\ \bar{\varphi}_2 & \bar{\varphi}_4 \end{bmatrix} = \begin{bmatrix} \bar{w}_1 & \bar{\theta}_{y1} \\ \bar{w}_2 & \bar{\theta}_{y2} \end{bmatrix} = \frac{1}{2} \begin{bmatrix} 1 & 1 \\ 1 & -1 \end{bmatrix} \begin{bmatrix} w_1 & \theta_{y1} \\ w_2 & \theta_{y2} \end{bmatrix}$$

$$= \frac{1}{2} \begin{bmatrix} w_1 + w_2 & \theta_{y1} + \theta_{y2} \\ w_1 - w_2 & \theta_{y1} - \theta_{y2} \end{bmatrix}$$

or $\bar{\Phi} = T\Phi$, with $T^{-1} = 2T$.

Conversely, the relation of the sets of nodal functions $\Phi^{(i)}$ to the sets of basis vectors $\bar{\Phi}^{(i)}$ is

$$\begin{bmatrix} \Phi^{(1)} \\ \Phi^{(2)} \end{bmatrix} = \begin{bmatrix} \varphi_1 & \varphi_3 \\ \varphi_2 & \varphi_4 \end{bmatrix} = \begin{bmatrix} w_1 & \theta_{y1} \\ w_2 & \theta_{y2} \end{bmatrix} = \begin{bmatrix} 1 & 1 \\ 1 & -1 \end{bmatrix} \begin{bmatrix} \bar{w}_1 & \bar{\theta}_{y1} \\ \bar{w}_2 & \bar{\theta}_{y2} \end{bmatrix}$$

$$= \begin{bmatrix} \bar{w}_1 + \bar{w}_2 & \bar{\theta}_{y1} + \bar{\theta}_{y2} \\ \bar{w}_1 - \bar{w}_2 & \bar{\theta}_{y1} - \bar{\theta}_{y2} \end{bmatrix}$$

or $\Phi = T^{-1}\bar{\Phi} = 2T\bar{\Phi}$.

The sets $\bar{\Phi}_l$ containing the basis vectors $\bar{\varphi}_j$ ($l = 1, 2; j = 1, 2, 3, 4$), representing $\bar{w}_1$, $\bar{w}_2$, $\bar{\theta}_{y1}$, $\bar{\theta}_{y2}$, and ordered according to step (2) in section 4.1, are given by

$$[\bar{\Phi}_1 \quad \bar{\Phi}_2] = \begin{bmatrix} \bar{\varphi}_1 & \bar{\varphi}_3 \\ \bar{\varphi}_2 & \bar{\varphi}_4 \end{bmatrix} = \begin{bmatrix} \bar{w}_1 & \bar{\theta}_{y1} \\ \bar{w}_2 & \bar{\theta}_{y2} \end{bmatrix}.$$

The relation of the sets of basis vectors $\bar{\Phi}_l$ to the sets of nodal functions Φ_l is expressed by the following system of equations with the supermatrix in diagonal form containing two transformation matrices T of the group C_2:

$$\begin{bmatrix} \bar{\Phi}_1 \\ \bar{\Phi}_2 \end{bmatrix} = \begin{bmatrix} T & \\ & T \end{bmatrix} \begin{bmatrix} \Phi_1 \\ \Phi_2 \end{bmatrix} \quad \text{or} \quad \bar{\Phi}_l = \bar{T}\Phi_l,$$

with

$$T = \frac{1}{2} \begin{bmatrix} 1 & 1 \\ 1 & -1 \end{bmatrix} \quad \text{and} \quad [\Phi_1 \quad \Phi_2] = \begin{bmatrix} \varphi_1 & \varphi_3 \\ \varphi_2 & \varphi_4 \end{bmatrix} = \begin{bmatrix} w_1 & \theta_{y1} \\ w_2 & \theta_{y2} \end{bmatrix}.$$

Conversely, the relation of the set of nodal functions Φ_I to the sets of basis vectors $\bar{\Phi}_I$ is

$$\begin{bmatrix} \Phi_1 \\ \Phi_2 \end{bmatrix} = 2 \begin{bmatrix} T & \\ & T \end{bmatrix} \begin{bmatrix} \bar{\Phi}_1 \\ \bar{\Phi}_2 \end{bmatrix} \quad \text{or} \quad \Phi_I = 2\bar{T}\bar{\Phi}_I,$$

since $T^{-1} = 2T$.

The displacement functions for the beam element are usually assumed to be

$$w = \alpha_1 + \alpha_2 x + \alpha_3 x^2 + \alpha_4 x^3$$

$$\theta_y = \frac{dw}{dx} = \alpha_2 + 2\alpha_3 x + 3\alpha_4 x^2$$

or

$$\delta(x, y) = \begin{bmatrix} w \\ \theta_y \end{bmatrix} = \begin{bmatrix} 1 & x & x^2 & x^3 \\ 0 & 1 & 2x & 3x^2 \end{bmatrix} [\alpha_1 \ \alpha_2 \ \alpha_3 \ \alpha_4]^T.$$

For the group supermatrix procedure, as in the case of the four-node bar element in section 5.2, the terms in the above polynomial will be allocated by the following expression to their respective G-invariant subspaces, with the unique order of the terms that gives the block diagonal form of the matrix of the system of equations relating displacements $\bar{U}$ to coefficients A:

$$\begin{bmatrix} F_1 \\ F_2 \end{bmatrix} = \begin{bmatrix} 1 \\ x \end{bmatrix} [1 \ x^2] = \begin{bmatrix} 1 & x^2 \\ x & x^3 \end{bmatrix} \quad \begin{matrix} \text{Subspace} \\ U_1 \\ U_2 \end{matrix}$$

As in section 5.2, the column vector and the row vector correspond to the Cartesian sets given in addition to the character table of the group C_2, pertaining to the symmetry types of group representations A, B.

Thus, with

$$\theta_y = -\frac{dw}{dx},$$

the displacement function

$$\begin{bmatrix} w \\ \theta_y \end{bmatrix} = \begin{bmatrix} 1 & x^2 & \vdots & x & x^3 \\ 0 & -2x & \vdots & -1 & -3x^2 \end{bmatrix} [\alpha_1 \ \alpha_2 \ \vdots \ \alpha_3 \ \alpha_4]^T$$

$$= [F_1 \ \vdots \ F_2][A_1 \ \vdots \ A_2]^T$$

is decomposed into two two-dimensional G-invariant subspaces U_1, U_2.

Substitution of the nodal coordinate of the first node in the nodal set $S(1, 2)$, i.e. $x = c$ for the node 1, into the function matrices F_1, F_2 of $[w \ \theta_y]^T$ pertaining to the

Sec. 5.3]   Stiffness equations in G-invariant subspaces for the beam element   269

subspaces U_1, U_2, provides the relations of the displacements $\bar{\Phi}$ to the coefficients A for each subspace separately:

$$\begin{bmatrix} \bar{w}_1 \\ \bar{\theta}_{y1} \\ \hdashline \bar{w}_2 \\ \bar{\theta}_{y2} \end{bmatrix} = \frac{1}{2} \begin{bmatrix} w_1 + w_2 \\ \theta_{y1} + \theta_{y2} \\ \hdashline w_1 - w_2 \\ \theta_{y1} - \theta_{y2} \end{bmatrix} = \begin{bmatrix} 1 & c^2 & & \\ 0 & -2c & & \\ \hdashline & & c & c^3 \\ & & -1 & -3c^2 \end{bmatrix} \begin{bmatrix} \alpha_1 \\ \alpha_2 \\ \hdashline \alpha_3 \\ \alpha_4 \end{bmatrix}$$

or

$$\begin{bmatrix} \bar{\Phi}^{(1)} \\ \bar{\Phi}^{(2)} \end{bmatrix} = \begin{bmatrix} \bar{C}_1 & \\ & \bar{C}_2 \end{bmatrix} \begin{bmatrix} A_1 \\ A_2 \end{bmatrix}$$

The four-dimensional G-vector space of the problem is decomposed now into two two-dimensional G-invariant subspaces U_1, U_2.

The coefficients A are determined by inverting the matrices $\bar{C}_1$, $\bar{C}_2$ in subspaces U_1, U_2

$$\begin{bmatrix} \alpha_1 \\ \alpha_2 \\ \hdashline \alpha_3 \\ \alpha_4 \end{bmatrix} = \begin{bmatrix} 1 & \dfrac{c}{2} & & \\ 0 & -\dfrac{1}{2c} & & \\ \hdashline & & \dfrac{3}{2c} & \dfrac{1}{2} \\ & & -\dfrac{1}{2c^3} & -\dfrac{1}{2c^2} \end{bmatrix} \begin{bmatrix} \bar{w}_1 \\ \bar{\theta}_{y1} \\ \hdashline \bar{w}_2 \\ \bar{\theta}_{y2} \end{bmatrix} \quad \text{or} \quad A = \bar{C}^{-1} \bar{\Phi}.$$

The four-dimensional G-vector space of the displacement field Δ, decomposed into two two-dimensional G-invariant subspaces U_1, U_2, is

$$\Delta = T_D \bar{C}^{-1} \bar{\Phi},$$

with

$$T_D = \text{Diag} \begin{bmatrix} 1 & x^2 & \vdots & x & x^3 \end{bmatrix}$$

$$\begin{bmatrix} \Delta_1 \\ \cdots \\ \Delta_2 \end{bmatrix} = \begin{bmatrix} 1 & \dfrac{c}{2} & & \\ 0 & -\dfrac{x^2}{2c} & & \\ \cdots & \cdots & \cdots & \cdots \\ & & \dfrac{3x}{2c} & \dfrac{x}{2} \\ & & -\dfrac{x^3}{2c^3} & -\dfrac{x^3}{2c^2} \end{bmatrix} \begin{bmatrix} \bar{w}_1 \\ \bar{\theta}_{y1} \\ \cdots \\ \bar{w}_2 \\ \bar{\theta}_{y2} \end{bmatrix} \quad \text{or} \quad \Delta = \bar{N}_\Delta \bar{\Phi}.$$

The shape functions $\bar{N}^{(i)}$ in G-invariant subspaces are obtained by
$$\bar{N}^{(i)} = S\bar{N}_\Delta,$$
with
$$S = [1 \quad 1 \ \vdots \ 1 \quad 1]$$

$$\begin{bmatrix} \bar{N}^{(1)} \\ \cdots \\ \bar{N}^{(2)} \end{bmatrix} = \begin{bmatrix} 1 & \dfrac{c}{2} - \dfrac{x^2}{2c} \\ \cdots & \cdots \\ \dfrac{3x}{2c} - \dfrac{x^3}{2c^3} & \dfrac{x}{2} - \dfrac{x^3}{2c^2} \end{bmatrix}.$$

In the subspace U_1

$$\bar{B}_1 = -y\frac{d^2}{dx^2}\bar{N}^{(1)} = -y\frac{d^2}{dx^2}\begin{bmatrix} 1 & \dfrac{c}{2} - \dfrac{x^2}{2c} \end{bmatrix} = -y\begin{bmatrix} 0 & -\dfrac{1}{c} \end{bmatrix}$$

$$\bar{K}_1 = E'\int_v \bar{B}_1^T \bar{B}_1 \, dV = E'\int_A y^2 \, dy \int_0^c \begin{bmatrix} 0 & 0 \\ 0 & \dfrac{1}{c^2} \end{bmatrix} dx = \frac{E'I}{c^3}\begin{bmatrix} 0 & 0 \\ 0 & c^2 \end{bmatrix}.$$

In the subspace U_2

$$\bar{B}_2 = -y\frac{d^2}{dx^2}\bar{N}^{(2)} = -y\frac{d^2}{dx^2}\begin{bmatrix} \dfrac{3x}{2c} - \dfrac{x^3}{2c^3} & \dfrac{x}{2} - \dfrac{x^3}{2c^2} \end{bmatrix}$$

$$= -y\begin{bmatrix} -\dfrac{3x}{c^3} & -\dfrac{3x}{c^2} \end{bmatrix}$$

Sec. 5.3] Stiffness equations in G-invariant subspaces for the beam element 271

$$\bar{K}_2 = E' \int_v \bar{B}_2^T \bar{B}_2 \, dV = E' \int_A y^2 \, dy \int_0^c \begin{bmatrix} \dfrac{9x^2}{c^6} & \dfrac{9x^2}{c^5} \\ \dfrac{9x^2}{c^5} & \dfrac{9x^2}{c^4} \end{bmatrix} dx = \dfrac{E'I}{c^3} \begin{bmatrix} 3 & 3c \\ 3c & 3c^2 \end{bmatrix}.$$

Thus, the stiffness equation with the stiffness matrix in block diagonal form, i.e. with stiffness matrices of G-invariant subspaces U_1, U_2, is obtained:

$$\begin{bmatrix} \bar{K}_1 & \\ & \bar{K}_2 \end{bmatrix} \begin{bmatrix} \bar{\Phi}^{(1)} \\ \bar{\Phi}^{(2)} \end{bmatrix} = \dfrac{E'I}{c^3} \begin{bmatrix} 0 & 0 & & \\ 0 & c^2 & & \\ & & 3 & 3c \\ & & 3c & 3c^2 \end{bmatrix} \begin{bmatrix} \bar{w}_1 \\ \bar{\theta}_{y1} \\ \bar{w}_2 \\ \bar{\theta}_{y2} \end{bmatrix}$$

$$= \dfrac{2E'I}{l^3} \begin{bmatrix} 0 & 0 & & \\ 0 & l^2 & & \\ & & 12 & 6l \\ & & 6l & 3l^2 \end{bmatrix} \begin{bmatrix} \bar{w}_1 \\ \bar{\theta}_{y1} \\ \bar{w}_2 \\ \bar{\theta}_{y2} \end{bmatrix} = \begin{bmatrix} \bar{P}^{(1)} \\ \bar{P}^{(2)} \end{bmatrix} = \begin{bmatrix} \bar{P}_{z1} \\ \bar{M}_{y1} \\ \bar{P}_{z2} \\ \bar{M}_{y2} \end{bmatrix}$$

or $\bar{K}\bar{\Phi} = \bar{P}$.

As shown in section 3.2, the above equation can be written as

$$\begin{bmatrix} A+B & \\ & A-B \end{bmatrix} \begin{bmatrix} \bar{\Phi}^{(1)} \\ \bar{\Phi}^{(2)} \end{bmatrix} = \begin{bmatrix} \bar{P}^{(1)} \\ \bar{P}^{(2)} \end{bmatrix}$$

and transformed into the stiffness equation with the stiffness group supermatrix in normal form by the group supermatrix transformation

$$T^{-1}\bar{G}TT^{-1}\bar{\Phi} = T^{-1}\bar{P}$$

derived in section 3.1 and applied to $\bar{K}\bar{\Phi} = \bar{P}$, giving

$$\begin{bmatrix} E & E \\ E & -E \end{bmatrix} \begin{bmatrix} A+B & \\ & A-B \end{bmatrix} \dfrac{1}{2} \begin{bmatrix} E & E \\ E & -E \end{bmatrix} \begin{bmatrix} E & E \\ E & -E \end{bmatrix} \dfrac{1}{2} \begin{bmatrix} \Phi^{(1)} + \Phi^{(2)} \\ \Phi^{(1)} - \Phi^{(2)} \end{bmatrix}$$

$$= \begin{bmatrix} E & E \\ E & -E \end{bmatrix} \dfrac{1}{2} \begin{bmatrix} P^{(1)} + P^{(2)} \\ P^{(1)} - P^{(2)} \end{bmatrix}$$

$$\begin{bmatrix} A & B \\ B & A \end{bmatrix} \begin{bmatrix} \Phi^{(1)} \\ \Phi^{(2)} \end{bmatrix} = \begin{bmatrix} P^{(1)} \\ P^{(2)} \end{bmatrix}$$

and

$$A = \tfrac{1}{2}[(A+B)+(A-B)], \quad B = \tfrac{1}{2}[(A+B)-(A-B)]$$

$$A = \frac{1}{2}\frac{2E'I}{l^3}\left\{\begin{bmatrix} 0 & 0 \\ 0 & l^2 \end{bmatrix} + \begin{bmatrix} 12 & 6l \\ 6l & 3l^2 \end{bmatrix}\right\} = \frac{E'I}{l^3}\begin{bmatrix} 12 & 6l \\ 6l & 4l^2 \end{bmatrix}$$

$$B = \frac{1}{2}\frac{2E'I}{l^3}\left\{\begin{bmatrix} 0 & 0 \\ 0 & l^2 \end{bmatrix} - \begin{bmatrix} 12 & 6l \\ 6l & 3l^2 \end{bmatrix}\right\} = \frac{E'I}{l^3}\begin{bmatrix} -12 & -6l \\ -6l & -2l^2 \end{bmatrix},$$

so that the stiffness equation with the stiffness group supermatrix in normal form is

$$\frac{E'I}{l^3}\begin{bmatrix} 12 & 6l & \vdots & -12 & -6l \\ 6l & 4l^2 & \vdots & -6l & -2l^2 \\ \cdots & \cdots & & \cdots & \cdots \\ -12 & -6l & \vdots & 12 & 6l \\ -6l & -2l^2 & \vdots & 6l & 4l^2 \end{bmatrix}\begin{bmatrix} w_1 \\ \theta_{y1} \\ \cdots \\ w_2 \\ \theta_{y2} \end{bmatrix} = \begin{bmatrix} P_{z1} \\ M_{y1} \\ \cdots \\ P_{z2} \\ M_{y2} \end{bmatrix} \quad \text{or} \quad K\Phi = P.$$

In order to obtain the stiffness equation in the conventional form, positive directions of the nodal rotation θ_{y2} and the nodal moment M_{y2} must be changed to suit the conventional positive directions, which is performed by

$$T_D K T_D T_D \Phi = T_D P, \quad \text{where} \quad T = \text{Diag}\,[1 \quad 1 \;\vdots\; 1 \quad \bar{1}],$$

producing $\theta'_{y2} = -\theta_{y2}$, $M'_{y2} = -M_{y2}$ and the conventional stiffness equation of the beam element

$$\frac{E'I}{l^3}\begin{bmatrix} 12 & 6l & -12 & 6l \\ 6l & 4l^2 & -6l & 2l^2 \\ -12 & -6l & 12 & -6l \\ 6l & 2l^2 & -6l & 4l^2 \end{bmatrix}\begin{bmatrix} w_1 \\ \theta_{y1} \\ w_2 \\ \theta'_{y2} \end{bmatrix} = \begin{bmatrix} P_{z1} \\ M_{y1} \\ P_{z2} \\ M'_{y2} \end{bmatrix} \quad \text{or} \quad K_c\Phi' = P'.$$

Conversely, by $T_D K_c T_D T_D \Phi' = T_D P'$, the above stiffness equation is transformed back into the stiffness equation with the stiffness group supermatrix in normal form

$$K\Phi = P \quad \text{or} \quad \begin{bmatrix} A & B \\ B & A \end{bmatrix}\begin{bmatrix} \Phi^{(1)} \\ \Phi^{(2)} \end{bmatrix} = \begin{bmatrix} P^{(1)} \\ P^{(2)} \end{bmatrix}.$$

The stiffness equation with the stiffness group supermatrix in diagonal form is obtained by application of the group supermatrix transformation

$$TGT^{-1}T\Phi = TP,$$

derived in section 3.1, to $K\Phi = P$, so that

Sec. 5.4]   Stiffness equations/rectangular element for planar analysis   273

$$\frac{1}{2}\begin{bmatrix} E & E \\ E & -E \end{bmatrix}\begin{bmatrix} A & B \\ B & A \end{bmatrix}\begin{bmatrix} E & E \\ E & -E \end{bmatrix}\frac{1}{2}\begin{bmatrix} E & E \\ E & -E \end{bmatrix}\begin{bmatrix} \Phi^{(1)} \\ \Phi^{(2)} \end{bmatrix} = \frac{1}{2}\begin{bmatrix} E & E \\ E & -E \end{bmatrix}\begin{bmatrix} P^{(1)} \\ P^{(2)} \end{bmatrix}$$

$$\begin{bmatrix} A+B & \\ & A-B \end{bmatrix}\begin{bmatrix} \bar{\Phi}^{(1)} \\ \bar{\Phi}^{(2)} \end{bmatrix} = \begin{bmatrix} \bar{P}^{(1)} \\ \bar{P}^{(2)} \end{bmatrix}$$

or

$$\frac{2E'I}{l^3}\begin{bmatrix} 0 & 0 & & \\ 0 & l^2 & & \\ \hdashline & & 12 & 6l \\ & & 6l & 3l^2 \end{bmatrix}\begin{bmatrix} \bar{w}_1 \\ \bar{\theta}_{y1} \\ \hdashline \bar{w}_2 \\ \bar{\theta}_{y2} \end{bmatrix} = \begin{bmatrix} \bar{P}_{z1} \\ \bar{M}_{y1} \\ \hdashline \bar{P}_{z2} \\ \bar{M}_{y2} \end{bmatrix}.$$

Table 5.1 presents a comparison of derivations of the stiffness equation of the beam element in the conventional way and by the group supermatrix procedure.

5.4 STIFFNESS EQUATIONS IN G-INVARIANT SUBSPACES FOR THE RECTANGULAR ELEMENT FOR PLANAR ANALYSIS

The four-node 8-degrees-of-freedom rectangular element, as given in Fig. 5.4(b), with the unique nodal numbering derived in section 3.5, is described by the group C_{2v}, where positive directions of displacements u_1, u_2, u_3, u_4, v_1, v_2, v_3, v_4 suit the first symmetry type of the group, which differs from the conventional set of positive directions in Fig. 5.4(a).

The nodes of its nodal set $S(1, 2, 3, 4)$ are permuted by action of group elements, i.e. symmetry operations E (identity), C_2 (rotation through 180° about the z axis), σ_1 and σ_2 (reflections in xz and yz planes respectively).

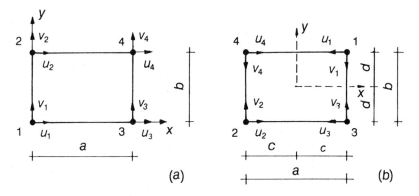

Fig. 5.4. The four-node rectangular element for planar analysis: (a) with the conventional set of positive directions of nodal displacements; (b) with the set of positive directions of nodal displacements that suit the symmetry type of the first irreducible representation of the group C_{2v}.

Table 5.1. Comparison of derivations of the stiffness equation of the beam element in the conventional way and by the group supermatrix procedure

	Derivation of the stiffness equation of the beam element with four degrees of freedom	
	Conventional way	Group supermatrix procedure
Displacement functions spanning the space of the deflection field	$\Phi' = \begin{bmatrix} w_1 \\ \theta_{y1} \\ w_2 \\ \theta'_{y2} \end{bmatrix}$ Space V	$\bar{\Phi}^{(1)} = \begin{bmatrix} \bar{w}_1 \\ \bar{\theta}_{y1} \end{bmatrix} = \frac{1}{2}\begin{bmatrix} w_1 + w_2 \\ \theta_{y1} + \theta_{y2} \end{bmatrix}$ Subspace U_1 $\bar{\Phi}^{(2)} = \begin{bmatrix} \bar{w}_2 \\ \bar{\theta}_{y2} \end{bmatrix} = \frac{1}{2}\begin{bmatrix} w_1 - w_2 \\ \theta_{y1} - \theta_{y2} \end{bmatrix}$ U_1
Matrix that is inverted to yield the coefficients of the polynomial displacement function	$\begin{bmatrix} 1 & 0 & 0 & 0 \\ 0 & 1 & 0 & 0 \\ 1 & l & l^2 & l^3 \\ 0 & 1 & 2l & 3l^2 \end{bmatrix}$	$\begin{bmatrix} 1 & c^2 & \vdots & & \\ 0 & -2c & \vdots & 0 & \\ \cdots & \cdots & \vdots & \cdots & \cdots \\ & & \vdots & c & c^3 \\ & 0 & \vdots & -1 & -3c^2 \end{bmatrix}$ Subspace U_1 U_2 $(c = l/2)$

Table 5.1. (continued)

Derivation of the stiffness equation of the beam element with four degrees of freedom

Polynomial displacement function	Conventional way	Group supermatrix procedure
	$w = \left[1 - 3\left(\dfrac{x}{l}\right)^2 + 2\left(\dfrac{x}{l}\right)^3\right] w_1$ $\quad + \left(x - 2\dfrac{x^2}{l} + \dfrac{x^3}{l^2}\right) \theta_{y1}$ $\quad + \left[3\left(\dfrac{x}{l}\right)^2 - 2\left(\dfrac{x}{l}\right)^3\right] w_2$ $\quad + \left(-\dfrac{x^2}{l} + \dfrac{x^3}{l^2}\right) \theta_{y2}$	Subspace U_1 $$\Delta_1 = \begin{bmatrix} 1 & \dfrac{c}{2} \\ 0 & -\dfrac{x^2}{2c} \end{bmatrix} \begin{matrix} \bar{w}_1 \\ \bar{\theta}_{y1} \end{matrix}$$ Subspace U_2 $$\Delta_2 = \begin{bmatrix} \dfrac{3x}{2c} & \dfrac{x}{2} \\ -\dfrac{x^3}{2c^3} & -\dfrac{x^3}{2c^2} \end{bmatrix} \begin{matrix} \bar{w}_2 \\ \bar{\theta}_{y2} \end{matrix}$$

Table 5.1. (continued)

	Derivation of the stiffness equation of the beam element with four degrees of freedom	
	Conventional way	Group supermatrix procedure
Matrix B in the product $B^T B$	$B^T = -y \begin{bmatrix} -\dfrac{6}{l^2} + \dfrac{12}{l^3} \\ -\dfrac{4}{l} + \dfrac{6}{l^2} \\ \dfrac{6}{l^2} - \dfrac{12}{l^3} \\ -\dfrac{2}{l} - \dfrac{6}{l^2} \end{bmatrix}$	Subspace U_1: $\bar{B}_1 = -y \begin{bmatrix} 0 & -\dfrac{1}{c} \end{bmatrix}$ Subspace U_2: $\bar{B}_2 = -y \begin{bmatrix} -\dfrac{3x}{c^3} & -\dfrac{3x}{c^2} \end{bmatrix}$

Table 5.1. (continued)

Derivation of the stiffness equation of the beam element with four degrees of freedom

	Conventional way	Group supermatrix procedure
Integration of polynomials	$\displaystyle\int_0^l \begin{bmatrix} \# & \# & \# & \# \\ & \# & \# & \# \\ & & \# & \# \\ \text{Sym.} & & & \# \end{bmatrix} dx$ Integration of 10 three-term polynomials (# non-zero element)	$\displaystyle\int_0^{l/2} \begin{bmatrix} 0 & 0 & 0 & 0 \\ 0 & \# & 0 & \# \\ \hline 0 & 0 & \# & \# \\ 0 & \# & \text{Sym.} & \# \end{bmatrix} dx$ Integration of 4 one-term polynomials
Stiffness equation	$\dfrac{E'I}{l^3}\begin{bmatrix} 12 & 6l & -12 & 6l \\ 6l & 4l^2 & -6l & 2l^2 \\ -12 & -6l & 12 & -6l \\ 6l & 2l^2 & -6l & 4l^2 \end{bmatrix} \begin{bmatrix} w_1 \\ \theta_{y1} \\ w_2 \\ \theta'_{y2} \end{bmatrix} = \begin{bmatrix} P_{z1} \\ M_{y1} \\ P_{z2} \\ M'_{y2} \end{bmatrix}$	$\dfrac{2E'I}{l^3}\begin{bmatrix} 0 & 0 & 0 & 0 \\ 0 & l^2 & 0 & 0 \\ \hline 0 & 0 & 12 & 6l \\ 0 & 0 & 6l & 3l^2 \end{bmatrix} \begin{bmatrix} \bar{w}_1 \\ \bar{\theta}_{y1} \\ \bar{w}_2 \\ \bar{\theta}_{y2} \end{bmatrix} = \begin{bmatrix} \bar{P}_{z1} \\ \bar{M}_{y1} \\ \bar{P}_{z2} \\ \bar{M}_{y2} \end{bmatrix}$ (in block diagonal form)

278 Stiffness equations in G-invariant subspaces [Ch. 5

The character table of the group C_{2v} (with Cartesian sets and products) is

C_{2v}	E	C_2	σ_1 (xz)	σ_2 (yz)		
A_1	1	1	1	1	z	x^2, y^2, z^2
A_2	1	1	-1	-1		xy
B_1	1	-1	1	-1	x	xz
B_2	1	-1	-1	1	y	yz

while the idempotents of the centre of group algebra are

$$\begin{bmatrix} \pi_1 \\ \pi_2 \\ \pi_3 \\ \pi_4 \end{bmatrix} = \frac{1}{4} \begin{bmatrix} 1 & 1 & 1 & 1 \\ 1 & 1 & -1 & -1 \\ 1 & -1 & 1 & -1 \\ 1 & -1 & -1 & 1 \end{bmatrix} \begin{bmatrix} E \\ C_2 \\ \sigma_1 \\ \sigma_2 \end{bmatrix}$$

or $\Pi = T\Sigma$, with $T^{-1} = 4T$ as obtained in section 4.2.

By application of the idempotents π_i ($i = 1, 2, 3, 4$) to the nodal functions φ_j ($j = 1, 2, \ldots, 8$) the basis vectors $\bar{\varphi}_j$ of G-invariant subspaces U_i are obtained:

$$\begin{bmatrix} \bar{\Phi}^{(1)} \\ \bar{\Phi}^{(2)} \\ \bar{\Phi}^{(3)} \\ \bar{\Phi}^{(4)} \end{bmatrix} = \begin{bmatrix} \bar{\varphi}_1 & \bar{\varphi}_5 \\ \bar{\varphi}_2 & \bar{\varphi}_6 \\ \bar{\varphi}_3 & \bar{\varphi}_7 \\ \bar{\varphi}_4 & \bar{\varphi}_8 \end{bmatrix} = \begin{bmatrix} \bar{u}_1 & \bar{v}_1 \\ \bar{u}_2 & \bar{v}_2 \\ \bar{u}_3 & \bar{v}_3 \\ \bar{u}_4 & \bar{v}_4 \end{bmatrix} = \frac{1}{4} \begin{bmatrix} 1 & 1 & 1 & 1 \\ 1 & 1 & -1 & -1 \\ 1 & -1 & 1 & -1 \\ 1 & -1 & -1 & 1 \end{bmatrix} \begin{bmatrix} u_1 & v_1 \\ u_2 & v_2 \\ u_3 & v_3 \\ u_4 & v_4 \end{bmatrix}$$

$$= \frac{1}{4} \begin{bmatrix} u_1 + u_2 + u_3 + u_4 & v_1 + v_2 + v_3 + v_4 \\ u_1 + u_2 - u_3 - u_4 & v_1 + v_2 - v_3 - v_4 \\ u_1 - u_2 + u_3 - u_4 & v_1 - v_2 + v_3 - v_4 \\ u_1 - u_2 - u_3 + u_4 & v_1 - v_2 - v_3 + v_4 \end{bmatrix}$$

or $\bar{\Phi} = T\Phi$, with $T^{-1} = 4T$.

Conversely, the relation of the nodal functions $\Phi^{(i)}$ to the basis vectors $\bar{\Phi}^{(i)}$ is

$$\begin{bmatrix} \Phi^{(1)} \\ \Phi^{(2)} \\ \Phi^{(3)} \\ \Phi^{(4)} \end{bmatrix} = \begin{bmatrix} \varphi_1 & \varphi_5 \\ \varphi_2 & \varphi_6 \\ \varphi_3 & \varphi_7 \\ \varphi_4 & \varphi_8 \end{bmatrix} = \begin{bmatrix} u_1 & v_1 \\ u_2 & v_2 \\ u_3 & v_3 \\ u_4 & v_4 \end{bmatrix} = \begin{bmatrix} 1 & 1 & 1 & 1 \\ 1 & 1 & -1 & -1 \\ 1 & -1 & 1 & -1 \\ 1 & -1 & -1 & 1 \end{bmatrix} \begin{bmatrix} \bar{u}_1 & \bar{v}_1 \\ \bar{u}_2 & \bar{v}_2 \\ \bar{u}_3 & \bar{v}_3 \\ \bar{u}_4 & \bar{v}_4 \end{bmatrix}$$

Sec. 5.4] Stiffness equations/rectangular element for planar analysis 279

$$= \begin{bmatrix} \bar{u}_1 + \bar{u}_2 + \bar{u}_3 + \bar{u}_4 & \bar{v}_1 + \bar{v}_2 + \bar{v}_3 + \bar{v}_4 \\ \bar{u}_1 + \bar{u}_2 - \bar{u}_3 - \bar{u}_4 & \bar{v}_1 + \bar{v}_2 - \bar{v}_3 - \bar{v}_4 \\ \bar{u}_1 - \bar{u}_2 + \bar{u}_3 - \bar{u}_4 & \bar{v}_1 - \bar{v}_2 + \bar{v}_3 - \bar{v}_4 \\ \bar{u}_1 - \bar{u}_2 - \bar{u}_3 + \bar{u}_4 & \bar{v}_1 - \bar{v}_2 - \bar{v}_3 + \bar{v}_4 \end{bmatrix}.$$

The sets $\bar{\Phi}_l$ ($l = 1, 2$) containing basis vectors $\bar{\varphi}_j$ ($j = 1, 2, \ldots, 8$), representing $\bar{u}_1, \bar{u}_2, \bar{u}_3, \bar{u}_4, \bar{v}_1, \bar{v}_2, \bar{v}_3, \bar{v}_4$ and ordered according to the sequence explained in section 4.1 (step (2)), are given by

$$[\bar{\Phi}_1 \quad \bar{\Phi}_2] = \begin{bmatrix} \bar{\varphi}_1 & \bar{\varphi}_5 \\ \bar{\varphi}_2 & \bar{\varphi}_6 \\ \bar{\varphi}_3 & \bar{\varphi}_7 \\ \bar{\varphi}_4 & \bar{\varphi}_8 \end{bmatrix} = \begin{bmatrix} \bar{u}_1 & \bar{v}_1 \\ \bar{u}_2 & \bar{v}_2 \\ \bar{u}_3 & \bar{v}_3 \\ \bar{u}_4 & \bar{v}_4 \end{bmatrix}.$$

The relation of the sets of basis vectors $\bar{\Phi}_l$ to the sets of nodal functions Φ_l is expressed by the following system of equations with the supermatrix in diagonal form containing two transformation matrices T of the group C_{2v}:

$$\begin{bmatrix} \bar{\Phi}_1 \\ \bar{\Phi}_2 \end{bmatrix} = \begin{bmatrix} T & \\ & T \end{bmatrix} \begin{bmatrix} \Phi_1 \\ \Phi_2 \end{bmatrix} \quad \text{or} \quad \bar{\Phi}_l = T\Phi_l,$$

with

$$T = \frac{1}{4} \begin{bmatrix} 1 & 1 & 1 & 1 \\ 1 & 1 & -1 & -1 \\ 1 & -1 & 1 & -1 \\ 1 & -1 & -1 & 1 \end{bmatrix}$$

and

$$[\Phi_1 \quad \Phi_2] = \begin{bmatrix} \varphi_1 & \varphi_5 \\ \varphi_2 & \varphi_6 \\ \varphi_3 & \varphi_7 \\ \varphi_4 & \varphi_8 \end{bmatrix} = \begin{bmatrix} u_1 & v_1 \\ u_2 & v_2 \\ u_3 & v_3 \\ u_4 & v_4 \end{bmatrix}.$$

Conversely, the relation of the sets of nodal functions Φ_l to the sets of basis vectors $\bar{\Phi}_l$ is

$$\begin{bmatrix} \Phi_1 \\ \Phi_2 \end{bmatrix} = 4 \begin{bmatrix} T & \\ & T \end{bmatrix} \begin{bmatrix} \bar{\Phi}_1 \\ \bar{\Phi}_2 \end{bmatrix} \quad \text{or} \quad \Phi_l = 4T\bar{\Phi}_l,$$

since $T^{-1} = 4T$.

The displacement function for the rectangular element for planar analysis is usually assumed to be

$$u = \alpha_1 + \alpha_2 x + \alpha_3 y + \alpha_4 xy$$

$$v = \alpha_5 + \alpha_6 x + \alpha_7 y + \alpha_8 xy.$$

For the group supermatrix procedure the terms in the above polynomial will be allocated by the following expression to their respective G-invariant subspaces with the order of the terms that provides the diagonal form of the supermatrix of the system of equations relating the displacements $\bar{\Phi}$ to the coefficients A:

$$[x \quad y] \begin{bmatrix} 1 & 0 & xy & 0 & x & 0 & y & 0 \\ 0 & 1 & 0 & xy & 0 & x & 0 & y \end{bmatrix} = \begin{bmatrix} x & 0 & x^2y & 0 & x^2 & 0 & xy & 0 \\ 0 & y & 0 & xy^2 & 0 & xy & 0 & y^2 \end{bmatrix},$$

where $1, xy, x, y$ correspond to the symmetry types of group representations A_1, A_2, B_1, B_2 respectively, as given in addition to the character table of the group C_{2v}. Since the term $x^2 y$ possesses the same symmetry type as y, x^2 as 1, xy^2 as x, y^2 as 1, the following displacement function is obtained:

$$\begin{bmatrix} u \\ v \end{bmatrix} = \begin{bmatrix} x & & y & & 1 & & xy & \\ & y & & x & & xy & & 1 \end{bmatrix} [\alpha_1 \quad \alpha_2 \mid \alpha_3 \quad \alpha_4$$

$$\alpha_5 \quad \alpha_6 \mid \alpha_7 \quad \alpha_8]^T$$

$$= [F_1 \mid F_2 \mid F_3 \mid F_4] [A_1 \mid A_2 \mid A_3 \mid A_4]^T.$$

Thus, the space is decomposed into four two-dimensional G-invariant subspaces U_1, U_2, U_3, U_4.

Substitution of the nodal coordinates of the first node in the nodal set $S(1, 2, 3, 4)$, i.e. $x = c$, $y = d$ for the node 1, into the function matrices F_1, F_2, F_3, F_4 of $[u \ v]^T$ pertaining to the subspaces U_1, U_2, U_3, U_4 will give the relations of the displacements $\bar{\Phi}$ to the coefficients A for each subspace separately:

$$\begin{bmatrix} \bar{u}_1 \\ \bar{v}_1 \\ --- \\ \bar{u}_2 \\ \bar{v}_2 \\ --- \\ \bar{u}_3 \\ \bar{v}_3 \\ --- \\ \bar{u}_4 \\ \bar{v}_4 \end{bmatrix} = \begin{bmatrix} c & & & & & & & \\ & d & & & & & & \\ & & d & & & & & \\ & & & c & & & & \\ & & & & 1 & & & \\ & & & & & cd & & \\ & & & & & & cd & \\ & & & & & & & 1 \end{bmatrix} \begin{bmatrix} \alpha_1 \\ \alpha_2 \\ --- \\ \alpha_3 \\ \alpha_4 \\ --- \\ \alpha_5 \\ \alpha_6 \\ --- \\ \alpha_7 \\ \alpha_8 \end{bmatrix}$$

or $\bar{\Phi} = \bar{C} A$.

Sec. 5.4] **Stiffness equations/rectangular element for planar analysis** 281

The eight-dimensional G-vector space of the problem is decomposed now into four two-dimensional G-invariant subspaces U_1, U_2, U_3, U_4. The matrix of the system of equations of the above relation is obtained in block diagonal form containing four diagonal 2×2 matrices. Thus, the inversion of the matrix $\bar{C}$ is reduced to writing reciprocal values of c, d, d, c, 1, cd, cd, 1, so that

$$[\alpha_1 \; \alpha_2 \mid \alpha_3 \; \alpha_4 \mid \alpha_5 \; \alpha_6 \mid \alpha_7 \; \alpha_8]^T$$

$$= \text{Diag} \begin{bmatrix} \dfrac{1}{c} & \dfrac{1}{d} \mid \dfrac{1}{d} & \dfrac{1}{c} \mid 1 & \dfrac{1}{cd} \mid \dfrac{1}{cd} & 1 \end{bmatrix}$$

$$[\bar{u}_1 \; \bar{v}_1 \mid \bar{u}_2 \; \bar{v}_2 \mid \bar{u}_3 \; \bar{v}_3 \mid \bar{u}_4 \; \bar{v}_4]^T,$$

or $A = \bar{C}^{-1} \bar{\Phi}$.

The eight-dimensional G-vector space of the displacement field Δ, decomposed into four two-dimensional G-invariant subspaces U_1, U_2, U_3, U_4, is

$$\Delta = T_D \bar{C}^{-1} \bar{\Phi},$$

with

$$T_D = \text{Diag} \, [x \; y \mid y \; x \mid 1 \; xy \mid xy \; 1],$$

$$\begin{bmatrix} \Delta_1 \\ \cdots \\ \Delta_2 \\ \cdots \\ \Delta_3 \\ \cdots \\ \Delta_4 \end{bmatrix} = \begin{bmatrix} \dfrac{x}{c} & & & & & & & \\ & \dfrac{y}{d} & & & & & & \\ \hline & & \dfrac{y}{d} & & & & & \\ & & & \dfrac{x}{c} & & & & \\ \hline & & & & 1 & & & \\ & & & & & \dfrac{xy}{cd} & & \\ \hline & & & & & & \dfrac{xy}{cd} & \\ & & & & & & & 1 \end{bmatrix} \begin{bmatrix} \bar{u}_1 \\ \bar{v}_1 \\ \cdots \\ \bar{u}_2 \\ \bar{v}_2 \\ \cdots \\ \bar{u}_3 \\ \bar{v}_3 \\ \cdots \\ \bar{u}_4 \\ \bar{v}_4 \end{bmatrix}$$

or $\Delta = \bar{N}_\Delta \bar{\Phi}$.

282 Stiffness equations in G-invariant subspaces [Ch. 5]

The shape functions $\bar{N}^{(i)}$ in G-invariant subspaces are

$$[\bar{N}^{(1)} \,\vdots\, \bar{N}^{(2)} \,\vdots\, \bar{N}^{(3)} \,\vdots\, \bar{N}^{(4)}] = \begin{bmatrix} \dfrac{x}{c} & 0 & \vdots & \dfrac{y}{d} & 0 & \vdots & 1 & 0 & \vdots & \dfrac{xy}{cd} & 0 \\ 0 & \dfrac{y}{d} & \vdots & 0 & \dfrac{x}{c} & \vdots & 0 & \dfrac{xy}{cd} & \vdots & 0 & 1 \end{bmatrix}.$$

Now it is possible to derive 2×2 stiffness matrices $\bar{K}_1, \bar{K}_2, \bar{K}_3, \bar{K}_4$ for each subspace U_1, U_2, U_3, U_4 separately.

Subspace U_1:

$$\bar{N}^{(1)} = \begin{bmatrix} \dfrac{x}{c} & 0 \\ 0 & \dfrac{y}{d} \end{bmatrix}, \quad \bar{B}_1 = \begin{bmatrix} \partial/\partial x & 0 \\ 0 & \partial/\partial y \\ \partial/\partial y & \partial/\partial x \end{bmatrix} \begin{bmatrix} \dfrac{x}{c} & 0 \\ 0 & \dfrac{y}{d} \end{bmatrix} = \begin{bmatrix} \dfrac{1}{c} & 0 \\ 0 & \dfrac{1}{d} \\ 0 & 0 \end{bmatrix}$$

and with

$$D = \begin{bmatrix} d_{11} & d_{12} & 0 \\ d_{21} & d_{22} & 0 \\ 0 & 0 & d_{33} \end{bmatrix}$$

for plane stress:

$$d_{11} = d_{22} = \frac{E'}{1-\nu^2}, \quad d_{21} = d_{12} = \frac{\nu E'}{1-\nu^2}, \quad d_{33} = \frac{E'}{2(1+\nu)},$$

for plane strain:

$$d_{11} = d_{22} = \frac{(1-\nu)E'}{(1+\nu)(1-2\nu)}, \quad d_{21} = d_{12} = \frac{\nu E'}{(1+\nu)(1-2\nu)},$$

$$d_{33} = \frac{E'}{2(1+\nu)}$$

$$\bar{B}_1^T D \bar{B}_1 = \begin{bmatrix} \dfrac{1}{c} & 0 & 0 \\ 0 & \dfrac{1}{d} & 0 \end{bmatrix} \begin{bmatrix} d_{11} & d_{12} & 0 \\ d_{21} & d_{22} & 0 \\ 0 & 0 & d_{33} \end{bmatrix} \begin{bmatrix} \dfrac{1}{c} & 0 \\ 0 & \dfrac{1}{d} \\ 0 & 0 \end{bmatrix} = \begin{bmatrix} \dfrac{1}{c^2} d_{11} & \dfrac{1}{cd} d_{12} \\ \dfrac{1}{cd} d_{21} & \dfrac{1}{d^2} d_{22} \end{bmatrix},$$

Sec. 5.4] Stiffness equations/rectangular element for planar analysis 283

$$\bar{K}_1 = \int_0^c \int_0^d \begin{bmatrix} \frac{1}{c^2}d_{11} & \frac{1}{cd}d_{12} \\ \frac{1}{cd}d_{21} & \frac{1}{d^2}d_{22} \end{bmatrix} dx\,dy = \begin{bmatrix} d_{11}p^{-1} & d_{21} \\ d_{21} & d_{22}p \end{bmatrix}$$

with

$$p = \frac{c}{d}.$$

Subspace U_2:

$$\bar{N}^{(2)} = \begin{bmatrix} \frac{y}{d} & 0 \\ 0 & \frac{x}{c} \end{bmatrix}, \quad \bar{B}_2 = \begin{bmatrix} \partial/\partial x & 0 \\ 0 & \partial/\partial y \\ \partial/\partial y & \partial/\partial x \end{bmatrix} \begin{bmatrix} \frac{y}{d} & 0 \\ 0 & \frac{x}{c} \end{bmatrix} = \begin{bmatrix} 0 & 0 \\ 0 & 0 \\ \frac{1}{d} & \frac{1}{c} \end{bmatrix}$$

$$\bar{B}_2^T D \bar{B}_2 = \begin{bmatrix} 0 & 0 & \frac{1}{d} \\ 0 & 0 & \frac{1}{c} \end{bmatrix} \begin{bmatrix} d_{11} & d_{12} & 0 \\ d_{21} & d_{22} & 0 \\ 0 & 0 & d_{33} \end{bmatrix} \begin{bmatrix} 0 & 0 \\ 0 & 0 \\ \frac{1}{d} & \frac{1}{c} \end{bmatrix} = \begin{bmatrix} \frac{1}{d^2}d_{33} & \frac{1}{cd}d_{33} \\ \frac{1}{cd}d_{33} & \frac{1}{c^2}d_{33} \end{bmatrix}$$

$$\bar{K}_2 = \int_0^c \int_0^d \begin{bmatrix} \frac{1}{d^2}d_{33} & \frac{1}{cd}d_{33} \\ \frac{1}{cd}d_{33} & \frac{1}{c^2}d_{33} \end{bmatrix} dx\,dy = \begin{bmatrix} d_{33}p & d_{33} \\ d_{33} & d_{33}p^{-1} \end{bmatrix}.$$

Subspace U_3:

$$\bar{N}^{(3)} = \begin{bmatrix} 1 & 0 \\ 0 & \frac{xy}{cd} \end{bmatrix}, \quad \bar{B}_3 = \begin{bmatrix} \partial/\partial x & 0 \\ 0 & \partial/\partial x \\ \partial/\partial y & \partial/\partial x \end{bmatrix} \begin{bmatrix} 1 & 0 \\ 0 & \frac{xy}{cd} \end{bmatrix} = \begin{bmatrix} 0 & 0 \\ 0 & \frac{x}{cd} \\ 0 & \frac{y}{cd} \end{bmatrix}$$

$$\bar{B}_3^T D \bar{B}_3 = \begin{bmatrix} 0 & 0 & 0 \\ 0 & \dfrac{x}{cd} & \dfrac{y}{cd} \end{bmatrix} \begin{bmatrix} d_{11} & d_{12} & 0 \\ d_{21} & d_{22} & 0 \\ 0 & 0 & d_{33} \end{bmatrix} \begin{bmatrix} 0 & 0 \\ 0 & \dfrac{x}{cd} \\ 0 & \dfrac{y}{cd} \end{bmatrix}$$

$$= \begin{bmatrix} 0 & 0 \\ 0 & \dfrac{x^2}{c^2 d^2} d_{22} + \dfrac{y^2}{c^2 d^2} d_{33} \end{bmatrix}$$

$$\bar{K}_3 = \int_0^c \int_0^d \begin{bmatrix} 0 & 0 \\ 0 & \dfrac{x^2}{c^2 d^2} d_{22} + \dfrac{y^2}{c^2 d^2} d_{33} \end{bmatrix} dx\, dy = \begin{bmatrix} 0 & 0 \\ 0 & \tfrac{1}{3} d_{22} p + \tfrac{1}{3} d_{33} p^{-1} \end{bmatrix}.$$

Subspace U_4:

$$\bar{N}^{(4)} = \begin{bmatrix} \dfrac{xy}{cd} & 0 \\ 0 & 1 \end{bmatrix}, \quad \bar{B}_4 = \begin{bmatrix} \partial/\partial x & 0 \\ 0 & \partial/\partial y \\ \partial/\partial y & \partial/\partial x \end{bmatrix} \begin{bmatrix} \dfrac{xy}{cd} & 0 \\ 0 & 1 \end{bmatrix} = \begin{bmatrix} \dfrac{y}{cd} & 0 \\ 0 & 0 \\ \dfrac{x}{cd} & 0 \end{bmatrix}$$

$$\bar{B}_4 D \bar{B}_4 = \begin{bmatrix} \dfrac{y}{cd} & 0 & \dfrac{x}{cd} \\ 0 & 0 & 0 \end{bmatrix} \begin{bmatrix} d_{11} & d_{12} & 0 \\ d_{21} & d_{22} & 0 \\ 0 & 0 & d_{33} \end{bmatrix} \begin{bmatrix} \dfrac{y}{cd} & 0 \\ 0 & 0 \\ \dfrac{x}{cd} & 0 \end{bmatrix}$$

$$= \begin{bmatrix} \dfrac{y^2}{c^2 d^2} d_{11} + \dfrac{x^2}{c^2 d^2} d_{33} & 0 \\ 0 & 0 \end{bmatrix}$$

$$\bar{K}_4 = \int_0^c \int_0^d \begin{bmatrix} \dfrac{y^2}{c^2 d^2} d_{11} + \dfrac{x^2}{c^2 d^2} d_{33} & 0 \\ 0 & 0 \end{bmatrix} dx\, dy = \begin{bmatrix} \tfrac{1}{3} d_{11} p^{-1} + \tfrac{1}{3} d_{33} p & 0 \\ 0 & 0 \end{bmatrix}.$$

Thus the stiffness equation with the stiffness group supermatrix in diagonal form is obtained

Sec. 5.4] Stiffness equations/rectangular element for planar analysis 285

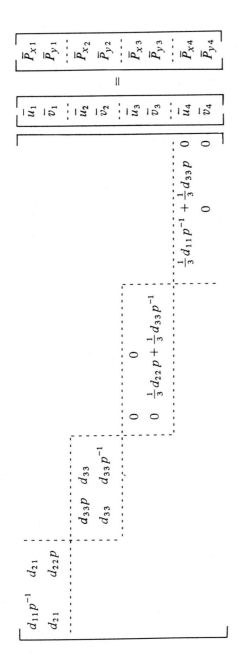

or

$$\begin{bmatrix} \bar{K}_1 & & & \\ & \bar{K}_2 & & \\ & & \bar{K}_3 & \\ & & & \bar{K}_4 \end{bmatrix} \begin{bmatrix} \bar{\Phi}^{(1)} \\ \bar{\Phi}^{(2)} \\ \bar{\Phi}^{(3)} \\ \bar{\Phi}^{(4)} \end{bmatrix} = \begin{bmatrix} \bar{P}^{(1)} \\ \bar{P}^{(2)} \\ \bar{P}^{(3)} \\ \bar{P}^{(4)} \end{bmatrix} \quad \text{or} \quad \bar{K}\bar{\Phi} = \bar{P},$$

or with substitutions $A = 2d_{11}p^{-1}$, $B = 2d_{22}p$, $C = 3d_{21}$, $F = 2d_{33}p$, $G = 2d_{33}p^{-1}$, $H = 3d_{33}$,

$$\frac{1}{6}\begin{bmatrix} 3A & 2C & & & & & & \\ 2C & 3B & & & & & & \\ & & 3F & 2H & & & & \\ & & 2H & 3G & & & & \\ & & & & 0 & 0 & & \\ & & & & 0 & B+G & & \\ & & & & & & A+F & 0 \\ & & & & & & 0 & 0 \end{bmatrix} \begin{bmatrix} \bar{u}_1 \\ \bar{v}_1 \\ \bar{u}_2 \\ \bar{v}_2 \\ \bar{u}_3 \\ \bar{v}_3 \\ \bar{u}_4 \\ \bar{v}_4 \end{bmatrix} = \begin{bmatrix} \bar{P}_{x1} \\ \bar{P}_{y1} \\ \bar{P}_{x2} \\ \bar{P}_{y2} \\ \bar{P}_{x3} \\ \bar{P}_{y3} \\ \bar{P}_{x4} \\ \bar{P}_{y4} \end{bmatrix}.$$

Transformation of the stiffness equation $\bar{K}\bar{\Phi} = \bar{P}$ into the stiffness equation $K\Phi = P$ with the stiffness group supermatrix in normal form is performed by the group supermatrix transformation

$$T^{-1}\bar{K}TT^{-1}\bar{\Phi} = T^{-1}\bar{P},$$

derived in section 3.1, producing

$$\begin{bmatrix} E & E & E & E \\ E & E & -E & -E \\ E & -E & E & -E \\ E & -E & -E & E \end{bmatrix} \begin{bmatrix} \bar{K}_1 & & & \\ & \bar{K}_2 & & \\ & & \bar{K}_3 & \\ & & & \bar{K}_4 \end{bmatrix} \frac{1}{4}\begin{bmatrix} E & E & E & E \\ E & E & -E & -E \\ E & -E & E & -E \\ E & -E & -E & E \end{bmatrix} \cdot$$

$$\cdot \begin{bmatrix} E & E & E & E \\ E & E & -E & -E \\ E & -E & E & -E \\ E & -E & -E & E \end{bmatrix} \begin{bmatrix} \bar{\Phi}^{(1)} \\ \bar{\Phi}^{(2)} \\ \bar{\Phi}^{(3)} \\ \bar{\Phi}^{(4)} \end{bmatrix} = \begin{bmatrix} E & E & E & E \\ E & E & -E & -E \\ E & -E & E & -E \\ E & -E & -E & E \end{bmatrix} \begin{bmatrix} \bar{P}^{(1)} \\ \bar{P}^{(2)} \\ \bar{P}^{(3)} \\ \bar{P}^{(4)} \end{bmatrix}$$

and

Sec. 5.4]   Stiffness equations/rectangular element for planar analysis   287

$$\begin{bmatrix} A_1 & A_2 & A_3 & A_4 \\ A_2 & A_1 & A_4 & A_3 \\ A_3 & A_4 & A_1 & A_2 \\ A_4 & A_3 & A_2 & A_1 \end{bmatrix} \begin{bmatrix} u_1 \\ v_1 \\ --- \\ u_2 \\ v_2 \\ --- \\ u_3 \\ v_3 \\ --- \\ u_4 \\ v_4 \end{bmatrix} = \begin{bmatrix} P_{x1} \\ P_{y1} \\ ---- \\ P_{x2} \\ P_{y2} \\ ---- \\ P_{x3} \\ P_{y3} \\ ---- \\ P_{x4} \\ P_{y4} \end{bmatrix}$$

with

$$\begin{bmatrix} A_1 \\ A_2 \\ A_3 \\ A_4 \end{bmatrix} = \frac{1}{4} \begin{bmatrix} E & E & E & E \\ E & E & -E & -E \\ E & -E & E & -E \\ E & -E & -E & E \end{bmatrix} \begin{bmatrix} \bar{K}_1 \\ \bar{K}_2 \\ \bar{K}_3 \\ \bar{K}_4 \end{bmatrix} .$$

After evaluation of A_1, A_2, A_3, A_4 the stiffness equation with the stiffness group supermatrix in normal form, with the nodal numbering of the group supermatrix procedure in Fig. 5.4(b), is obtained: (see (1) p. 288) or $K\Phi = P$.

In order to obtain the stiffness equation $K'\Phi' = P'$ with the conventional nodal numbering given in Fig. 5.4(a), it is necessary to change the numbering of the nodes from (1, 2, 3, 4) to (4, 1, 3, 2), which is achieved by

$$\begin{bmatrix} & & & E \\ & E & & \\ & & E & \\ E & & & \end{bmatrix} \begin{bmatrix} A_1 & A_2 & A_3 & A_4 \\ A_2 & A_1 & A_4 & A_3 \\ A_3 & A_4 & A_1 & A_2 \\ A_4 & A_3 & A_2 & A_1 \end{bmatrix} \begin{bmatrix} & & & E \\ & E & & \\ & & E & \\ E & & & \end{bmatrix} = \begin{bmatrix} A_1 & A_3 & A_4 & A_2 \\ A_3 & A_1 & A_2 & A_4 \\ A_4 & A_2 & A_1 & A_3 \\ A_2 & A_4 & A_3 & A_1 \end{bmatrix}$$

and (see (2) p. 288) or $K'\Phi' = P'$.

This stiffness equation is transformed into the conventional stiffness equation by changing positive directions of $v_2, u_3, u_4, v_4, P_{y2}, P_{x3}, P_{x4}, P_{y4}$ in the set of positive directions that suit the first symmetry type of the group C_{2v} given in Fig. 5.4(b). In this way the conventional set of positive directions of nodal displacements and forces according to Fig. 5.4(a) is obtained. This operation is performed by

$$T_D K' T_D T_D \Phi' = T_D P',$$

288   Stiffness equations in G-invariant subspaces   [Ch. 5

(1)
$$\frac{1}{12}\left\{\begin{bmatrix} 2A & C & A & C & A & C & 2A & C \\ C & 2B & C & B & C & 2B & C & B \\ \hline A & C & 2A & C & 2A & C & A & C \\ C & B & C & 2B & C & B & C & 2B \\ \hline A & C & 2A & C & 2A & C & A & C \\ C & 2B & C & B & C & 2B & C & B \\ \hline 2A & C & A & C & A & C & 2A & C \\ C & B & C & 2B & C & B & C & 2B \end{bmatrix}\right.$$

(2)
$$\frac{1}{12}\left\{\begin{bmatrix} 2A & C & A & C & 2A & C & A & C \\ C & 2B & C & 2B & C & B & C & B \\ \hline A & C & 2A & C & A & C & 2A & C \\ C & 2B & C & 2B & C & B & C & B \\ \hline 2A & C & A & C & 2A & C & A & C \\ C & B & C & B & C & 2B & C & 2B \\ \hline A & C & 2A & C & A & C & 2A & C \\ C & B & C & B & C & 2B & C & 2B \end{bmatrix}\right.$$

where

$$T_D = \text{Diag}\ [1 \quad 1 \mid 1 \quad \bar{1} \mid \bar{1} \quad 1 \mid \bar{1} \quad \bar{1}],$$

producing

$$v'_2 = -v_2, \quad u'_3 = -u_3, \quad u'_4 = -u_4, \quad v'_4 = -v_4, \quad P'_{y2} = -P_{y2},$$
$$P'_{x3} = -P_{x3}, \quad P'_{x4} = -P_{x4}, \quad P'_{y4} = -P_{y4}.$$

Thus, the conventional stiffness equation is obtained: (see p. 290).

or $K_c \Phi_c = P_c$, pertaining to the rectangular element for planar analysis according to Fig. 5.4(a), with substitutions for A, B, C, F, G, H given earlier.

The conventional stiffness equation $K_c \Phi_c = P_c$, with the nodal numbering and the set of positive directions of nodal displacements and forces given in Fig. 5.4(a), is transformed by

$$T_D K_c T_D T_D \Phi_c = T_D P_c,$$

Sec. 5.4]   Stiffness equations/rectangular element for planar analysis   289

$$+ \begin{bmatrix} 2F & H & F & H & -2F & -H & -F & -H \\ H & 2G & H & G & -H & -G & -H & -2G \\ F & H & 2F & H & -F & -H & -2F & -H \\ H & G & H & 2G & -H & -2G & -H & -G \\ -2F & -H & -F & -H & 2F & H & F & H \\ -H & -G & -H & -2G & H & 2G & H & G \\ -F & -H & -2F & -H & F & H & 2F & H \\ -H & -2G & -H & -G & H & G & H & 2G \end{bmatrix} \begin{Bmatrix} u_1 \\ v_1 \\ u_2 \\ v_2 \\ u_3 \\ v_3 \\ u_4 \\ v_4 \end{Bmatrix} = \begin{Bmatrix} P_{x1} \\ P_{y1} \\ P_{x2} \\ P_{y2} \\ P_{x3} \\ P_{y3} \\ P_{x4} \\ P_{y4} \end{Bmatrix}$$

$$+ \begin{bmatrix} 2F & H & -2F & -H & -F & -H & F & H \\ H & 2G & -H & -G & -H & -2G & H & G \\ -2F & -H & 2F & H & F & H & -F & -H \\ -H & -G & H & 2G & H & G & -H & -2G \\ -F & -H & F & H & 2F & H & -2F & -H \\ -H & -2G & H & G & H & 2G & -H & -G \\ F & H & -F & -H & -2F & -H & 2F & H \\ H & G & -H & -2G & -H & -G & H & 2G \end{bmatrix} \begin{Bmatrix} u_1 \\ v_1 \\ u_2 \\ v_2 \\ u_3 \\ v_3 \\ u_4 \\ v_4 \end{Bmatrix} \begin{Bmatrix} P_{x1} \\ P_{y1} \\ P_{x2} \\ P_{y2} \\ P_{x3} \\ P_{y3} \\ P_{x4} \\ P_{y4} \end{Bmatrix}$$

with

$$T_D = \text{Diag}\,[1 \quad 1 \quad 1 \quad \bar{1} \quad \bar{1} \quad 1 \quad \bar{1} \quad \bar{1}]$$

into the stiffness equation $K'\Phi' = P'$ with the stiffness group supermatrix in normal form.

In order to obtain the stiffness equation $K\Phi = P$ with the nodal numbering of the group supermatrix procedure as given in Fig. 5.4(b), it is necessary to change the numbering of nodes from (1, 2, 3, 4) to (2, 4, 3, 1), which is achieved by

$$\begin{bmatrix} & & & E \\ E & & & \\ & & E & \\ & E & & \end{bmatrix} \begin{bmatrix} H_1 & H_2 & H_3 & H_4 \\ H_2 & H_1 & H_4 & H_3 \\ H_3 & H_4 & H_1 & H_2 \\ H_4 & H_3 & H_2 & H_1 \end{bmatrix} \begin{bmatrix} & & & E \\ & & E & \\ & E & & \\ E & & & \end{bmatrix}$$

$$\frac{1}{12}\left\{\begin{bmatrix} 2A & C & A & -C & -2A & C & -A & -C \\ C & 2B & C & -2B & -C & B & -C & -B \\ A & C & 2A & -C & -A & C & -2A & -C \\ -C & -2B & -C & 2B & C & -B & C & B \\ -2A & -C & -A & C & 2A & -C & A & C \\ C & B & C & -B & -C & 2B & -C & -2B \\ -A & -C & -2A & C & A & -C & 2A & C \\ -C & -B & -C & B & C & -2B & C & 2B \end{bmatrix}\right.$$

$$= \begin{bmatrix} H_1 & H_4 & H_2 & H_3 \\ H_4 & H_1 & H_3 & H_2 \\ H_2 & H_3 & H_1 & H_4 \\ H_3 & H_2 & H_4 & H_1 \end{bmatrix}.$$

Transformation of the stiffness equation $K\Phi = P$ into the stiffness equation $\bar{K}\bar{\Phi} = \bar{P}$ with the stiffness group supermatrix in diagonal form is accomplished by

$$TKT^{-1}T\Phi = TP,$$

producing

$$\begin{bmatrix} \bar{K}_1 & & & \\ & \bar{K}_2 & & \\ & & \bar{K}_3 & \\ & & & \bar{K}_4 \end{bmatrix} \begin{bmatrix} \bar{\Phi}^{(1)} \\ \bar{\Phi}^{(2)} \\ \bar{\Phi}^{(3)} \\ \bar{\Phi}^{(4)} \end{bmatrix} = \begin{bmatrix} \bar{P}^{(1)} \\ \bar{P}^{(2)} \\ \bar{P}^{(3)} \\ \bar{P}^{(4)} \end{bmatrix}.$$

Table 5.2 presents a comparison of derivations of the stiffness matrix of the rectangular element for planar analysis in the conventional way and by the group supermatrix procedure.

5.5 STIFFNESS EQUATIONS IN G-INVARIANT SUBSPACES FOR THE RECTANGULAR ELEMENT FOR PLATE FLEXURE

The four node 12-degrees-of-freedom rectangular plate bending element, as given in Fig. 5.5(b) with the unique nodal numbering derived in section 3.5, is described by the group C_{2v}, where positive directions of generalized displacements $w_1, \theta_{x1}, \theta_{y1}, w_2, \theta_{x2}, \theta_{y2}, w_3, \theta_{x3}, \theta_{y3}, w_4, \theta_{x4}, \theta_{y4}$ suit the first symmetry type of the group, which differs from the conventional set of positive directions in Fig. 5.5(a).

Sec. 5.5] Stiffness equations/rectangular element for plate flexure 291

$$+ \begin{bmatrix} 2F & H & -2F & H & F & -H & -F & -H \\ H & 2G & -H & G & H & -2G & -H & -G \\ -2F & -H & 2F & -H & -F & H & F & H \\ H & G & -H & 2G & H & -G & -H & -2G \\ F & H & -F & H & 2F & -H & -2F & -H \\ -H & -2G & H & -G & -H & 2G & H & G \\ -F & -H & F & -H & -2F & H & 2F & H \\ -H & -G & H & -2G & -H & G & H & 2G \end{bmatrix} \begin{Bmatrix} u_1 \\ v_1 \\ u_2 \\ v'_2 \\ u'_3 \\ v_3 \\ u'_4 \\ v'_4 \end{Bmatrix} \begin{Bmatrix} P_{x1} \\ P_{y1} \\ P_{x2} \\ P'_{y2} \\ P'_{x3} \\ P_{y3} \\ P'_{x4} \\ P'_{y4} \end{Bmatrix}$$

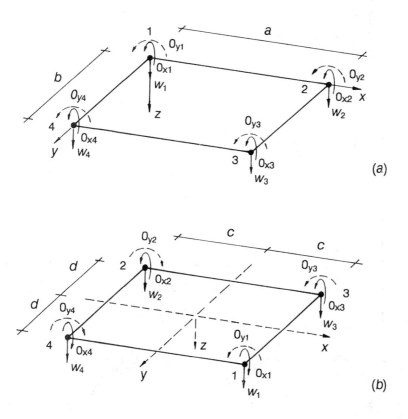

Fig. 5.5. The four-node rectangular element for plate flexure: (a) with the conventional set of positive directions of generalized nodal displacements; (b) with the set of positive directions of generalized nodal displacements that suit the symmetry type of the first irreducible representation of the group C_{2v}.

Table 5.2. Comparison of derivations of the stiffness matrix of the rectangular element for planar analysis in the conventional way and by the group supermatrix procedure

For planar analysis with eight degrees of freedom

Conventional procedure	Group supermatrix procedure

Functions spanning the space of the displacement field

$$\Phi' = \begin{bmatrix} u_1 \\ v_1 \\ u_2 \\ v_2' \\ u_3' \\ v_3 \\ u_4' \\ v_4' \end{bmatrix}$$

$$\bar{\Phi} = \begin{bmatrix} \bar{u}_1 \\ \bar{v}_1 \\ \hdashline \bar{u}_2 \\ \bar{v}_2 \\ \hdashline \bar{u}_3 \\ \bar{v}_3 \\ \hdashline \bar{u}_4 \\ \bar{v}_4 \end{bmatrix} = \frac{1}{4} \begin{bmatrix} u_1 + u_2 + u_3 + u_4 \\ v_1 + v_2 + v_3 + v_4 \\ \hdashline u_1 + u_2 - u_3 - u_4 \\ v_1 + v_2 - v_3 - v_4 \\ \hdashline u_1 - u_2 + u_3 - u_4 \\ v_1 - v_2 + v_3 - v_4 \\ \hdashline u_1 - u_2 - u_3 + u_4 \\ v_1 - v_2 - v_3 + v_4 \end{bmatrix} \begin{matrix} \left.\vphantom{\begin{matrix}a\\b\end{matrix}}\right\} U_1 \\ \left.\vphantom{\begin{matrix}a\\b\end{matrix}}\right\} U_2 \\ \left.\vphantom{\begin{matrix}a\\b\end{matrix}}\right\} U_3 \\ \left.\vphantom{\begin{matrix}a\\b\end{matrix}}\right\} U_4 \end{matrix}$$

Table 5.2. (continued)

For planar analysis with eight degrees of freedom

	Conventional procedure	Group supermatrix procedure
Matrix that is inverted to obtain the coefficients of the polynomial displacement function	(8×8 matrix with scattered # non-zero elements)	(8×8 block-diagonal matrix with # elements grouped into blocks U_1, U_2, U_3, U_4)

\# non-zero element   · zero elements

Table 5.2. (continued)

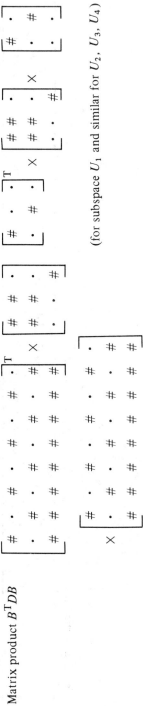

Table 5.2. (continued)

For planar analysis with eight degrees of freedom

Stiffness matrix

Conventional procedure

$$\begin{bmatrix} \# & \# & \# & \# & \# & \# & \# & \# \\ & \# & \# & \# & \# & \# & \# & \# \\ & & \# & \# & \# & \# & \# & \# \\ & & & \# & \# & \# & \# & \# \\ & & & & \# & \# & \# & \# \\ & & & \text{Sym.} & & \# & \# & \# \\ & & & & & & \# & \# \\ & & & & & & & \# \end{bmatrix}$$

Group supermatrix procedure

$\underbrace{}_{U_1} \underbrace{}_{U_2} \underbrace{}_{U_3} \underbrace{}_{U_4}$

in block diagonal form (symmetric)

As given in section 5.4 for the rectangular element for plane analysis, the nodes of its nodal set $S(1, 2, 3, 4)$ are permuted by action of group elements, i.e. symmetry operations $E, C_2, \sigma_1, \sigma_2$.

The character table of the group C_{2v} (with Cartesian sets and products) is

C_{2v}	E	C_2	σ_1 (xz)	σ_2 (yz)		
A_1	1	1	1	1	z	x^2, y^2, z^2
A_2	1	1	-1	-1		xy
B_1	1	-1	1	-1	x	xz
B_2	1	-1	-1	1	y	yz

and the idempotents of the centre of the group algebra are

$$\begin{bmatrix} \pi_1 \\ \pi_2 \\ \pi_3 \\ \pi_4 \end{bmatrix} = \frac{1}{4} \begin{bmatrix} 1 & 1 & 1 & 1 \\ 1 & 1 & -1 & -1 \\ 1 & -1 & 1 & -1 \\ 1 & -1 & -1 & 1 \end{bmatrix} \begin{bmatrix} E \\ C_2 \\ \sigma_1 \\ \sigma_2 \end{bmatrix}$$

or
$$\Pi = T\Sigma,$$

with
$$T^{-1} = 4T,$$

as obtained in section 4.2.

By application of the idempotents π_i ($i = 1, 2, 3, 4$) to the nodal functions φ_j ($j = 1, 2, \ldots, 12$) the basis vectors $\bar{\varphi}_j$ of G-invariant subspaces U_i are derived:

$$\begin{bmatrix} \bar{\Phi}^{(1)} \\ \bar{\Phi}^{(2)} \\ \bar{\Phi}^{(3)} \\ \bar{\Phi}^{(4)} \end{bmatrix} = \begin{bmatrix} \bar{\varphi}_1 & \bar{\varphi}_5 & \bar{\varphi}_9 \\ \bar{\varphi}_2 & \bar{\varphi}_6 & \bar{\varphi}_{10} \\ \bar{\varphi}_3 & \bar{\varphi}_7 & \bar{\varphi}_{11} \\ \bar{\varphi}_4 & \bar{\varphi}_8 & \bar{\varphi}_{12} \end{bmatrix} = \begin{bmatrix} \bar{w}_1 & \bar{\theta}_{x1} & \bar{\theta}_{y1} \\ \bar{w}_2 & \bar{\theta}_{x2} & \bar{\theta}_{y2} \\ \bar{w}_3 & \bar{\theta}_{x3} & \bar{\theta}_{y3} \\ \bar{w}_4 & \bar{\theta}_{x4} & \bar{\theta}_{y4} \end{bmatrix}$$

$$= \frac{1}{4} \begin{bmatrix} 1 & 1 & 1 & 1 \\ 1 & 1 & -1 & -1 \\ 1 & -1 & 1 & -1 \\ 1 & -1 & -1 & 1 \end{bmatrix} \begin{bmatrix} w_1 & \theta_{x1} & \theta_{y1} \\ w_2 & \theta_{x2} & \theta_{y2} \\ w_3 & \theta_{x3} & \theta_{y3} \\ w_4 & \theta_{x4} & \theta_{y4} \end{bmatrix}$$

Sec. 5.5] Stiffness equations/rectangular element for plate flexure

$$= \frac{1}{4} \begin{bmatrix} w_1 + w_2 + w_3 + w_4 & \theta_{x1} + \theta_{x2} + \theta_{x3} + \theta_{x4} \\ w_1 + w_2 - w_3 - w_4 & \theta_{x1} + \theta_{x2} - \theta_{x3} - \theta_{x4} \\ w_1 - w_2 + w_3 - w_4 & \theta_{x1} - \theta_{x2} + \theta_{x3} - \theta_{x4} \\ w_1 - w_2 - w_3 + w_4 & \theta_{x1} - \theta_{x2} - \theta_{x3} + \theta_{x4} \end{bmatrix}$$

$$\begin{bmatrix} \theta_{y1} + \theta_{y2} + \theta_{y3} + \theta_{y4} \\ \theta_{y1} + \theta_{y2} - \theta_{y3} - \theta_{y4} \\ \theta_{y1} - \theta_{y2} + \theta_{y3} - \theta_{y4} \\ \theta_{y1} - \theta_{y2} - \theta_{y3} + \theta_{y4} \end{bmatrix}$$

or $\bar{\Phi} = T\Phi$ with $T^{-1} = 4T$.

Conversely, the relation of the nodal functions $\Phi^{(i)}$ to the basis vectors $\bar{\Phi}^{(i)}$ is

$$\begin{bmatrix} \Phi^{(1)} \\ \Phi^{(2)} \\ \Phi^{(3)} \\ \Phi^{(4)} \end{bmatrix} = \begin{bmatrix} \varphi_1 & \varphi_5 & \varphi_9 \\ \varphi_2 & \varphi_6 & \varphi_{10} \\ \varphi_3 & \varphi_7 & \varphi_{11} \\ \varphi_4 & \varphi_8 & \varphi_{12} \end{bmatrix} = \begin{bmatrix} w_1 & \theta_{x1} & \theta_{y1} \\ w_2 & \theta_{x2} & \theta_{y2} \\ w_3 & \theta_{x3} & \theta_{y3} \\ w_4 & \theta_{x4} & \theta_{y4} \end{bmatrix}$$

$$= \begin{bmatrix} 1 & 1 & 1 & 1 \\ 1 & 1 & -1 & -1 \\ 1 & -1 & 1 & -1 \\ 1 & -1 & -1 & 1 \end{bmatrix} \begin{bmatrix} \bar{w}_1 & \bar{\theta}_{x1} & \bar{\theta}_{y1} \\ \bar{w}_2 & \bar{\theta}_{x2} & \bar{\theta}_{y2} \\ \bar{w}_3 & \bar{\theta}_{x3} & \bar{\theta}_{y3} \\ \bar{w}_4 & \bar{\theta}_{x4} & \bar{\theta}_{y4} \end{bmatrix}$$

$$= \begin{bmatrix} \bar{w}_1 + \bar{w}_2 + \bar{w}_3 + \bar{w}_4 & \bar{\theta}_{x1} + \bar{\theta}_{x2} + \bar{\theta}_{x3} + \bar{\theta}_{x4} \\ \bar{w}_1 + \bar{w}_2 - \bar{w}_3 - \bar{w}_4 & \bar{\theta}_{x1} + \bar{\theta}_{x2} - \bar{\theta}_{x3} - \bar{\theta}_{x4} \\ \bar{w}_1 - \bar{w}_2 + \bar{w}_3 - \bar{w}_4 & \bar{\theta}_{x1} - \bar{\theta}_{x2} + \bar{\theta}_{x3} - \bar{\theta}_{x4} \\ \bar{w}_1 - \bar{w}_2 - \bar{w}_3 + \bar{w}_4 & \bar{\theta}_{x1} - \bar{\theta}_{x2} - \bar{\theta}_{x3} + \bar{\theta}_{x4} \end{bmatrix}$$

$$\begin{bmatrix} \bar{\theta}_{y1} + \bar{\theta}_{y2} + \bar{\theta}_{y3} + \bar{\theta}_{y4} \\ \bar{\theta}_{y1} + \bar{\theta}_{y2} - \bar{\theta}_{y3} - \bar{\theta}_{y4} \\ \bar{\theta}_{y1} - \bar{\theta}_{y2} + \bar{\theta}_{y3} - \bar{\theta}_{y4} \\ \bar{\theta}_{y1} - \bar{\theta}_{y2} - \bar{\theta}_{y3} + \bar{\theta}_{y4} \end{bmatrix}$$

or $\Phi = 4T\bar{\Phi}$ with $T^{-1} = 4T$.

The sets $\bar{\Phi}_l$ ($l = 1, 2, 3$) of basis vectors $\bar{\varphi}_j$ ($j = 1, 2, \ldots, 12$) that represent $\bar{w}_1$, $\bar{\theta}_{x1}$, $\bar{\theta}_{y1}$, $\bar{w}_2$, $\bar{\theta}_{x2}$, $\bar{\theta}_{y2}$, $\bar{w}_3$, $\bar{\theta}_{x3}$, $\bar{\theta}_{y3}$, $\bar{w}_4$, $\bar{\theta}_{x4}$, $\bar{\theta}_{y4}$, ordered according to the sequence explained in section 4.1 (step (2)), are given by

$$[\bar{\Phi}_1 \quad \bar{\Phi}_2 \quad \bar{\Phi}_3] = \begin{bmatrix} \bar{\varphi}_1 & \bar{\varphi}_5 & \bar{\varphi}_9 \\ \bar{\varphi}_2 & \bar{\varphi}_6 & \bar{\varphi}_{10} \\ \bar{\varphi}_3 & \bar{\varphi}_7 & \bar{\varphi}_{11} \\ \bar{\varphi}_4 & \bar{\varphi}_8 & \bar{\varphi}_{12} \end{bmatrix} = \begin{bmatrix} \bar{w}_1 & \bar{\theta}_{x1} & \bar{\theta}_{y1} \\ \bar{w}_2 & \bar{\theta}_{x2} & \bar{\theta}_{y2} \\ \bar{w}_3 & \bar{\theta}_{x3} & \bar{\theta}_{y3} \\ \bar{w}_4 & \bar{\theta}_{x4} & \bar{\theta}_{y4} \end{bmatrix}.$$

The relation of the sets of basis vectors $\bar{\Phi}_l$ to the sets of nodal functions Φ_l is expressed by the following system of equations with the supermatrix in diagonal form containing three transformation matrices T of the group C_{2v}

$$\begin{bmatrix} \bar{\Phi}_1 \\ \bar{\Phi}_2 \\ \bar{\Phi}_3 \end{bmatrix} = \begin{bmatrix} T & & \\ & T & \\ & & T \end{bmatrix} \begin{bmatrix} \Phi_1 \\ \Phi_2 \\ \Phi_3 \end{bmatrix} \quad \text{or} \quad \bar{\Phi}_l = \bar{T}\Phi_l,$$

with

$$T = \frac{1}{4} \begin{bmatrix} 1 & 1 & 1 & 1 \\ 1 & 1 & -1 & -1 \\ 1 & -1 & 1 & -1 \\ 1 & -1 & -1 & 1 \end{bmatrix}$$

and

$$[\Phi_1 \quad \Phi_2 \quad \Phi_3] = \begin{bmatrix} \varphi_1 & \varphi_5 & \varphi_9 \\ \varphi_2 & \varphi_6 & \varphi_{10} \\ \varphi_3 & \varphi_7 & \varphi_{11} \\ \varphi_4 & \varphi_8 & \varphi_{12} \end{bmatrix} = \begin{bmatrix} w_1 & \theta_{x1} & \theta_{y1} \\ w_2 & \theta_{x2} & \theta_{y2} \\ w_3 & \theta_{x3} & \theta_{y3} \\ w_4 & \theta_{x4} & \theta_{y4} \end{bmatrix}.$$

Conversely, the relation of the sets of nodal functions Φ_l to the sets of basis vectors $\bar{\Phi}_l$ is

$$\begin{bmatrix} \Phi_1 \\ \Phi_2 \\ \Phi_3 \end{bmatrix} = 4 \begin{bmatrix} T & & \\ & T & \\ & & T \end{bmatrix} \begin{bmatrix} \bar{\Phi}_1 \\ \bar{\Phi}_2 \\ \bar{\Phi}_3 \end{bmatrix} \quad \text{or} \quad \Phi_l = 4\bar{T}\bar{\Phi}_l,$$

since $T^{-1} = 4T$.

The displacement function for the 12-d.-o,-f. rectangular plate bending element is usually assumed as

$$w = \alpha_1 + \alpha_2 x + \alpha_3 y + \alpha_4 x^2 + \alpha_5 xy + \alpha_6 y^2 + \alpha_7 x^3 + \alpha_8 x^2 y + \alpha_9 xy^2$$
$$+ \alpha_{10} y^3 + \alpha_{11} x^3 y + \alpha_{12} xy^3,$$

Sec. 5.5]   Stiffness equations/rectangular element for plate flexure   299

with

$$\theta_x = \partial w/\partial y, \quad \theta_y = -\partial w/\partial x$$

according to Fig. 5.5(a).

For the group supermatrix procedure by the following expression the terms of the above polynomial will be allocated into respective G-invariant subspaces where they belong and with the unique order of the terms that provides the diagonal form of the supermatrix of the system of equations relating generalized displacements $\overline{\Phi}$ to coefficients A:

$$\begin{bmatrix} F_1 \\ F_2 \\ F_3 \\ F_4 \end{bmatrix} = \begin{bmatrix} 1 \\ xy \\ x \\ y \end{bmatrix} \begin{bmatrix} 1 & x^2 & y^2 \end{bmatrix} = \begin{bmatrix} 1 & x^2 & y^2 \\ xy & x^3y & xy^3 \\ x & x^3 & xy^2 \\ y & x^2y & y^3 \end{bmatrix} \quad \begin{array}{l} \text{Subspace} \\ U_1 \\ U_2 \\ U_3 \\ U_4 \end{array}$$

The column vector and the row vector are taken from the Cartesian sets and products as they stand in addition to the character table of the group C_{2v} according to the pertinence of the terms to the symmetry types of group representations A_1, A_2, B_1, B_2.

In Fig. 5.5(b), where positive directions of nodal displacements and rotations possess the symmetry type of the first irreducible group representation, the rotation θ_x in the first quadrant has the positive direction that is opposite to the positive direction of the corresponding rotation in Fig. 5.5(a) that presents the conventional set of positive directions. It is therefore necessary to change the conventional set w, $\theta_x = \partial w/\partial y$, $\theta_y = -\partial w/\partial x$ into w, $\theta_x = -\partial w/\partial y$, $\theta_y = -\partial w/\partial x$, so that the functions w, θ_x, θ_y are

$$\begin{bmatrix} w \\ \theta_x \\ \theta_y \end{bmatrix} = \begin{bmatrix} 1 & x^2 & y^2 & \vdots & xy & x^3y & xy^3 \\ & & -2y & \vdots & -x & -x^3 & -3xy^2 \\ -2x & & & \vdots & -y & -3x^2y & -y^3 \end{bmatrix}$$

$$\begin{bmatrix} \vdots & x & x^3 & xy^2 & \vdots & y & x^2y & y^3 \\ \vdots & & & -2xy & \vdots & -1 & -x^2 & -3y^2 \\ \vdots & -1 & -3x^2 & -y^2 & \vdots & & -2xy & \end{bmatrix} \cdot$$

$$\circ [\alpha_1 \quad \alpha_2 \quad \alpha_3 \,\vdots\, \alpha_4 \quad \alpha_5 \quad \alpha_6 \,\vdots\, \alpha_7 \quad \alpha_8 \quad \alpha_9 \,\vdots\, \alpha_{10} \quad \alpha_{11} \quad \alpha_{12}]^T$$

$$= [F_1 \,\vdots\, F_2 \,\vdots\, F_3 \,\vdots\, F_4][A_1 \,\vdots\, A_2 \,\vdots\, A_3 \,\vdots\, A_4]^T.$$

Thus, the above function is decomposed into four three-dimensional subspaces U_1, U_2, U_3, U_4.

Substitution of the nodal coordinates of the first node in the nodal set $S(1, 2, 3, 4)$, i.e. $x = c$, $y = d$ for the node 1, into the function matrices F_1, F_2, F_3, F_4 of $[w \; \theta_x \; \theta_y]^T$ pertaining to the subspaces U_1, U_2, U_3, U_4, gives relations of generalized displacements $\overline{\Phi}$ to coefficients A for each subspace separately:

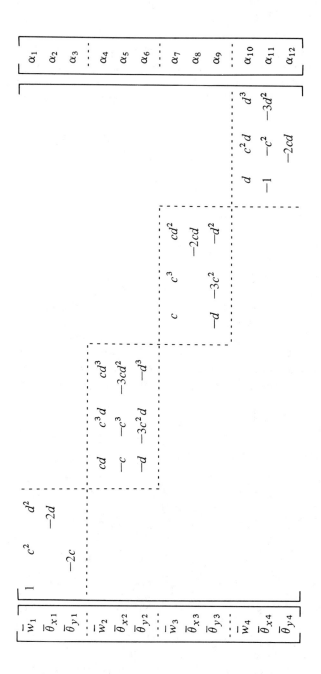

Sec. 5.5]   Stiffness equations/rectangular element for plate flexure   301

or $\bar{\Phi} = \bar{C}A$.

The twelve-dimensional G-vector space of the problem is decomposed now into four three-dimensional G-invariant subspaces U_1, U_2, U_3, U_4. The matrix of the system of equations of the above relation is obtained in block diagonal form containing four 3 × 3 matrices.

The coefficients A are determined by inverting the matrices $\bar{C}_1, \bar{C}_2, \bar{C}_3, \bar{C}_4$ in the subspaces U_1, U_2, U_3, U_4: (see p. 302) or $A = \bar{C}^{-1}\bar{\Phi}$.

The twelve-dimensional G-vector space of the displacement field Δ is decomposed into four three-dimensional subspaces U_1, U_2, U_3, U_4

$$\Delta = T_D \bar{C}^{-1} \bar{\Phi}$$

with

$$T_D = \text{Diag } [1 \quad x^2 \quad y^2 \,\vdots\, xy \quad x^3 y \quad xy^3 \,\vdots\, x \quad x^3 \quad xy^2 \,\vdots\, y \quad x^2 y \quad y^3],$$

(see p. 303).

Thus, in the subspaces U_1, U_2, U_3, U_4 one obtains

$$U_1: \quad \Delta_1 = \begin{bmatrix} 1 & \dfrac{d}{2} - \dfrac{y^2}{2d} & \dfrac{c}{2} - \dfrac{x^2}{2c} \end{bmatrix} [\bar{w}_1 \quad \bar{\theta}_{x1} \quad \bar{\theta}_{y1}]^T = \bar{N}^{(1)} \bar{\Phi}^{(1)}$$

$$U_2: \quad \Delta_2 = \begin{bmatrix} \dfrac{2xy}{cd} - \dfrac{x^3 y}{2c^3 d} - \dfrac{xy^3}{2cd^3} & \dfrac{xy}{2c} - \dfrac{xy^3}{2cd^2} & \dfrac{xy}{2d} - \dfrac{x^3 y}{2c^2 d} \end{bmatrix}$$
$$[\bar{w}_2 \quad \bar{\theta}_{x2} \quad \bar{\theta}_{y2}]^T = \bar{N}^{(2)} \bar{\Phi}^{(2)}$$

$$U_3: \quad \Delta_3 = \begin{bmatrix} \dfrac{3x}{2c} - \dfrac{x^3}{2c^3} & \dfrac{dx}{2c} - \dfrac{xy^2}{2cd} & \dfrac{x}{2} - \dfrac{x^3}{2c^2} \end{bmatrix}$$
$$[\bar{w}_3 \quad \bar{\theta}_{x3} \quad \bar{\theta}_{y3}]^T = \bar{N}^{(3)} \bar{\Phi}^{(3)}$$

$$U_4: \quad \Delta_4 = \begin{bmatrix} \dfrac{3y}{2d} - \dfrac{y^3}{2d^3} & \dfrac{y}{2} - \dfrac{y^3}{2d^2} & \dfrac{cy}{2d} - \dfrac{x^2 y}{2cd} \end{bmatrix}$$
$$[\bar{w}_4 \quad \bar{\theta}_{x4} \quad \bar{\theta}_{y4}]^T = \bar{N}^{(4)} \bar{\Phi}^{(4)}.$$

The 3 × 3 stiffness matrices $\bar{K}_1, \bar{K}_2, \bar{K}_3, \bar{K}_4$ of the subspaces U_1, U_2, U_3, U_4 will be derived using the above shape functions $\bar{N}^{(1)}, \bar{N}^{(2)}, \bar{N}^{(3)}, \bar{N}^{(4)}$, so that the stiffness group supermatrix in diagonal form will be obtained:

$$
\begin{bmatrix} \alpha_1 \\ \alpha_2 \\ \alpha_3 \\ \hdashline \alpha_4 \\ \alpha_5 \\ \alpha_6 \\ \hdashline \alpha_7 \\ \alpha_8 \\ \alpha_9 \\ \hdashline \alpha_{10} \\ \alpha_{11} \\ \alpha_{12} \end{bmatrix}
=
\left[\begin{array}{ccc:ccc:ccc:ccc}
1 & & & & & & & & & & & \\
\dfrac{d}{2} & \dfrac{c}{2} & & & & & & & & & & \\
-\dfrac{1}{2d} & -\dfrac{1}{2c} & & & & & & & & & & \\
\hdashline
 & & & \dfrac{2}{cd} & \dfrac{1}{2c} & \dfrac{1}{2d} & & & & & & \\
 & & & -\dfrac{1}{2c^3 d} & & -\dfrac{1}{2c^2 d} & & & & & & \\
 & & & -\dfrac{1}{2cd^3} & -\dfrac{1}{2cd^2} & & & & & & & \\
\hdashline
 & & & & & & \dfrac{3}{2c} & \dfrac{d}{2c} & \dfrac{1}{2} & & & \\
 & & & & & & -\dfrac{1}{2c^3} & & -\dfrac{1}{2c^2} & & & \\
 & & & & & & & -\dfrac{1}{2cd} & & & & \\
\hdashline
 & & & & & & & & & \dfrac{3}{2d} & \dfrac{1}{2} & \dfrac{c}{2d} \\
 & & & & & & & & & & & -\dfrac{1}{2cd} \\
 & & & & & & & & & -\dfrac{1}{2d^3} & -\dfrac{1}{2d^2} & \\
\end{array}\right]
\begin{bmatrix} \bar{w}_1 \\ \bar{\theta}_{x1} \\ \bar{\theta}_{y1} \\ \hdashline \bar{w}_2 \\ \bar{\theta}_{x2} \\ \bar{\theta}_{y2} \\ \hdashline \bar{w}_3 \\ \bar{\theta}_{x3} \\ \bar{\theta}_{y3} \\ \hdashline \bar{w}_4 \\ \bar{\theta}_{x4} \\ \bar{\theta}_{y4} \end{bmatrix}
$$

$$
\begin{bmatrix}
1 & \dfrac{d}{2} & \dfrac{c}{2} & & & & & & & & & \\
-\dfrac{y^2}{2d} & & -\dfrac{x^2}{2c} & & & & & & & & & \\
\dfrac{2xy}{cd} & \dfrac{xy}{2c} & \dfrac{xy}{2d} & & & & & & & & & \\
-\dfrac{x^3y}{2c^3d}-\dfrac{xy^3}{2cd^3} & -\dfrac{xy^3}{2cd^2} & -\dfrac{x^3y}{2c^2d} & \dfrac{3x}{2c}-\dfrac{x^3}{2c^3} & \dfrac{dx}{2c} & \dfrac{x}{2}-\dfrac{x^3}{2c^2} & & & & & & \\
 & & & & & -\dfrac{xy^2}{2cd} & \dfrac{3y}{2d} & \dfrac{y}{2} & \dfrac{cy}{2d} & & & \\
 & & & & & & -\dfrac{y^3}{2d^3} & -\dfrac{y^3}{2d^2} & -\dfrac{x^2y}{2cd} & & &
\end{bmatrix}
\begin{bmatrix} \bar{w}_1 \\ \bar{\theta}_{x1} \\ \bar{\theta}_{y1} \\ \bar{w}_2 \\ \bar{\theta}_{x2} \\ \bar{\theta}_{y2} \\ \bar{w}_3 \\ \bar{\theta}_{x3} \\ \bar{\theta}_{y3} \\ \bar{w}_4 \\ \bar{\theta}_{x4} \\ \bar{\theta}_{y4} \end{bmatrix}
=
\begin{bmatrix} \Delta_1 \\ \Delta_2 \\ \Delta_3 \\ \Delta_4 \end{bmatrix}
$$

$$\bar{K} = \begin{bmatrix} \bar{K}_1 & & & \\ & \bar{K}_2 & & \\ & & \bar{K}_3 & \\ & & & \bar{K}_4 \end{bmatrix} = \mathrm{Diag}\,[\bar{K}_1 \quad \bar{K}_2 \quad \bar{K}_3 \quad \bar{K}_4].$$

Subspace U_1:

$$\bar{N}^{(1)} = \begin{bmatrix} 1 & \dfrac{d}{2} - \dfrac{y^2}{2d} & \dfrac{c}{2} - \dfrac{x^2}{2c} \end{bmatrix},$$

$$\bar{B}_1 = -\begin{bmatrix} \partial^2/\partial x^2 \\ \partial^2/\partial y^2 \\ 2\partial^2/\partial x \partial y \end{bmatrix} \bar{N}^{(1)} = \begin{bmatrix} 0 & 0 & -\dfrac{1}{c} \\ 0 & -\dfrac{1}{d} & 0 \\ 0 & 0 & 0 \end{bmatrix}$$

$$\bar{B}_1^{\mathrm{T}} D \bar{B}_1 = \begin{bmatrix} 0 & 0 & 0 \\ 0 & -\dfrac{1}{d} & 0 \\ -\dfrac{1}{c} & 0 & 0 \end{bmatrix} \begin{bmatrix} D_x & D_1 & 0 \\ D_1 & D_y & 0 \\ 0 & 0 & D_{xy} \end{bmatrix} \begin{bmatrix} 0 & 0 & -\dfrac{1}{c} \\ 0 & -\dfrac{1}{d} & 0 \\ 0 & 0 & 0 \end{bmatrix}$$

$$= \begin{bmatrix} 0 & 0 & 0 \\ 0 & \dfrac{D_y}{d^2} & \dfrac{D_1}{cd} \\ 0 & \dfrac{D_1}{cd} & \dfrac{D_x}{c^2} \end{bmatrix}$$

$$\bar{K}_1 = \int_0^c \int_0^d \bar{B}_1^{\mathrm{T}} D \bar{B}_1 \, \mathrm{d}x \, \mathrm{d}y = \begin{bmatrix} 0 & 0 & 0 \\ & \dfrac{c}{d} D_y & D_1 \\ \mathrm{sym.} & & \dfrac{d}{c} D_x \end{bmatrix}.$$

Sec. 5.5] Stiffness equations/rectangular element for plate flexure

Subspace U_2:

$$\bar{N}^{(2)} = \left[\frac{2xy}{cd} - \frac{x^3y}{2c^3d} - \frac{xy^3}{2cd^3} \quad \frac{xy}{2c} - \frac{xy^3}{2cd^2} \quad \frac{xy}{2d} - \frac{x^3y}{2c^2d} \right]$$

$$\bar{B}_2 = - \begin{bmatrix} \partial^2/\partial x^2 \\ \partial^2/\partial y^2 \\ 2\partial^2/\partial x \partial y \end{bmatrix} \bar{N}^{(2)}$$

$$= \begin{bmatrix} \dfrac{3xy}{c^3d} & 0 & \dfrac{3xy}{c^2d} \\[6pt] \dfrac{3xy}{cd^3} & \dfrac{3xy}{cd^2} & 0 \\[6pt] -\dfrac{4}{cd} + \dfrac{3x^2}{c^3d} + \dfrac{3y^2}{cd^3} & -\dfrac{1}{c} + \dfrac{3y^2}{cd^2} & -\dfrac{1}{d} + \dfrac{3x^2}{c^2d} \end{bmatrix}$$

$$\bar{K}_2 = \int_0^c \int_0^d \bar{B}_2^T D \bar{B}_2 \, dx \, dy$$

$$= \begin{bmatrix} \dfrac{d}{c^3}D_x + \dfrac{c}{d^3}D_y + & \dfrac{c}{d^2}D_y + \dfrac{1}{c}D_1 + & \dfrac{d}{c^2}D_x + \dfrac{1}{d}D_1 + \\[4pt] + \dfrac{2}{cd}D_1 + \dfrac{28}{5}\dfrac{1}{cd}D_{xy} & + \dfrac{4}{5}\dfrac{1}{c}D_{xy} & + \dfrac{4}{5}\dfrac{1}{d}D_{xy} \\[6pt] \hdashline & \dfrac{c}{d}D_y + \dfrac{4}{5}\dfrac{d}{c}D_{xy} & D_1 \\[6pt] \hdashline \text{sym.} & & \dfrac{d}{c}D_x + \dfrac{4}{5}\dfrac{c}{d}D_{xy} \end{bmatrix}$$

Subspace U_3:

$$\bar{N}^{(3)} = \left[\frac{3x}{2c} - \frac{x^3}{2c^3} \quad \frac{dx}{2c} - \frac{xy^2}{2cd} \quad \frac{x}{2} - \frac{x^3}{2c^2} \right]$$

$$\bar{B}_3 = - \begin{bmatrix} \partial^2/\partial x^2 \\ \partial^2/\partial y^2 \\ 2\partial^2/\partial x \partial y \end{bmatrix} N^{(3)} = \begin{bmatrix} \dfrac{3x}{c^3} & 0 & \dfrac{3x}{c^2} \\ 0 & \dfrac{x}{cd} & 0 \\ 0 & \dfrac{2y}{cd} & 0 \end{bmatrix}$$

$$\bar{K}_3 = \int_0^c \int_0^d \bar{B}_3^T D \bar{B}_3 \, dx \, dy = \begin{bmatrix} 3\dfrac{d}{c^3} D_x & \dfrac{1}{c} D_1 & 3\dfrac{d}{c^2} D_x \\ & \dfrac{1}{3}\dfrac{c}{d} D_y + \dfrac{4}{3}\dfrac{d}{c} D_{xy} & D_1 \\ \text{sym.} & & 3\dfrac{d}{c} D_x \end{bmatrix}.$$

Subspace U_4:

$$\bar{N}^{(4)} = \begin{bmatrix} \dfrac{3y}{2d} - \dfrac{y^3}{2d^3} & \dfrac{y}{2} - \dfrac{y^3}{2d^2} & \dfrac{cy}{2d} - \dfrac{x^2 y}{2cd} \end{bmatrix}$$

$$\bar{B}_4 = - \begin{bmatrix} \partial^2/\partial x^2 \\ \partial^2/\partial y^2 \\ 2\partial^2/\partial x \partial y \end{bmatrix} \bar{N}^{(4)} = \begin{bmatrix} 0 & 0 & \dfrac{y}{cd} \\ \dfrac{3y}{d^3} & \dfrac{3y}{d^2} & 0 \\ 0 & 0 & \dfrac{2x}{cd} \end{bmatrix}$$

$$\bar{K}_4 = \int_0^c \int_0^d \bar{B}_4^T D \bar{B}_4 \, dx \, dy \begin{bmatrix} 3\dfrac{c}{d^3} D_y & 3\dfrac{c}{d^2} D_y & \dfrac{1}{d} D_1 \\ & 3\dfrac{c}{d} D_y & D_1 \\ \text{sym.} & & \dfrac{1}{3}\dfrac{d}{c} D_x + \dfrac{4}{3}\dfrac{c}{d} D_{xy} \end{bmatrix}.$$

Sec. 5.5]   Stiffness equations/rectangular element for plate flexure

The stiffness equation with the stiffness group supermatrix in diagonal form is given in Table 5.3. It can be also written as

$$\begin{bmatrix} \bar{K}_1 & & & \\ & \bar{K}_2 & & \\ & & \bar{K}_3 & \\ & & & \bar{K}_4 \end{bmatrix} \begin{bmatrix} \bar{\Phi}^{(1)} \\ \bar{\Phi}^{(2)} \\ \bar{\Phi}^{(3)} \\ \bar{\Phi}^{(4)} \end{bmatrix} = \begin{bmatrix} \bar{P}^{(1)} \\ \bar{P}^{(2)} \\ \bar{P}^{(3)} \\ \bar{P}^{(4)} \end{bmatrix} \quad \text{or} \quad \bar{K}\bar{\Phi} = \bar{P}.$$

Transformation of the stiffness equation $\bar{K}\bar{\Phi} = \bar{P}$ into the stiffness equation $K\Phi = P$ with the stiffness group supermatrix in normal form is performed by the group supermatrix transformation

$$T^{-1}\bar{K}TT^{-1}\bar{\Phi} = T^{-1}\bar{P},$$

derived in section 3.1.

As shown in section 5.4 this transformation results in

$$\begin{bmatrix} C_1 & C_2 & C_3 & C_4 \\ C_2 & C_1 & C_4 & C_3 \\ C_3 & C_4 & C_1 & C_2 \\ C_4 & C_3 & C_2 & C_1 \end{bmatrix} \begin{bmatrix} \Phi^{(1)} \\ \Phi^{(2)} \\ \Phi^{(3)} \\ \Phi^{(4)} \end{bmatrix} = \begin{bmatrix} P^{(1)} \\ P^{(2)} \\ P^{(3)} \\ P^{(4)} \end{bmatrix}$$

with

$$\begin{bmatrix} C_1 \\ C_2 \\ C_3 \\ C_4 \end{bmatrix} = \frac{1}{4} \begin{bmatrix} E & E & E & E \\ E & E & -E & -E \\ E & -E & E & -E \\ E & -E & -E & E \end{bmatrix} \begin{bmatrix} \bar{K}_1 \\ \bar{K}_2 \\ \bar{K}_3 \\ \bar{K}_4 \end{bmatrix}.$$

By using the last relation the elements $c^{(i)}_{jk}$ of the matrices C_i are determined (with substitutions $a = 2c$, $b = 2d$, $p = a/b$):

$$c^{(1)}_{11} = \frac{1}{15ab}(60p^{-2}D_x + 60p^2 D_y + 30D_1 + 84D_{xy}) = \frac{1}{15ab}A$$

$$c^{(1)}_{12} = \frac{1}{15ab}(30apD_y + 15dD_1 + 6bD_{xy}) = \frac{1}{15ab}B$$

$$c^{(1)}_{13} = \frac{1}{15ab}(30p^{-1}D_x + 15aD_1 + 6aD_{xy}) = \frac{1}{15ab}D$$

$$c^{(1)}_{22} = \frac{1}{15ab}(20a^2 D_y + 8b^2 D_{xy}) = \frac{1}{15ab}C$$

Table 5.3. The stiffness equation with the stiffness group supermatrix in diagonal form for the rectangular element for plate flexure

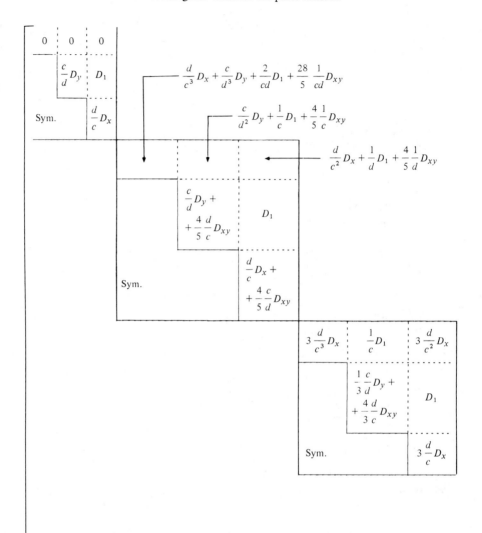

Sec. 5.5] Stiffness equations/rectangular element for plate flexure 309

$$
\begin{bmatrix}
3\dfrac{c}{d^3}D_y & 3\dfrac{c}{d^2}D_y & \dfrac{1}{d}D_1 \\
& 3\dfrac{c}{d}D_y & D_1 \\
& & \dfrac{1}{3}\dfrac{d}{c}D_x + \dfrac{4}{3}\dfrac{c}{d}D_{xy} \\
\text{Sym.} & & &
\end{bmatrix}
\begin{bmatrix}
\bar{w}_1 = \tfrac{1}{4}(w_1 + w_2 + w_3 + w_4) \\
\bar{\theta}_{x1} = \tfrac{1}{4}(\theta_{x1} + \theta_{x2} + \theta_{x3} + \theta_{x4}) \\
\bar{\theta}_{y1} = \tfrac{1}{4}(\theta_{y1} + \theta_{y2} + \theta_{y3} + \theta_{y4}) \\
\hline
\bar{w}_2 = \tfrac{1}{4}(w_1 + w_2 - w_3 + w_4) \\
\bar{\theta}_{x2} = \tfrac{1}{4}(\theta_{x1} + \theta_{x2} - \theta_{x3} - \theta_{x4}) \\
\bar{\theta}_{y2} = \tfrac{1}{4}(\theta_{y1} + \theta_{y2} - \theta_{y3} - \theta_{y4}) \\
\hline
\bar{w}_3 = \tfrac{1}{4}(w_1 - w_2 + w_3 - w_4) \\
\bar{\theta}_{x3} = \tfrac{1}{4}(\theta_{x1} - \theta_{x2} + \theta_{x3} - \theta_{x4}) \\
\bar{\theta}_{y3} = \tfrac{1}{4}(\theta_{y1} - \theta_{y2} + \theta_{y3} - \theta_{y4}) \\
\hline
\bar{w}_4 = \tfrac{1}{4}(w_1 - w_2 - w_3 + w_4) \\
\bar{\theta}_{x4} = \tfrac{1}{4}(\theta_{x1} - \theta_{x2} - \theta_{x3} + \theta_{x4}) \\
\bar{\theta}_{y4} = \tfrac{1}{4}(\theta_{y1} - \theta_{y2} - \theta_{y3} + \theta_{y4})
\end{bmatrix}
=
\begin{bmatrix}
\bar{P}_1 = \tfrac{1}{4}(P_1 + P_2 + P_3 + P_4) \\
\bar{M}_{x1} = \tfrac{1}{4}(M_{x1} + M_{x2} + M_{x3} + M_{x4}) \\
\bar{M}_{y1} = \tfrac{1}{4}(M_{y1} + M_{y2} + M_{y3} + M_{y4}) \\
\hline
\bar{P}_2 = \tfrac{1}{4}(P_1 + P_2 - P_3 - P_4) \\
\bar{M}_{x2} = \tfrac{1}{4}(M_{x1} + M_{x2} - M_{x3} - M_{x4}) \\
\bar{M}_{y2} = \tfrac{1}{4}(M_{y1} + M_{y2} - M_{y3} - M_{y4}) \\
\hline
\bar{P}_3 = \tfrac{1}{4}(P_1 - P_2 + P_3 - P_4) \\
\bar{M}_{x3} = \tfrac{1}{4}(M_{x1} - M_{y2} + M_{x3} - M_{x4}) \\
\bar{M}_{y3} = \tfrac{1}{4}(M_{y1} - M_{y2} + M_{y3} - M_{y4}) \\
\hline
\bar{P}_4 = \tfrac{1}{4}(P_1 - P_2 - P_3 + P_4) \\
\bar{M}_{x4} = \tfrac{1}{4}(M_{x1} - M_{x2} - M_{x3} + M_{x4}) \\
\bar{M}_{y4} = \tfrac{1}{4}(M_{y1} - M_{y2} - M_{y3} + M_{y4})
\end{bmatrix}
$$

$$c_{23}^{(1)} = \frac{1}{15ab} 15abD_1 = \frac{1}{15ab} \text{``}E\text{''}$$

$$c_{33}^{(1)} = \frac{1}{15ab} (20b^2 D_x + 8a^2 D_{xy}) = \frac{1}{15ab} E$$

$$c_{11}^{(2)} = \frac{1}{15ab} (-30p^{-2} D_x - 30p^2 D_y + 30D_1 + 84D_{xy}) = \frac{1}{15ab} M$$

$$c_{12}^{(2)} = \frac{1}{15ab} (-15ap D_y + 6b D_{xy}) = \frac{1}{15ab} (-N)$$

$$c_{13}^{(2)} = \frac{1}{15ab} (-15bp^{-1} D_x + 6a D_{xy}) = \frac{1}{15ab} (-\text{``}O\text{''})$$

$$c_{22}^{(2)} = \frac{1}{15ab} (-5a^2 D_y - 2b^2 D_{xy}) = \frac{1}{15ab} (-P)$$

$$c_{23}^{(2)} = 0$$

$$c_{33}^{(2)} = \frac{1}{15ab} (-5b^2 D_x - 2a^2 D_{xy}) = \frac{1}{15ab} (-Q)$$

$$c_{11}^{(3)} = \frac{1}{15ab} (30p^{-2} D_x - 60p^2 D_y - 30D_1 - 84D_{xy}) = \frac{1}{15ab} R$$

$$c_{12}^{(3)} = \frac{1}{15ab} (-30ap D_y - 6b D_{xy}) = \frac{1}{15ab} S$$

$$c_{13}^{(3)} = \frac{1}{15ab} (15bp^{-1} D_x - 15a D_1 - 6a D_{xy}) = \frac{1}{15ab} T$$

$$c_{22}^{(3)} = \frac{1}{15ab} (-10a^2 D_y + 2b^2 D_{xy}) = \frac{1}{15ab} (-U)$$

$$c_{23}^{(3)} = 0$$

$$c_{33}^{(3)} = \frac{1}{15ab} (10b^2 D_x - 8a^2 D_{xy}) = \frac{1}{15ab} V$$

$$c_{11}^{(4)} = \frac{1}{15ab} (-60p^{-2} D_x + 30p^2 D_y - 30D_1 - 84D_{xy}) = \frac{1}{15ab} G$$

$$c_{12}^{(4)} = \frac{1}{15ab} (15a^2 D_y - 15b D_1 - 6b D_{xy}) = \frac{1}{15ab} (-H)$$

$$c_{13}^{(4)} = \frac{1}{15ab} (-30p^{-1} b D_x - 6a D_{xy}) = \frac{1}{15ab} (-I)$$

Sec. 5.5]   Stiffness equations/rectangular element for plate flexure   311

$$c_{22}^{(4)} = \frac{1}{15ab}(10a^2 D_y - 8b^2 D_{xy}) = \frac{1}{15ab} J$$

$$c_{23}^{(4)} = 0$$

$$c_{33}^{(4)} = \frac{1}{15ab}(-10b^2 D_x + 2a^2 D_{xy}) = \frac{1}{15ab}(-L).$$

Thus, the stiffness equation with the stiffness group supermatrix in normal form is obtained (with the nodal numbering of the group supermatrix procedure in Fig. 5.5 (b)) (see p. 312) or $K\Phi = P$.

In order to obtain the stiffness equation $K'\Phi' = P'$ with the conventional nodal numbering given in Fig. 5.5(a), it is necessary to change the numbering of nodes from (1, 2, 3, 4) to (3, 1, 2, 4), which is performed by

$$\begin{bmatrix} & & & E \\ & & E & \\ & E & & \\ E & & & \end{bmatrix} \begin{bmatrix} H_1 & H_2 & H_3 & H_4 \\ H_2 & H_1 & H_4 & H_3 \\ H_3 & H_4 & H_1 & H_2 \\ H_4 & H_3 & H_2 & H_1 \end{bmatrix} \begin{bmatrix} & & & E \\ & & E & \\ & E & & \\ E & & & \end{bmatrix} = \begin{bmatrix} H_1 & H_4 & H_2 & H_3 \\ H_4 & H_1 & H_3 & H_2 \\ H_2 & H_3 & H_1 & H_4 \\ H_3 & H_2 & H_4 & H_1 \end{bmatrix}$$

and (see p. 313) or $K'\Phi' = P'$.

This equation is transformed into the conventional stiffness equation by changing positive directions of $\theta_{y1}, \theta_{x3}, \theta_{x4}, \theta_{y4}, M_{y1}, M_{x3}, M_{x4}, M_{y4}$ in the set of positive directions that suit the first symmetry type of the group C_{2v} given in Fig. 5.5(b). In this way the conventional set of positive directions of nodal generalized displacements and forces according to Fig. 5.5(a) is obtained. This is performed by

$$T_D K' T_D T_D \Phi' = T_D P',$$

where

$$T_D = \text{Diag } [1 \quad 1 \quad \bar{1} \mid 1 \quad 1 \quad 1 \mid 1 \quad \bar{1} \quad 1 \mid 1 \quad \bar{1} \quad \bar{1}],$$

producing

$$\theta'_{y1} = -\theta_{y1}, \quad \theta'_{x3} = -\theta_{x3}, \quad \theta'_{x4} = -\theta_{x4}, \quad \theta'_{y4} = -\theta_{y4},$$

$$M'_{y1} = -M_{y1}, \quad M'_{x3} = -M_{x3}, \quad M'_{x4} = -M_{x4}, \quad M'_{y4} = -M_{y4}.$$

Thus, the conventional stiffness equation is obtained: (see p. 314) or $K_c\Phi_c = P_c$, where K_c is the conventional stiffness matrix of the rectangular element for plate flexure according to Fig. 5.5(a) with substitutions for $A, B, C, \ldots$ given earlier.

The conventional stiffness equation $K_c\Phi_c = P_c$, with the nodal numbering and the set of positive directions of nodal generalized displacements and forces given in Fig. 5.5(a), is transformed by

$$T_D K_c T_D T_D \Phi_c = T_D P_c,$$

312 Stiffness equations in G-invariant subspaces [Ch. 5

$$
\begin{bmatrix} P_1 \\ M_{x1} \\ M_{y1} \\ P_2 \\ M_{x2} \\ M_{y2} \\ P_3 \\ M_{x3} \\ M_{y3} \\ P_4 \\ M_{x4} \\ M_{y4} \end{bmatrix}
= \frac{1}{15ab}
\begin{bmatrix}
A & B & D & M & -N & -\text{``}O\text{''} & R & S & T & G & -H & -I \\
B & C & \text{``}E\text{''} & -N & -P & -\text{``}E\text{''} & S & -U & -\text{``}E\text{''} & -H & J & -\text{``}E\text{''} \\
D & \text{``}E\text{''} & F & -\text{``}O\text{''} & -\text{``}E\text{''} & -Q & T & -\text{``}E\text{''} & V & -I & -\text{``}E\text{''} & -L \\
M & -N & -\text{``}O\text{''} & A & B & D & G & -H & -I & R & S & T \\
-N & -P & -\text{``}E\text{''} & B & C & \text{``}E\text{''} & -H & J & -\text{``}E\text{''} & S & -U & -\text{``}E\text{''} \\
-\text{``}O\text{''} & -\text{``}E\text{''} & -Q & D & \text{``}E\text{''} & F & -I & -\text{``}E\text{''} & -L & T & -\text{``}E\text{''} & V \\
R & S & T & G & -H & -I & A & B & D & M & -N & -\text{``}O\text{''} \\
S & -U & -\text{``}E\text{''} & -H & J & -\text{``}E\text{''} & B & C & \text{``}E\text{''} & -N & -P & -\text{``}E\text{''} \\
T & -\text{``}E\text{''} & V & -I & -\text{``}E\text{''} & -L & D & \text{``}E\text{''} & F & -\text{``}O\text{''} & -\text{``}E\text{''} & -Q \\
G & -H & -I & R & S & T & M & -N & -\text{``}O\text{''} & A & B & D \\
-H & J & -\text{``}E\text{''} & S & -U & -\text{``}E\text{''} & -N & -P & -\text{``}E\text{''} & B & C & \text{``}E\text{''} \\
-I & -\text{``}E\text{''} & -L & T & -\text{``}E\text{''} & V & -\text{``}O\text{''} & -\text{``}E\text{''} & -Q & D & \text{``}E\text{''} & F
\end{bmatrix}
\begin{bmatrix} w_1 \\ \theta_{x1} \\ \theta_{y1} \\ w_2 \\ \theta_{x2} \\ \theta_{y2} \\ w_3 \\ \theta_{x3} \\ \theta_{y3} \\ w_4 \\ \theta_{x4} \\ \theta_{y4} \end{bmatrix}
$$

Sec. 5.5] Stiffness equations/rectangular element for plate flexure

$$
\begin{Bmatrix} P_1 \\ M_{x1} \\ M_{y1} \\ P_2 \\ M_{x2} \\ M_{y2} \\ P_3 \\ M_{x3} \\ M_{y3} \\ P_4 \\ M_{x4} \\ M_{y4} \end{Bmatrix}
=
\frac{1}{15ab}
\begin{bmatrix}
A & B & D & G & -H & -I & M & -N & -\text{``}O\text{''} & R & S & T \\
B & C & \text{``}E\text{''} & -H & J & -L & -N & -P & -Q & S & -U & V \\
D & \text{``}E\text{''} & F & -I & -L & \text{``}E\text{''} & -\text{``}O\text{''} & -Q & F & T & V & F \\
G & -H & -I & A & -B & -D & R & -S & T & M & N & -\text{``}O\text{''} \\
-H & J & -L & -B & C & -\text{``}E\text{''} & -S & -U & V & N & -P & Q \\
-I & -L & \text{``}E\text{''} & -D & -\text{``}E\text{''} & F & -T & V & F & \text{``}O\text{''} & Q & F \\
M & -N & -\text{``}O\text{''} & R & -S & -T & A & -B & -D & G & H & -I \\
-N & -P & -Q & -S & -U & V & -B & C & \text{``}E\text{''} & H & J & -L \\
-\text{``}O\text{''} & -Q & F & T & V & F & -D & \text{``}E\text{''} & F & I & L & \text{``}E\text{''} \\
R & S & T & M & N & -\text{``}O\text{''} & G & H & -I & A & B & D \\
S & -U & V & N & -P & Q & H & J & -L & B & C & \text{``}E\text{''} \\
T & V & F & -\text{``}O\text{''} & Q & F & I & L & \text{``}E\text{''} & D & \text{``}E\text{''} & F
\end{bmatrix}
\begin{Bmatrix} w_1 \\ \theta_{x1} \\ \theta_{y1} \\ w_2 \\ \theta_{x2} \\ \theta_{y2} \\ w_3 \\ \theta_{x3} \\ \theta_{y3} \\ w_4 \\ \theta_{x4} \\ \theta_{y4} \end{Bmatrix}
$$

$$\frac{1}{15ab}\begin{bmatrix} A & B & -D & G & -H & -I & M & -N & -\text{``}O\text{''} & R & -S & -T \\ B & C & -\text{``}E\text{''} & -H & J & L & N & P & Q & -S & U & V \\ -D & -\text{``}E\text{''} & F & I & L & \text{``}O\text{''} & \text{``}O\text{''} & Q & -T & -T & V & \text{``}O\text{''} \\ G & -H & I & A & B & D & R & S & T & M & N & -\text{``}O\text{''} \\ -H & J & L & B & C & \text{``}E\text{''} & -S & U & V & -N & P & Q \\ -I & L & \text{``}O\text{''} & D & \text{``}E\text{''} & F & T & V & \text{``}O\text{''} & -\text{``}O\text{''} & Q & -T \\ M & -N & -\text{``}O\text{''} & R & S & T & A & -B & -D & G & H & I \\ N & P & Q & -S & U & V & -B & C & -\text{``}E\text{''} & H & J & L \\ -\text{``}O\text{''} & Q & -T & T & V & \text{``}O\text{''} & D & -\text{``}E\text{''} & F & -I & L & \text{``}O\text{''} \\ R & S & T & M & -N & -\text{``}O\text{''} & G & H & -I & A & -B & -D \\ -S & U & V & N & P & Q & H & J & L & -B & C & \text{``}E\text{''} \\ -T & V & \text{``}O\text{''} & \text{``}O\text{''} & Q & -T & -I & L & \text{``}O\text{''} & -D & \text{``}E\text{''} & F \end{bmatrix} \begin{bmatrix} w_1 \\ \theta'_{x1} \\ \theta'_{y1} \\ w_2 \\ \theta_{x2} \\ \theta_{y2} \\ w_3 \\ \theta'_{x3} \\ \theta_{y3} \\ w_4 \\ \theta'_{x4} \\ \theta'_{y4} \end{bmatrix} = \begin{bmatrix} P_1 \\ M_{x1} \\ M'_{y1} \\ P_2 \\ M_{x2} \\ M_{y2} \\ P_3 \\ M'_{x3} \\ M_{y3} \\ P_4 \\ M'_{x4} \\ M'_{y4} \end{bmatrix}$$

with

$$T_D = \text{Diag } [1\ 1\ \bar{1} \mathrel{\vdots} 1\ 1\ 1 \mathrel{\vdots} 1\ \bar{1}\ 1 \mathrel{\vdots} 1\ \bar{1}\ \bar{1}],$$

into the stiffness equation $K'\Phi' = P'$ with the stiffness supermatrix in normal form.

In order to obtain the stiffness equation $K\Phi = P$ with the nodal numbering of the group supermatrix procedure, as given in Fig. 5.5(b), it is necessary to change the number of nodes from $(1, 2, 3, 4)$ to $(2, 3, 1, 4)$, which is achieved by

$$\begin{bmatrix} & E & & \\ E & & & \\ & & E & \\ & & & E \end{bmatrix} \begin{bmatrix} A_1 & A_2 & A_3 & A_4 \\ A_2 & A_1 & A_4 & A_3 \\ A_3 & A_4 & A_1 & A_2 \\ A_4 & A_3 & A_2 & A_1 \end{bmatrix} \begin{bmatrix} E & & & \\ & E & & \\ & & E & \\ & & & E \end{bmatrix} = \begin{bmatrix} A_1 & A_3 & A_4 & A_2 \\ A_3 & A_1 & A_2 & A_4 \\ A_4 & A_2 & A_1 & A_3 \\ A_2 & A_4 & A_3 & A_1 \end{bmatrix}$$

Transformation of the stiffness equation $K\Phi = P$ into the stiffness equation $\bar{K}\bar{\Phi} = \bar{P}$ with the stiffness group supermatrix in diagonal form is accomplished by

$$TKT^{-1} T\Phi = TP,$$

producing

$$\begin{bmatrix} \bar{K}_1 & & & \\ & \bar{K}_2 & & \\ & & \bar{K}_3 & \\ & & & \bar{K}_4 \end{bmatrix} \begin{bmatrix} \bar{\Phi}^{(1)} \\ \bar{\Phi}^{(2)} \\ \bar{\Phi}^{(3)} \\ \bar{\Phi}^{(4)} \end{bmatrix} = \begin{bmatrix} \bar{P}^{(1)} \\ \bar{P}^{(2)} \\ \bar{P}^{(3)} \\ \bar{P}^{(4)} \end{bmatrix}.$$

Table 5.4 presents a comparative display of derivations of the stiffness matrix of the rectangular element for plate flexure in the conventional way and by the group supermatrix procedure.

The four-node 16-d.-o.-f. rectangular plate bending element is obtained by adding the second-order twist derivative terms $\omega = \partial^2 w/\partial x\, \partial y$ to the 12-d,-o.-f. element having w, $\theta_x = -\partial w/\partial y$, $\theta_y = -\partial w/\partial x$ at each corner node, as shown in Fig. 5.5.

As the 12-d.-o.-f. element given in Fig. 5.5(b), the 16-d.-o.-f. element is described by the group C_{2v} and it has the same unique nodal numbering. By using the character table of the group C_{2v} and the idempotents Π of the centre of the group algebra, as previously in this section, the basis vectors $\bar{\varphi}_j$ ($j = 1, 2, \ldots, 16$) of G-invariant subspaces U_i ($i = 1, 2, 3, 4$) are derived:

$$\begin{bmatrix} \bar{\Phi}^{(1)} \\ \bar{\Phi}^{(2)} \\ \bar{\Phi}^{(3)} \\ \bar{\Phi}^{(4)} \end{bmatrix} = \begin{bmatrix} \bar{\varphi}_1 & \bar{\varphi}_5 & \bar{\varphi}_9 & \bar{\varphi}_{13} \\ \bar{\varphi}_2 & \bar{\varphi}_6 & \bar{\varphi}_{10} & \bar{\varphi}_{14} \\ \bar{\varphi}_3 & \bar{\varphi}_7 & \bar{\varphi}_{11} & \bar{\varphi}_{15} \\ \bar{\varphi}_4 & \bar{\varphi}_8 & \bar{\varphi}_{12} & \bar{\varphi}_{16} \end{bmatrix} = \begin{bmatrix} \bar{w}_1 & \bar{\theta}_{x1} & \bar{\theta}_{y1} & \bar{\omega}_1 \\ \bar{w}_2 & \bar{\theta}_{x2} & \bar{\theta}_{y2} & \bar{\omega}_2 \\ \bar{w}_3 & \bar{\theta}_{x3} & \bar{\theta}_{y3} & \bar{\omega}_3 \\ \bar{w}_4 & \bar{\theta}_{x4} & \bar{\theta}_{y4} & \bar{\omega}_4 \end{bmatrix}$$

$$= \frac{1}{4} \begin{bmatrix} 1 & 1 & 1 & 1 \\ 1 & 1 & -1 & -1 \\ 1 & -1 & 1 & -1 \\ 1 & -1 & -1 & 1 \end{bmatrix} \begin{bmatrix} w_1 & \theta_{x1} & \theta_{y1} & \omega_1 \\ w_2 & \theta_{x2} & \theta_{y2} & \omega_2 \\ w_3 & \theta_{x3} & \theta_{y3} & \omega_3 \\ w_4 & \theta_{x4} & \theta_{y4} & \omega_4 \end{bmatrix}$$

$$= \frac{1}{4} \begin{bmatrix} w_1 + w_2 + w_3 + w_4 & \theta_{x1} + \theta_{x2} + \theta_{x3} + \theta_{x4} \\ w_1 + w_2 - w_3 - w_4 & \theta_{x1} + \theta_{x2} - \theta_{x3} - \theta_{x4} \\ w_1 - w_2 + w_3 - w_4 & \theta_{x1} - \theta_{x2} + \theta_{x3} - \theta_{x4} \\ w_1 - w_2 - w_3 + w_4 & \theta_{x1} - \theta_{x2} - \theta_{x3} + \theta_{x4} \end{bmatrix}$$

$$\begin{bmatrix} \theta_{y1} + \theta_{y2} + \theta_{y3} + \theta_{y4} & \omega_1 + \omega_2 + \omega_3 + \omega_4 \\ \theta_{y1} + \theta_{y2} - \theta_{y3} - \theta_{y4} & \omega_1 + \omega_2 - \omega_3 - \omega_4 \\ \theta_{y1} - \theta_{y2} + \theta_{y3} - \theta_{y4} & \omega_1 - \omega_2 + \omega_3 - \omega_4 \\ \theta_{y1} - \theta_{y2} - \theta_{y3} + \theta_{y4} & \omega_1 - \omega_2 - \omega_3 + \omega_4 \end{bmatrix}$$

or $\bar{\Phi} = T\Phi$, with $T^{-1} = 4T$.

Conversely, the relation of the nodal functions $\Phi^{(i)}$ to the basis vectors $\bar{\Phi}^{(i)}$ is

$$\begin{bmatrix} \Phi^{(1)} \\ \Phi^{(2)} \\ \Phi^{(3)} \\ \Phi^{(4)} \end{bmatrix} = \begin{bmatrix} \varphi_1 & \varphi_5 & \varphi_9 & \varphi_{13} \\ \varphi_2 & \varphi_6 & \varphi_{10} & \varphi_{14} \\ \varphi_3 & \varphi_7 & \varphi_{11} & \varphi_{15} \\ \varphi_4 & \varphi_8 & \varphi_{12} & \varphi_{16} \end{bmatrix} \begin{bmatrix} w_1 & \theta_{x1} & \theta_{y1} & \omega_1 \\ w_2 & \theta_{x2} & \theta_{y2} & \omega_2 \\ w_3 & \theta_{x3} & \theta_{y3} & \omega_3 \\ w_4 & \theta_{x4} & \theta_{y4} & \omega_4 \end{bmatrix}$$

$$= \begin{bmatrix} 1 & 1 & 1 & 1 \\ 1 & 1 & -1 & -1 \\ 1 & -1 & 1 & -1 \\ 1 & -1 & -1 & 1 \end{bmatrix} \begin{bmatrix} \bar{w}_1 & \bar{\theta}_{x1} & \bar{\theta}_{y1} & \bar{\omega}_1 \\ \bar{w}_2 & \bar{\theta}_{x2} & \bar{\theta}_{y2} & \bar{\omega}_2 \\ \bar{w}_3 & \bar{\theta}_{x3} & \bar{\theta}_{y3} & \bar{\omega}_3 \\ \bar{w}_4 & \bar{\theta}_{x4} & \bar{\theta}_{y4} & \bar{\omega}_4 \end{bmatrix}$$

$$= \begin{bmatrix} \bar{w}_1 + \bar{w}_2 + \bar{w}_3 + \bar{w}_4 & \bar{\theta}_{x1} + \bar{\theta}_{x2} + \bar{\theta}_{x3} + \bar{\theta}_{x4} \\ \bar{w}_1 + \bar{w}_2 - \bar{w}_3 - \bar{w}_4 & \bar{\theta}_{x1} + \bar{\theta}_{x2} - \bar{\theta}_{x3} - \bar{\theta}_{x4} \\ \bar{w}_1 - \bar{w}_2 + \bar{w}_3 - \bar{w}_4 & \bar{\theta}_{x1} - \bar{\theta}_{x2} + \bar{\theta}_{x3} - \bar{\theta}_{x4} \\ \bar{w}_1 - \bar{w}_2 - \bar{w}_3 + \bar{w}_4 & \bar{\theta}_{x1} - \bar{\theta}_{x2} - \bar{\theta}_{x3} + \bar{\theta}_{x4} \end{bmatrix}$$

$$\begin{bmatrix} \bar{\theta}_{y1} + \bar{\theta}_{y2} + \bar{\theta}_{y3} + \bar{\theta}_{y4} & \bar{\omega}_1 + \bar{\omega}_2 + \bar{\omega}_3 + \bar{\omega}_4 \\ \bar{\theta}_{y1} + \bar{\theta}_{y2} - \bar{\theta}_{y3} - \bar{\theta}_{y4} & \bar{\omega}_1 + \bar{\omega}_2 - \bar{\omega}_3 - \bar{\omega}_4 \\ \bar{\theta}_{y1} - \bar{\theta}_{y2} + \bar{\theta}_{y3} - \bar{\theta}_{y4} & \bar{\omega}_1 - \bar{\omega}_2 + \bar{\omega}_3 - \bar{\omega}_4 \\ \bar{\theta}_{y1} - \bar{\theta}_{y2} - \bar{\theta}_{y3} + \bar{\theta}_{y4} & \bar{\omega}_1 - \bar{\omega}_2 - \bar{\omega}_3 + \bar{\omega}_4 \end{bmatrix}$$

Sec. 5.5]   Stiffness equations/rectangular element for plate flexure   317

or $\Phi = 4T\bar{\Phi}$, with $T^{-1} = 4T$.

The sets $\bar{\Phi}_l$ ($l = 1, 2, 3, 4$) containing basis vectors $\bar{\varphi}_j$ ($j = 1, 2, \ldots, 16$) that represent $w_1, \theta_{x1}, \theta_{y1}, \omega_1, w_2, \theta_{x2}, \theta_{y2}, \omega_2, w_3, \theta_{x3}, \theta_{y3}, \omega_3, w_4, \theta_{x4}, \theta_{y4}, \omega_4$, ordered according to the sequence explained in section 4.1 (step (2)), are given by

$$[\bar{\Phi}_1 \ \bar{\Phi}_2 \ \bar{\Phi}_3 \ \bar{\Phi}_4] = \begin{bmatrix} \bar{\varphi}_1 & \bar{\varphi}_5 & \bar{\varphi}_9 & \bar{\varphi}_{13} \\ \bar{\varphi}_2 & \bar{\varphi}_6 & \bar{\varphi}_{10} & \bar{\varphi}_{14} \\ \bar{\varphi}_3 & \bar{\varphi}_7 & \bar{\varphi}_{11} & \bar{\varphi}_{15} \\ \bar{\varphi}_4 & \bar{\varphi}_8 & \bar{\varphi}_{12} & \bar{\varphi}_{16} \end{bmatrix} = \begin{bmatrix} w_1 & \theta_{x1} & \theta_{y1} & \omega_1 \\ w_2 & \theta_{x2} & \theta_{y2} & \omega_2 \\ w_3 & \theta_{x3} & \theta_{y3} & \omega_3 \\ w_4 & \theta_{x4} & \theta_{y4} & \omega_4 \end{bmatrix}.$$

The relation of the sets of basis vectors $\bar{\Phi}_l$ to the sets of nodal functions Φ_l is expressed by the following system of equations with the supermatrix in diagonal form containing four transformation matrices T of the group C_{2v}:

$$\begin{bmatrix} \bar{\Phi}_1 \\ \bar{\Phi}_2 \\ \bar{\Phi}_3 \\ \bar{\Phi}_4 \end{bmatrix} = \begin{bmatrix} T & & & \\ & T & & \\ & & T & \\ & & & T \end{bmatrix} \begin{bmatrix} \Phi_1 \\ \Phi_2 \\ \Phi_3 \\ \Phi_4 \end{bmatrix} \quad \text{or} \quad \bar{\Phi}_l = T\Phi_l,$$

with

$$T = \frac{1}{4} \begin{bmatrix} 1 & 1 & 1 & 1 \\ 1 & 1 & -1 & -1 \\ 1 & -1 & 1 & -1 \\ 1 & -1 & -1 & 1 \end{bmatrix}$$

and

$$[\Phi_1 \ \Phi_2 \ \Phi_3 \ \Phi_4] = \begin{bmatrix} \varphi_1 & \varphi_5 & \varphi_9 & \varphi_{13} \\ \varphi_2 & \varphi_6 & \varphi_{10} & \varphi_{14} \\ \varphi_3 & \varphi_7 & \varphi_{11} & \varphi_{15} \\ \varphi_4 & \varphi_8 & \varphi_{12} & \varphi_{16} \end{bmatrix} = \begin{bmatrix} w_1 & \theta_{x1} & \theta_{y1} & \omega_1 \\ w_2 & \theta_{x2} & \theta_{y2} & \omega_2 \\ w_3 & \theta_{x3} & \theta_{y3} & \omega_3 \\ w_4 & \theta_{x4} & \theta_{y4} & \omega_4 \end{bmatrix}.$$

Conversely, the relation of the sets of nodal functions Φ_l to the sets of basis vectors $\bar{\Phi}_l$ is

$$\begin{bmatrix} \Phi_1 \\ \Phi_2 \\ \Phi_3 \\ \Phi_4 \end{bmatrix} = 4 \begin{bmatrix} T & & & \\ & T & & \\ & & T & \\ & & & T \end{bmatrix} \begin{bmatrix} \bar{\Phi}_1 \\ \bar{\Phi}_2 \\ \bar{\Phi}_3 \\ \bar{\Phi}_4 \end{bmatrix} \quad \text{or} \quad \Phi_l = 4T\bar{\Phi}_l, \quad \text{with} \quad T^{-1} = 4T.$$

Table 5.4. Comparison of derivations of the stiffness matrix of the rectangular element for plate flexure in the conventional way and by the group supermatrix procedure

Derivation of the stiffness matrix of the rectangular element for plate flexure with twelve degrees of freedom

	Conventional way	Group supermatrix procedure	Subspace
[1] Functions spanning the space of the displacement field	$\begin{bmatrix} w \\ \theta_x \\ \theta_y \end{bmatrix} = \begin{bmatrix} w_1 \\ \theta_{x1} \\ \theta_{y1} \\ \hdashline w_2 \\ \theta_{x2} \\ \theta_{y2} \\ \hdashline w_3 \\ \theta_{x3} \\ \theta_{y3} \\ \hdashline w_4 \\ \theta_{x4} \\ \theta_{y4} \end{bmatrix}$	$\begin{bmatrix} \bar{w}_1 \\ \bar{\theta}_{x1} \\ \bar{\theta}_{y1} \\ \hdashline \bar{w}_2 \\ \bar{\theta}_{x2} \\ \bar{\theta}_{y2} \\ \hdashline \bar{w}_3 \\ \bar{\theta}_{x3} \\ \bar{\theta}_{y3} \\ \hdashline \bar{w}_4 \\ \bar{\theta}_{x4} \\ \bar{\theta}_{y4} \end{bmatrix} = \frac{1}{4} \begin{bmatrix} w_1 + w_2 + w_3 + w_4 \\ \theta_{x1} + \theta_{x2} + \theta_{x3} + \theta_{x4} \\ \theta_{y1} + \theta_{y2} + \theta_{y3} + \theta_{y4} \\ \hdashline w_1 + w_2 - w_3 - w_4 \\ \theta_{x1} + \theta_{x2} - \theta_{x3} - \theta_{x4} \\ \theta_{y1} + \theta_{y2} - \theta_{y3} - \theta_{y4} \\ \hdashline w_1 - w_2 + w_3 - w_4 \\ \theta_{x1} - \theta_{x2} + \theta_{x3} - \theta_{x4} \\ \theta_{y1} - \theta_{y2} + \theta_{y3} - \theta_{y4} \\ \hdashline w_1 - w_2 - w_3 + w_4 \\ \theta_{x1} - \theta_{x2} - \theta_{x3} + \theta_{x4} \\ \theta_{y1} - \theta_{y2} - \theta_{y3} + \theta_{y4} \end{bmatrix}$	U_1 U_2 U_3 U_4

Table 5.4. (continued)

Derivation of the stiffness matrix of the rectangular element for plate flexure with twelve degrees of freedom

Conventional way	Group supermatrix procedure	Subspace

[2] Matrix that is inverted to produce the coefficients of the polynomial displacement function

$$C = \begin{bmatrix} \cdots \end{bmatrix} \qquad \bar{C} = \begin{bmatrix} \cdots \end{bmatrix}$$

with subspaces U_1, U_2, U_3, U_4.

\# non-zero element   • zero element

Table 5.4. (continued)

Derivation of the stiffness matrix of the rectangular element for plate flexure with twelve degrees of freedom

Conventional way	Group supermatrix procedure

[3]
Derivation of the stiffness matrix

$$R^T DR = \begin{bmatrix} \cdot & \cdot & \cdot & \# & \cdot & \cdot & \# & \# & \cdot & \# & \# \\ \cdot & \cdot & \cdot & \cdot & \# & \cdot & \cdot & \# & \# & \cdot & \# \\ \cdot & \cdot & \# & \cdot & \cdot & \# & \# & \cdot & \cdot & \# & \cdot \end{bmatrix} \times \begin{bmatrix} \# & \# & \cdot \\ \# & \# & \cdot \\ \cdot & \cdot & \# \end{bmatrix} \times \begin{bmatrix} \cdot & \cdot & \cdot & \# & \cdot & \cdot & \# & \# & \cdot & \# & \# \\ \cdot & \cdot & \cdot & \cdot & \# & \cdot & \cdot & \# & \# & \cdot & \# \\ \cdot & \cdot & \# & \cdot & \cdot & \# & \# & \cdot & \cdot & \# & \cdot \end{bmatrix}$$

$$\bar{B}_1^T D \bar{B}_1 = \begin{bmatrix} \cdot & \cdot & \cdot \\ \cdot & \# & \cdot \\ \cdot & \cdot & \# \end{bmatrix} \times \begin{bmatrix} \cdot & \cdot & \# \\ \# & \# & \cdot \\ \# & \# & \cdot \\ \cdot & \cdot & \# \end{bmatrix} \times \begin{bmatrix} \# & \cdot & \cdot \\ \cdot & \# & \cdot \\ \cdot & \cdot & \# \end{bmatrix}$$

$$\bar{B}_2^T D \bar{B}_2 = \begin{bmatrix} \cdot & \# & \# \\ \# & \# & \cdot \\ \cdot & \cdot & \# \end{bmatrix} \times \begin{bmatrix} \cdot & \cdot & \# \\ \# & \# & \cdot \\ \# & \# & \cdot \\ \cdot & \cdot & \# \end{bmatrix} \times \begin{bmatrix} \# & \cdot & \# \\ \cdot & \# & \# \\ \# & \# & \# \end{bmatrix}$$

$$\bar{B}_3^T D \bar{B}_3 = \begin{bmatrix} \# & \# & \cdot \\ \cdot & \cdot & \# \\ \cdot & \cdot & \# \end{bmatrix} \times \begin{bmatrix} \cdot & \cdot & \# \\ \# & \# & \cdot \\ \# & \# & \cdot \\ \cdot & \cdot & \# \end{bmatrix} \times \begin{bmatrix} \# & \cdot & \cdot \\ \cdot & \# & \# \\ \# & \# & \cdot \end{bmatrix}$$

$$\bar{B}_4^T D \bar{B}_4 = \begin{bmatrix} \cdot & \# & \cdot \\ \cdot & \# & \cdot \\ \# & \cdot & \# \end{bmatrix} \times \begin{bmatrix} \cdot & \cdot & \# \\ \# & \# & \cdot \\ \# & \# & \cdot \\ \cdot & \cdot & \# \end{bmatrix} \times \begin{bmatrix} \# & \cdot & \# \\ \cdot & \# & \# \\ \# & \# & \cdot \end{bmatrix}$$

$$\bar{K} = \text{Diag}\,[\bar{K}_1 \quad \bar{K}_2 \quad \bar{K}_3 \quad \bar{K}_4]$$

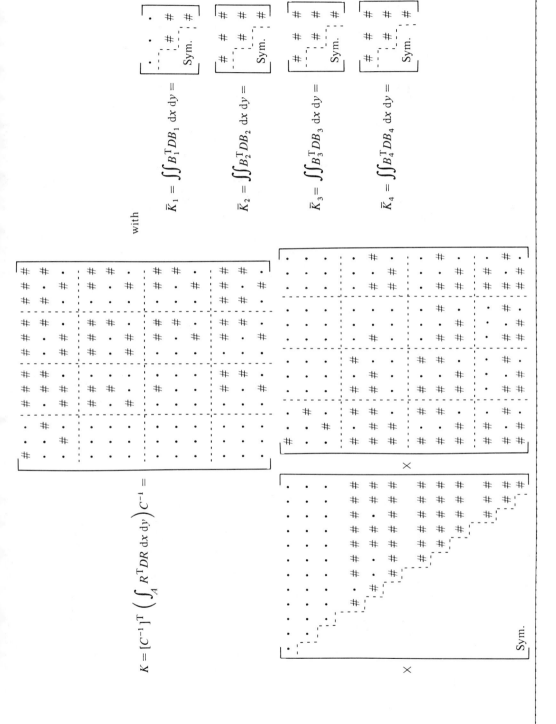

Table 5.4. (continued)

Derivation of the stiffness matrix of the rectangular element for plate flexure with twelve degrees of freedom

	Conventional way	Group supermatrix procedure	Subspace
[4] Stiffness matrix	$K=$ (full symmetric matrix with 12 dof shown in staircase block form, Sym.)	$\bar{K}=$ (block diagonal form with four symmetric blocks)	U_1 U_2 U_3 U_4

In block diagonal form

The displacement function for the four-node 16-d.-o.-f. rectangular plate bending element is usually assumed to be

$$w = \alpha_1 + \alpha_2 x + \alpha_3 y + \alpha_4 x^2 + \alpha_5 xy + \alpha_6 y^2 + \alpha_7 x^3 + \alpha_8 x^2 y + \alpha_9 xy^2$$
$$+ \alpha_{10} y^3 + \alpha_{11} x^3 y + \alpha_{12} x^2 y^2 + \alpha_{13} xy^3 + \alpha_{14} x^3 y^2 + \alpha_{15} x^2 y^3 + \alpha_{16} x^3 y^3.$$

For the group supermatrix procedure the terms in the above polynomial are allocated by the following expression to their respective G-invariant subspaces, with the unique order of the terms that provides the diagonal form of the supermatrix of the system of equations relating displacements $\bar{\Phi}$ to coefficients A:

$$\begin{bmatrix} F_1 \\ F_2 \\ F_3 \\ F_4 \end{bmatrix} = \begin{bmatrix} 1 \\ xy \\ x \\ y \end{bmatrix} \begin{bmatrix} 1 & x^2 & y^2 & x^2 y^2 \end{bmatrix}$$

$$= \begin{bmatrix} 1 & x^2 & y^2 & x^2 y^2 \\ xy & x^3 y & xy^3 & x^3 y^3 \\ x & x^3 & xy^2 & x^3 y^2 \\ y & x^2 y & y^3 & x^2 y^3 \end{bmatrix} \begin{matrix} \text{Subspace} \\ U_1 \\ U_2 \\ U_3 \\ U_4 \end{matrix}$$

The column vector and the row vector are taken from the Cartesian sets and products as they stand in addition to the character table of the group C_{2v} according to the pertinence of the terms to the symmetry types of group representations A_1, A_2, B_1, B_2.

The above expression is identical to the expression derived for the sixteen-node rectangular element in section 4.5.

In Fig. 5.5(b), where positive directions of displacements and rotations of the nodes possess the symmetry type of the first subspace, the derivative term $\partial^2 w/\partial x\, \partial y$ in the first quadrant has a positive direction opposite to that of the corresponding derivative term in Fig. 5.5(a), which presents the conventional set of positive directions. It is therefore necessary to add the twist $\omega = -\partial^2 w/\partial x\, \partial y$ to w, $\theta_x = -\partial w/\partial y$, $\theta_y = -\partial w/\partial x$, which were used for the 12-d.-o.-f. plate bending element.

Consequently, the displacement function w and its derivatives $\theta_x = -\partial w/\partial y$, $\theta_y = -\partial w/\partial x$, $\omega = -\partial^2 w/\partial x\, \partial y$ can be expressed as

$$\begin{bmatrix} w \\ \theta_x \\ \theta_y \\ \omega \end{bmatrix} = \begin{bmatrix} 1 & x^2 & y^2 & x^2y^2 & xy & x^3y & xy^3 & x^3y^3 \\ & -2y & -2x^2y & -x & -x^3 & -3xy^2 & -3x^3y^2 \\ -2x & & -2xy^2 & -y & -3x^2y & -y^3 & -3x^2y^3 \\ & & -4xy & -1 & -3x^2 & -3y^2 & -9x^2y^2 \end{bmatrix}$$

$$\begin{bmatrix} x & x^3 & xy^2 & x^3y^2 & y & x^2y & y^3 & x^2y^3 \\ & & -2xy & -2x^3y & -1 & -x^2 & -3y^2 & -3x^2y^2 \\ -1 & -3x^2 & -y^2 & -3x^2y^2 & & -2xy & & -2xy^3 \\ & & -2y & -6x^2y & & -2x & & -6xy^2 \end{bmatrix}$$

$$\cdot [\alpha_1 \ \alpha_2 \ \alpha_3 \ \alpha_4 \ \vdots \ \alpha_5 \ \alpha_6 \ \alpha_7 \ \alpha_8 \ \vdots \ \alpha_9 \ \alpha_{10} \ \alpha_{11} \ \alpha_{12}$$
$$\vdots \ \alpha_{13} \ \alpha_{14} \ \alpha_{15} \ \alpha_{16}]^T$$
$$= [F_1 \ \vdots \ F_2 \ \vdots \ F_3 \ \vdots \ F_4][A_1 \ \vdots \ A_2 \ \vdots \ A_3 \ \vdots \ A_4]^T.$$

Thus, the above function is decomposed into four four-dimensional G-invariant subspaces U_1, U_2, U_3, U_4.

Substitution of the nodal coordinates of the first node in the nodal set $S(1, 2, 3, 4)$, i.e. $x = c$, $y = d$ for the node 1, into the function matrices F_1, F_2, F_3, F_4 of $[w \ \theta_x \ \theta_y \ \omega]^T$ pertaining to the subspaces U_1, U_2, U_3, U_4 will give the relations of the generalized displacements $\bar{\Phi}$ to the coefficients A for each subspace separately (see p. 325).

or $\bar{\Phi} = \bar{C}A$.

The sixteen-dimensional G-vector space of the problem is decomposed now into four four-dimensional G-invariant subspaces U_1, U_2, U_3, U_4. The matrix of the system of equations of the above relation is obtained in block diagonal form containing four 4 × 4 matrices.

The coefficients A are determined by inverting the matrices $\bar{C}_1, \bar{C}_2, \bar{C}_3, \bar{C}_4$ in the subspaces U_1, U_2, U_3, U_4. Thus, the sixteen-dimensional G-vector space of the displacement field is decomposed into four four-dimensional subspaces by

$$\begin{bmatrix} \Delta_1 \\ \Delta_2 \\ \Delta_3 \\ \Delta_4 \end{bmatrix} = \text{Diag } [F_1 \ F_2 \ F_3 \ F_4] \begin{bmatrix} \bar{C}_1^{-1} & & & \\ & \bar{C}_2^{-1} & & \\ & & \bar{C}_3^{-1} & \\ & & & \bar{C}_4^{-1} \end{bmatrix} \begin{bmatrix} \bar{\Phi}^{(1)} \\ \bar{\Phi}^{(2)} \\ \bar{\Phi}^{(3)} \\ \bar{\Phi}^{(4)} \end{bmatrix}$$

or $\Delta = T_D \bar{C}^{-1} \bar{\Phi}$.

As in the case of the 12-d.o.-f. element the shape functions $\bar{N}^{(i)}$ ($i = 1, 2, 3, 4$) and stiffness matrices $\bar{K}_I$ are derived for each subspace U_i separately.

$$
\begin{bmatrix}
\bar{w}_1 \\
\bar{\theta}_{x1} \\
\bar{\theta}_{y1} \\
\bar{\omega}_1 \\
\hline
\bar{w}_2 \\
\bar{\theta}_{x2} \\
\bar{\theta}_{y2} \\
\bar{\omega}_2 \\
\hline
\bar{w}_3 \\
\bar{\theta}_{x3} \\
\bar{\theta}_{y3} \\
\bar{\omega}_3 \\
\hline
\bar{w}_4 \\
\bar{\theta}_{x4} \\
\bar{\theta}_{y4} \\
\bar{\omega}_4
\end{bmatrix}
=
\left[
\begin{array}{cccc|cccc|cccc|cccc}
1 & c^2 & d^2 & c^2d^2 & & & & & & & & & & & & \\
0 & 0 & -2d & -2c^2d & & & & & & & & & & & & \\
0 & -2c & 0 & -2cd^2 & & & & & & & & & & & & \\
0 & 0 & 0 & -4cd & & & & & & & & & & & & \\
\hline
 & & & & cd & c^3d & cd^3 & c^3d^3 & & & & & & & & \\
 & & & & -c & -c^3 & -3cd^2 & -3c^3d^2 & & & & & & & & \\
 & & & & -d & -3c^2d & -d^3 & -3c^2d^3 & & & & & & & & \\
 & & & & -1 & -3c^2 & -3d^2 & -9c^2d^2 & & & & & & & & \\
\hline
 & & & & & & & & c & c^3 & cd^2 & c^3d^2 & & & & \\
 & & & & & & & & 0 & 0 & -2cd & -2c^3d & & & & \\
 & & & & & & & & -1 & -3c^2 & -d^2 & -3c^2d^2 & & & & \\
 & & & & & & & & 0 & 0 & -2d & -6c^2d & & & & \\
\hline
 & & & & & & & & & & & & d & c^2d & d^3 & c^2d^3 \\
 & & & & & & & & & & & & -1 & -c^2 & -3d^2 & -3c^2d^2 \\
 & & & & & & & & & & & & 0 & -2cd & 0 & -2cd^3 \\
 & & & & & & & & & & & & 0 & -2c & 0 & -6cd^2
\end{array}
\right]
\begin{bmatrix}
\alpha_1 \\ \alpha_2 \\ \alpha_3 \\ \alpha_4 \\
\alpha_5 \\ \alpha_6 \\ \alpha_7 \\ \alpha_8 \\
\alpha_9 \\ \alpha_{10} \\ \alpha_{11} \\ \alpha_{12} \\
\alpha_{13} \\ \alpha_{14} \\ \alpha_{15} \\ \alpha_{16}
\end{bmatrix}
$$

6

Group supermatrices in formulation and assembly of stiffness equations

6.1 GROUP SUPERMATRIX PROCEDURE IN THE DIRECT STIFFNESS METHOD

The method of direct superposition, i.e. the direct stiffness method, can be used to assemble structure stiffness equations in finite element analysis.

The direct stiffness method permits the system stiffness matrix to be compiled by direct superpositions of stiffness matrices of individual structural elements. When a structure is defined, the system stiffness matrix is formulated with respect to

(1) the numbering of the nodes (nodal points) of the structure,
(2) the local coordinate systems of elements of the structure,
(3) the global coordinate system of the structure and
(4) the directions of displacements and rotations of the nodes.

It is the usual practice to employ a numbering of the nodes that provides an accumulation of non-zero matrix elements along the main diagonal of the stiffness matrix. A system of linear equations with such a matrix is more favourable for computation than systems with matrices where non-zero elements are scattered far from the main diagonal.

In the case of a structural element connecting two nodes, denoted by numbers m and n ($m < n$), the origin of the local coordinate system is usually positioned at node m, while the x axis coincides with the axis of the element.

The choice of the position of the origin of the global coordinate system is usually made with respect to the shape of the structure, for instance, at the corner of a rectangular configuration and with the coordinate axes coinciding with the sides of the rectangle.

It is a convention that positive directions of components of displacements coincide with positive directions of respective parallel coordinate axes, while positive directions of rotations about axes, parallel to coordinate axes, are counterclockwise rotations.

When a structure possesses symmetry properties described by a group G, for the group supermatrix procedure the above conventions concerning the numbering of nodes, positions of origins and directions of axes of the local and global coordinate systems, as well as the sets of positive directions of displacements and rotations of the nodes, must be modified to suit the first symmetry type of the group. Then it will be possible to formulate system stiffness equations with stiffness group supermatrices in normal and diagonal forms, giving system stiffness equations in G-invariant subspaces of the G-vector space of the problem.

G-vector analysis in the direct stiffness method was developed by Zloković (1989). In this present volume it is renamed 'the group supermatrix procedure in the direct stiffness method' because of the introduction of formulations of the group supermatrix transformations derived in Chapter 3.

The group supermatrix procedure in the direct stiffness method is systematized into seven consecutive steps.

(1) *Determination of the symmetry group of the structure*
When a structural configuration is transformed into itself by a set of symmetry operations that form a group of high order, it may be useful to consider applications of the subgroups of the group too and to compare their performances. As shown in section 2.1, a subgroup of the group G, with fewer symmetry elements and a fewer G-invariant subspaces than group G, may be more effective in the analysis than is group G.

It is also convenient when no nodes lie in reflection planes because this feature provides G-invariant subspaces of smaller dimensions than when nodes are in these planes, which can be confirmed by analysis of the expression for determination of dimensions of the subspaces

Dim $U_i = h_i(\text{Tr}, \chi_i)$,

as given in section 2.1.

Thus, in the case of a nodal pattern based on the square, it may be simpler and more efficient to apply the symmetry group C_{2v}, which describes the symmetry properties of the rectangle and possesses four symmetry operations and four one-dimensional irreducible group representations, instead of the symmetry group C_{4v} of the square with eight symmetry operations and five irreducible group representations (four one-dimensional and one two-dimensional ones).

The origin of the global coordinate system of the configuration must lie in the centre of symmetry of the structure. It remains only to consider various positions of the coordinate system, produced by rotations compatible with applications of symmetry operations of the group. When the coordinate system is in a compatible position and is such that no nodes lie in reflection planes, applications of all symmetry operations σ of the group on functions φ_j at nodes j produce the minimum number of sets of functions φ_j that permute between themselves. This relates also to the most favourable decomposition of the G-vector space V of the problem into the maximum number of G-invariant subspaces with smallest dimensions.

(2) *The numbering of the nodes and variables*

When the symmetry group G of the configuration is chosen and the global coordinate system established, it is necessary to select a numbering of the nodes and a numbering of variables (displacements and rotations) pertaining to individual nodes.

By applying all symmetry operations of the group G to functions φ_j at the node j for all nodes, one obtains the sets of functions φ_j (or nodes) that permute between themselves. When $S_1, S_2, \ldots, S_l$ are these sets of nodes, then each set represents an independent nodal pattern and the complete configuration can be regarded as an assembly of l nodal patterns pertaining to these sets. In Chapter 3 nodal numberings are derived for nodal patterns described by symmetry groups C_2, C_3, C_4, C_{2v}, C_{3v}, C_{4v} and D_{2h}.

The numbering of variables (generalized displacements) begins from node 1 with variables $u_1, v_1, w_1, \theta_{x1}, \theta_{y1}, \theta_{z1}$, numbering them from 1 to 6 (in a general case). The numbering of variables is continued in the same sequence at node 2, etc. In this way the compatibility for superpositions of submatrices in system stiffness group supermatrices is established. i.e. the generalized displacements of the node j will be superposed on the corresponding generalized displacements of all nodes which permute with the node j.

(3) *Determination of the basis vectors of G-invariant subspaces U_i*

By using the character table of G and the expression

$$\pi_i = \frac{h_i}{h} \sum_\sigma \chi_i(\sigma^{-1})\sigma,$$

the idempotents of the centre of group algebra are obtained. The basis vectors of G-invariant subspaces U_i are derived by applying π_i to nodal functions φ_j ($j = 1, 2, \ldots, n$).

(4) *Utilization of group supermatrices of G*

The group supermatrix G in normal form can be used for a straightforward formulation of the system stiffness group supermatrix in normal form. Thus, for instance, for the group C_{2v}, this supermatrix is

$$G = \begin{bmatrix} A & B & C & D \\ B & A & D & C \\ C & D & A & B \\ D & C & B & A \end{bmatrix}.$$

By the group supermatrix transformation TGT^{-1} the supermatrix G is transformed into the group supermatrix $\bar{G}$ in diagonal form

$$\bar{G} = \begin{bmatrix} A+B+C+D & & & \\ & A+B-C-D & & \\ & & A-B+C-D & \\ & & & A-B-C+D \end{bmatrix},$$

Sec. 6.1]   Group supermatrix procedure in the direct stiffness method   329

which is utilized for formulation of system stiffness matrices in G-invariant subspaces.

(5) *Establishing the correspondence of positive directions of displacements and rotations of the nodes with respective positive directions in the subspace U_1*

Since the configuration with displacements and rotations of the nodes in each subspace U_i must possess the symmetry type of U_i, it is necessary to form a two- or three-dimensional graphic diagram of the configuration with designated positive directions of displacements and rotations of the nodes that correspond to respective positive directions in the subspace U_1.

The correctness of the graphic diagram of the configuration having the symmetry type of U_1 may be tested by applying all symmetry operations σ of G with positive signs, each of which must transform the configuration into itself.

It is not necessary to draw graphic diagrams for displacements and rotations having the symmetry type of other subspaces when the graphic diagram of U_1 is drawn, since all derivations for these subspaces are performed algebraically by linear combinations which are based on the configuration in the subspace U_1.

(6) *Formulation of element stiffness matrices*

The configuration with the designated directions of generalized displacements of the symmetry type of the first subspace U_1 will serve to determine the origins of the local coordinate systems of structural elements and the directions of displacements and rotations of their nodes. It is necessary to formulate element stiffness matrices of all types of elements that appear in the configuration with respect to the combinations of directions of displacements and rotations of the nodes of the element.

(7) *Formulation of the system stiffness equations with group supermatrices in normal and diagonal forms*

Assembling element stiffness matrices yields the system stiffness supermatrix K in normal form, as explained in step (4).

The system stiffness group supermatrix $\bar{K}$ in diagonal form is obtained by transforming the group supermatrix K in normal form by using the group supermatrix transformation

$$\bar{K} = TKT^{-1},$$

where T is the transformation matrix of the group G.

Thus, the stiffness equation with the system stiffness supermatrix in diagonal form is

$$\bar{K} = \begin{bmatrix} \bar{K}_1 & & & \\ & \bar{K}_2 & & \\ & & \ddots & \\ & & & \bar{K}_k \end{bmatrix} \begin{bmatrix} \bar{\Phi}^{(1)} \\ \bar{\Phi}^{(2)} \\ \vdots \\ \bar{\Phi}^{(k)} \end{bmatrix} = \begin{bmatrix} \bar{P}^{(1)} \\ \bar{P}^{(2)} \\ \vdots \\ \bar{P}^{(k)} \end{bmatrix},$$

where system stiffness matrices $\bar{K}_i$, symmetry-adapted generalized displacements $\bar{\Phi}^{(i)}$ and symmetry-adapted generalized loads $\bar{P}^{(i)}$ pertain to G-invariant subspaces U_i ($i = 1, 2, \ldots, k$).

6.2 LINEAR BEAM ELEMENT ASSEMBLY

The linear beam element assembly, as given in Fig. 6.1, is composed of $n-1$ equal beam elements and n nodes, n being an even number. The nodal numbering on the positive branch of the x axis runs from the end node 1 to the node $\frac{1}{2}n$, the nearest one to the origin, while on the negative branch of the x axis the nodal numbering runs from the end node $\frac{1}{2}n + 1$ to the node n, the nearest one to the origin. This is the nodal numbering derined by $C_2(j) = (n/2) + j$, with $j = 1, 2, \ldots, \frac{1}{2}n$ and C_2 the rotation through 180° about the z axis, as given in section 3.2.

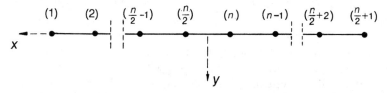

Fig. 6.1. The nodal numbering of the linear beam assembly described by the group C_2.

The above nodal numbering suits the symmetry properties of the assembly described by the group C_2, and it fulfils the requirements for the group supermatrices G and $\bar{G}$ of the group C_2 (derived in 3.2), i.e. the compatibility of columns pertaining to nodes and variables in

$$G = \begin{bmatrix} A & B \\ B & A \end{bmatrix} \quad \text{and} \quad \bar{G} = \begin{bmatrix} A+B & \\ & A-B \end{bmatrix}.$$

The above nodal numbering provides the numbering of nodal functions pertaining to the columns of the stiffness group supermatrix in normal form

$$K = \begin{array}{c} \\ \end{array} \begin{array}{cccccc} 1 & 2 & \cdots & \dfrac{n}{2} & \dfrac{n}{2}+1 & \cdots & n \end{array}$$

$$K = \left| \begin{array}{c|c} M & N \\ \hline N & M \end{array} \right|$$

the nodal numbering which runs continuously from 1 to n.

The nodal displacement v_j and the nodal rotation θ_j pertaining to the node j ($j = 1, 2, \ldots, n$) must have positive directions that suit the first symmetry type of the group C_2, as given in Fig. 6.2. This configuration must be turned into itself by action of all group elements, i.e. by E (identity) and C_2 (rotation through 180° about the z axis).

Sec. 6.2] Linear beam element assembly 331

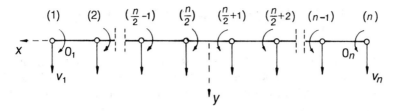

Fig. 6.2. The set of positive directions of displacements and rotations of the nodes that suit the first symmetry type of the group C_2 which describes the symmetry properties of the linear beam assembly.

The stiffness matrix of the beam element is derived with respect to the local coordinate system and the directions of positive displacements and rotations of the nodes given in Fig. 6.3. The beam member has uniform flexural rigidity, undergoes no shear deformation and is loaded only at its end points. Accordingly, the beam element stiffness matrix is

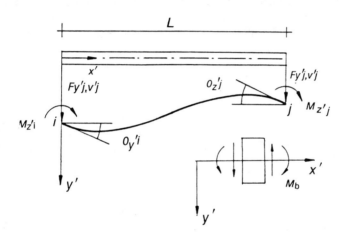

Fig. 6.3. Local coordinate system and directions of positive displacements and rotations of the nodes with respect to the axis of the beam element and the numbering of the nodes.

$$K'_{ij} = \frac{E'I_{z'}}{L^3} \begin{bmatrix} 12 & 6L & -12 & 6L \\ 6L & 4L^2 & -6L & 2L^2 \\ -12 & -6L & 12 & -6L \\ 6L & 2L^2 & -6L & 4L^2 \end{bmatrix} = E'I \begin{bmatrix} s & t & -s & t \\ t & 2v & -t & v \\ -s & -t & s & -t \\ t & v & -t & 2v \end{bmatrix},$$

with $s = 12/L^3$, $t = 6/L^2$ and $v = 2/L$. The displacements and rotations $v'_i, \theta_{z'i}, v'_j, \theta_{z'j}$ are denoted in a row above the matrix as a reminder of the meaning of the elements of K'_{ij}.

The transformation of K'_{ij} from the local to the global coordinates is accomplished by

$$K_{ij} = T^T K'_{ij} T,$$

where T is the transformation matrix. When the local and the global coordinate axes x' and x, y' and y, z' and z are parallel and have the same directions, the matrix T becomes an identity matrix E and the beam stiffness matrix is the same in both the local and the global coordinate systems:

$$K_{ij} = E^T K'_{ij} E = K'_{ij}.$$

The inspection of the linear beam element assembly in Fig. 6.2 shows that beside the conventional beam element, given in Fig. 6.3, there are beam elements with positive directions of nodal rotations that do not coincide with positive directions of the conventional beam element.

Therefore, the stiffness matrices will be determined for beam elements with all combinations of positive and negative directions of nodal rotations, as given in Fig. 6.4. Since positive displacements of all nodes have downward direction, they are not given in the figure.

The following beam element stiffness matrices correspond to the pairs of diagrams denoted by I, II, III, IV in Fig. 6.4:

$$K_{ij}^{I} = E'I \begin{bmatrix} s & t & -s & t \\ t & 2v & -t & v \\ -s & -t & s & -t \\ t & v & -t & 2v \end{bmatrix} \qquad K_{ij}^{II} = E'I \begin{bmatrix} s & t & -s & -t \\ t & 2v & -t & -v \\ -s & -t & s & t \\ -t & -v & t & 2v \end{bmatrix}$$

$$K_{ij}^{III} = E'I \begin{bmatrix} s & -t & -s & t \\ -t & 2v & t & -v \\ -s & t & s & -t \\ t & -v & -t & 2v \end{bmatrix} \qquad K_{ij}^{IV} = E'I \begin{bmatrix} s & -t & -s & -t \\ -t & 2v & t & v \\ -s & t & s & t \\ -t & v & t & 2v \end{bmatrix}.$$

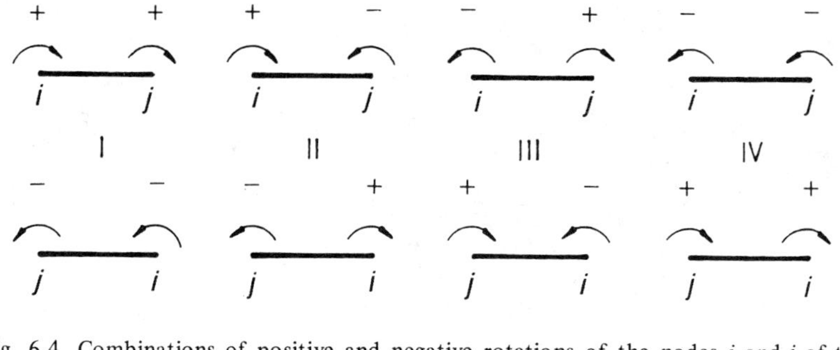

Fig. 6.4. Combinations of positive and negative rotations of the nodes i and j of the beam element.

Sec. 6.2]   Linear beam element assembly   333

The linear beam element assembly, as given in Fig. 6.5(b), with the nodal numbering derived in this section earlier, is described by the group C_2, where positive directions of the nodal displacements v_1, v_2, v_3, v_4 and the nodal rotations $\theta_1, \theta_2, \theta_3, \theta_4$ suit the first symmetry type of the group C_2, which differs from the conventional set of positive directions in Fig. 6.5(a).

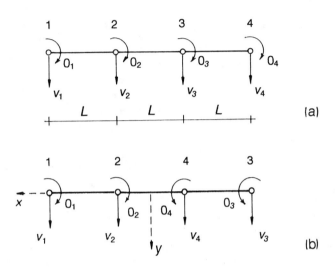

Fig. 6.5. The four-node linear beam assembly: (a) with the usual nodal numbering and the conventional set of positive directions of displacements and rotations of the nodes; (b) with the nodal numbering of the group supermatrix procedure and the set of positive directions of displacements and rotations of the nodes that suit the first symmetry type of the group C_2.

The nodes are grouped into two nodal sets $S_1(1, 3)$, $S_2(2, 4)$, where the nodes in each set are permuted by group elements, i.e. symmetry operations E (identity) and C_2 (rotation through 180° about the y axis).

The character table of the group C_2 and the idempotents of the centre of the group algebra are

C_2	E	C_2
A	1	1
B	1	-1

$$\pi_i = \frac{h_i}{h} \sum_\sigma \chi_i(\sigma^{-1})\sigma$$

$$\pi_1 = \tfrac{1}{2}(E + C_2), \quad \pi_2 = \tfrac{1}{2}(E - C_2).$$

Applying the idempotents to the functions of nodal displacements and rotations derives the basis vectors of G-invariant subspaces U_i ($i = 1, 2$):

U_1:

$$\bar{\Phi}^{(1)} = \tfrac{1}{2}(\Phi^{(1)} + \Phi^{(2)}) \quad \text{or} \quad \begin{bmatrix} \bar{v}_1 \\ \bar{\theta}_1 \\ \bar{v}_2 \\ \bar{\theta}_2 \end{bmatrix} = \frac{1}{2} \begin{bmatrix} v_1 + v_3 \\ \theta_1 + \theta_3 \\ v_2 + v_4 \\ \theta_2 + \theta_4 \end{bmatrix}$$

U_2:

$$\bar{\Phi}^{(2)} = \tfrac{1}{2}(\Phi^{(1)} - \Phi^{(2)}) \quad \text{or} \quad \begin{bmatrix} \bar{v}_3 \\ \bar{\theta}_3 \\ \bar{v}_4 \\ \bar{\theta}_4 \end{bmatrix} = \frac{1}{2} \begin{bmatrix} v_1 - v_3 \\ \theta_1 - \theta_3 \\ v_2 - v_4 \\ \theta_2 - \theta_4 \end{bmatrix}.$$

For the linear beam assembly, shown in Fig. 6.5(b), it is necessary to formulate the system stiffness group supermatrix K in normal form and the system stiffness group supermatrix $\bar{K}$ in diagonal form, as well as to solve the beam assembly with the boundary conditions $v_1 = v_3 = 0$ in the case of a concentrated downward load P at node 2.

By inspection of types of beam elements in Fig. 6.4 and the linear beam assembly in Fig. 6.5(b), one finds that type I corresponds to K'_{12} and K'_{34}, and type II to K'_{24} (the subscripts 12, 34, 24 denote respective beam elements). The following beam element stiffness matrices have designations of rows and columns from the set of numbers 1, 2, 3, 4, 5, 6, 7, 8 which denote variables $v_1, \theta_1, v_2, \theta_2, v_3, \theta_3, v_4, \theta_4$ respectively, with (1), (2), (3), (4) indicating the nodes:

$$K'_{12} = \begin{array}{c} \\ 1 \\ 2 \\ 3 \\ 4 \end{array} \begin{array}{cc} \overbrace{}^{(1)} & \overbrace{}^{(2)} \\ \begin{array}{cccc} 1 & 2 & 3 & 4 \end{array} & \\ \begin{bmatrix} s & t & -s & t \\ t & 2v & -t & v \\ -s & -t & s & -t \\ t & v & -t & 2v \end{bmatrix} & \end{array} E'I,$$

$$K'_{24} = \begin{array}{c} \\ 3 \\ 4 \\ 7 \\ 8 \end{array} \begin{array}{cc} \overbrace{}^{(2)} & \overbrace{}^{(4)} \\ \begin{array}{cccc} 3 & 4 & 7 & 8 \end{array} & \\ \begin{bmatrix} s & t & -s & -t \\ t & 2v & -t & -v \\ -s & -t & s & t \\ -t & -v & t & 2v \end{bmatrix} & \end{array} E'I$$

$$K'_{34} = \begin{array}{c} \\ 5 \\ 6 \\ 7 \\ 8 \end{array} \begin{array}{c} \\ \left[\begin{array}{cccc} 5 & \overset{(3)}{6} & 7 & \overset{(4)}{8} \\ s & t & -s & t \\ t & 2v & -t & v \\ -s & -t & s & -t \\ t & v & -t & 2v \end{array} \right] E'I. \end{array}$$

Assembling these beam element stiffness matrices produces the system stiffness group supermatrix in normal form, where positive directions of displacements and rotations of the nodes correspond to positive directions in the subspace U_1

$$K = \begin{array}{c} \\ 1 \\ 2 \\ 3 \\ 4 \\ \\ 5 \\ 6 \\ 7 \\ 8 \end{array} \left[\begin{array}{cccc|cccc} 1 & 2 & 3 & 4 & 5 & 6 & 7 & 8 \\ s & t & -s & t & & & & \\ t & 2v & -t & v & & & & \\ -s & -t & 2s & & & & -s & -t \\ t & v & & 4v & & & -t & -v \\ \hline & & & & s & t & -s & t \\ & & & & t & 2v & -t & v \\ & & -s & -t & -s & -t & 2s & \\ & & -t & -v & t & v & & 4v \end{array} \right] E'I.$$

Expressed in supermatrix form, this is

$$K = \begin{bmatrix} A & B \\ B & A \end{bmatrix} E'I.$$

This system stiffness group supermatrix in normal form is transformed into the system stiffness group supermatrix in diagonal form by the group supermatrix transformation TKT^{-1} (given in section 3.2)

$$\bar{K} = \frac{1}{2} \begin{bmatrix} E & E \\ E & -E \end{bmatrix} \begin{bmatrix} A & B \\ B & A \end{bmatrix} \begin{bmatrix} E & E \\ E & -E \end{bmatrix} E'I = \begin{bmatrix} A+B & \\ & A-B \end{bmatrix} E'I,$$

and

$$\bar{K} = \begin{bmatrix} s & t & -s & t & & & & \\ t & 2v & -t & v & & & & \\ -s & -t & s & -t & & & & \\ t & v & -t & 2v & & & & \\ \hdashline & & & & s & t & -s & t \\ & & & & t & 2v & -t & v \\ & & & & -s & -t & 3s & \\ & & & & t & v & & 5v \end{bmatrix} E'I.$$

The linear beam assembly with boundary conditions $v_1 = v_3 = 0$ (variables 1 and 5 in the stiffness group supermatrix K) and a vertical concentrated load P at node (2) is shown in Fig. 6.6(a).

The load P is decomposed into two pairs of symmetrical and antisymmetrical loads according to the coordinates of the basis vectors

$$\bar{\Phi}_2 = \tfrac{1}{2}(\Phi_2 + \Phi_4), \qquad \bar{\Phi}_4 = \tfrac{1}{2}(\Phi_2 - \Phi_4)$$

in the subspaces U_1 (Fig. 6.6b) and U_2 (Fig. 6.6c).

The boundary conditions $v_1 = v_3 = 0$ are introduced by deleting the first and the fifth rows and columns of the system stiffness group supermatrix K in normal form, so that for determination of unknown displacements and rotations the following system stiffness equation with the stiffness group supermatrix in normal form is obtained:

$$E'I \begin{bmatrix} 2v & -t & v & & & \\ -t & 2s & & & -s & -t \\ v & & 4v & & -t & -v \\ \hdashline & & & 2v & -t & v \\ & -s & -t & -t & 2s & \\ & -t & -v & v & & 4v \end{bmatrix} \begin{bmatrix} \Psi_1 \\ \Psi_2 \\ \Psi_3 \\ \Psi_4 \\ \Psi_5 \\ \Psi_6 \end{bmatrix} = \begin{bmatrix} 0 \\ P/2 \\ 0 \\ 0 \\ P/2 \\ 0 \end{bmatrix},$$

with column labels $\Psi_1, \Psi_2, \Psi_3, \Psi_4, \Psi_5, \Psi_6$ corresponding to $\theta_1, v_2, \theta_2, \theta_3, v_4, \theta_4$ (positions 2, 3, 4, 6, 7, 8).

where the variables $\theta_1, v_2, \theta_2, \theta_3, v_4, \theta_4$ are denoted by $\Psi_1, \Psi_2, \Psi_3, \Psi_4, \Psi_5, \Psi_6$ respectively.

In the subspace U_1 with

$$\bar{\Psi}_1 = \tfrac{1}{2}(\Psi_1 + \Psi_4), \qquad \bar{\Psi}_2 = \tfrac{1}{2}(\Psi_2 + \Psi_5), \qquad \bar{\Psi}_3 = \tfrac{1}{2}(\Psi_3 + \Psi_6)$$

one obtains

Sec. 6.2] Linear beam element assembly 337

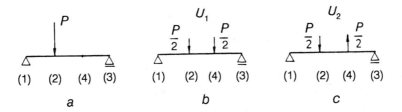

Fig. 6.6. Beam element assembly: (a) subjected to a vertical load P at node (2); (b) with a symmetric pair of loads $P/2$ in subspace U_1; (c) with an antisymmetric pair of loads $P/2$ in subspace U_2.

$$E'I \begin{bmatrix} 2v & -t & v \\ -t & s & -t \\ v & -t & 3v \end{bmatrix} \begin{bmatrix} \bar{\Psi}_1 \\ \bar{\Psi}_2 \\ \bar{\Psi}_3 \end{bmatrix} = \begin{bmatrix} 0 \\ P/2 \\ 0 \end{bmatrix}.$$

If the length of a beam elements equals one unit (metre), i.e. $L = 1$, then

$$s = 12/L^3 = 12, \qquad t = 6/L^2 = 6, \qquad v = 2/L = 2,$$

and after substitutions of s, t, v

$$E'I \begin{bmatrix} 4 & -6 & 2 \\ -6 & 12 & -6 \\ 2 & -6 & 6 \end{bmatrix} \begin{bmatrix} \bar{\Psi}_1 \\ \bar{\Psi}_2 \\ \bar{\Psi}_3 \end{bmatrix} = \begin{bmatrix} 0 \\ P/2 \\ 0 \end{bmatrix}.$$

The solution of this system of equations yields the symmetry-adapted generalized displacements in the subspace U_1:

$$\begin{bmatrix} \bar{\Psi}_1 \\ \bar{\Psi}_2 \\ \bar{\Psi}_3 \end{bmatrix} = \frac{1}{2} \begin{bmatrix} \Psi_1 + \Psi_4 \\ \Psi_2 + \Psi_5 \\ \Psi_3 + \Psi_6 \end{bmatrix} = \frac{1}{2} \begin{bmatrix} \theta_1 + \theta_3 \\ v_2 + v_4 \\ \theta_2 + \theta_4 \end{bmatrix} = \frac{P}{E'I} \begin{bmatrix} \frac{1}{2} \\ \frac{5}{12} \\ \frac{1}{4} \end{bmatrix}.$$

Similar to the previous analysis, in the subspace U_2 with

$$\bar{\Psi}_4 = \tfrac{1}{2}(\Psi_1 - \Psi_4), \qquad \bar{\Psi}_5 = \tfrac{1}{2}(\Psi_2 - \Psi_5), \qquad \bar{\Psi}_6 = \tfrac{1}{2}(\Psi_3 - \Psi_6),$$

one finds

$$E'I \begin{bmatrix} 2v & -t & v \\ -t & 3s & t \\ v & t & 5v \end{bmatrix} \begin{bmatrix} \bar{\Psi}_4 \\ \bar{\Psi}_5 \\ \bar{\Psi}_6 \end{bmatrix} = \frac{1}{2} \begin{bmatrix} 0 \\ P/2 \\ 0 \end{bmatrix}.$$

After substitutions for s, t, v by solving

$$E'I \begin{bmatrix} 4 & -6 & 2 \\ -6 & 36 & 6 \\ 2 & 6 & 10 \end{bmatrix} \begin{bmatrix} \bar{\Psi}_4 \\ \bar{\Psi}_5 \\ \bar{\Psi}_6 \end{bmatrix} = \frac{1}{2} \begin{bmatrix} 0 \\ P/2 \\ 0 \end{bmatrix}$$

one obtains the symmetry-adapted generalized displacements in the subspace U_2:

$$\begin{bmatrix} \bar{\Psi}_4 \\ \bar{\Psi}_5 \\ \bar{\Psi}_6 \end{bmatrix} = \frac{1}{2} \begin{bmatrix} \Psi_1 - \Psi_4 \\ \Psi_2 - \Psi_5 \\ \Psi_3 - \Psi_6 \end{bmatrix} = \frac{1}{2} \begin{bmatrix} \theta_1 - \theta_3 \\ v_2 - v_4 \\ \theta_2 - \theta_4 \end{bmatrix} = \frac{P}{36E'I} \begin{bmatrix} 2 \\ 1 \\ -1 \end{bmatrix}.$$

Evaluation of the shear forces and the bending moments (after substitutions for s, t, v) gives, in subspace U_1,

$$\begin{bmatrix} F_1^{(1)} \\ M_1^{(1)} \\ F_2^{(1)} \\ M_2^{(1)} \end{bmatrix} = \begin{bmatrix} 12 & 6 & -12 & 6 \\ 6 & 4 & -6 & 2 \\ -12 & -6 & 12 & -6 \\ 6 & 2 & -6 & 4 \end{bmatrix} \begin{bmatrix} 0 \\ \frac{1}{2} \\ \frac{5}{12} \\ \frac{1}{4} \end{bmatrix} P = \frac{P}{2} \begin{bmatrix} -1 \\ 0 \\ 1 \\ -1 \end{bmatrix}$$

$$\begin{bmatrix} F_2^{(1)} \\ M_2^{(1)} \\ F_4^{(1)} \\ M_4^{(1)} \end{bmatrix} = \begin{bmatrix} 12 & 6 & -12 & -6 \\ 6 & 4 & -6 & -2 \\ -12 & -6 & 12 & 6 \\ -6 & -2 & 6 & 4 \end{bmatrix} \begin{bmatrix} \frac{5}{12} \\ \frac{1}{4} \\ \frac{5}{12} \\ \frac{1}{4} \end{bmatrix} P = \frac{P}{2} \begin{bmatrix} 0 \\ 1 \\ 0 \\ 1 \end{bmatrix};$$

in subspace U_2,

$$\begin{bmatrix} F_1^{(2)} \\ M_1^{(2)} \\ F_2^{(2)} \\ M_2^{(2)} \end{bmatrix} = \begin{bmatrix} 12 & 6 & -12 & 6 \\ 6 & 4 & -6 & 2 \\ -12 & -6 & 12 & -6 \\ 6 & 2 & -6 & 4 \end{bmatrix} \begin{bmatrix} 0 \\ \frac{1}{18} \\ \frac{1}{36} \\ -\frac{1}{36} \end{bmatrix} P = \frac{P}{6} \begin{bmatrix} -1 \\ 0 \\ 1 \\ -1 \end{bmatrix}$$

$$\begin{bmatrix} F_2^{(2)} \\ M_2^{(2)} \\ F_4^{(2)} \\ M_4^{(2)} \end{bmatrix} = \begin{bmatrix} 12 & 6 & -12 & -6 \\ 6 & 4 & -6 & -2 \\ -12 & -6 & 12 & 6 \\ -6 & -2 & 6 & 4 \end{bmatrix} \begin{bmatrix} \frac{1}{36} \\ -\frac{1}{36} \\ -\frac{1}{36} \\ \frac{1}{36} \end{bmatrix} P = \frac{P}{6} \begin{bmatrix} 2 \\ 1 \\ -2 \\ -1 \end{bmatrix}.$$

By suming up the results in the subspaces U_1 and U_2 one obtains

$$\begin{bmatrix} F_1 \\ M_1 \\ F_2 \\ M_2 \end{bmatrix} = P \begin{bmatrix} -\frac{1}{2} - \frac{1}{6} \\ 0 + 0 \\ \frac{1}{2} + \frac{1}{6} \\ -\frac{1}{2} - \frac{1}{6} \end{bmatrix} = \frac{P}{3} \begin{bmatrix} -2 \\ 0 \\ 2 \\ -2 \end{bmatrix}$$

$$\begin{bmatrix} F_2 \\ M_2 \\ F_4 \\ M_4 \end{bmatrix} = P \begin{bmatrix} 0 + \frac{1}{3} \\ \frac{1}{2} + \frac{1}{6} \\ 0 - \frac{1}{3} \\ \frac{1}{2} - \frac{1}{6} \end{bmatrix} = \frac{P}{3} \begin{bmatrix} 1 \\ 2 \\ -1 \\ 1 \end{bmatrix}.$$

6.3 GIRDER GRILLAGE

For the girder grillage, given in Fig. 6.7, it is necessary to formulate the system stiffness equation with the group supermatrix in normal form and the system stiffness equation with the group supermatrix in diagonal form. Various boundary conditions, compatible with the G-vector analysis, will be considered.

The symmetry properties of this girder grillage will be described by the group C_{2v} using the nodal numbering derived in section 3.5.

The character table of the group C_{2v} and the idempotents of the centre of group algebra are

C_{2v}	E	C_2	σ_1	σ_2
A_1	1	1	1	1
A_2	1	1	-1	-1
B_1	1	-1	1	-1
B_2	1	-1	-1	1

$$\pi_i = \frac{h_i}{h} \sum_\sigma \chi_i(\sigma^{-1})\sigma$$

$$\pi_1 = \tfrac{1}{4}(E + C_2 + \sigma_1 + \sigma_2)$$
$$\pi_2 = \tfrac{1}{4}(E + C_2 - \sigma_1 - \sigma_2)$$
$$\pi_3 = \tfrac{1}{4}(E - C_2 + \sigma_1 - \sigma_2)$$
$$\pi_4 = \tfrac{1}{4}(E - C_2 - \sigma_1 + \sigma_2),$$

with the symmetry operations E (identity), C_2 (rotation through 180° about the z axis), σ_1 and σ_2 (reflections in the xz and yz planes respectively). Applying π_i ($i = 1, 2, 3, 4$) to the nodal functions φ_j ($j = 1, 2, \ldots, 12$) produces the basis vectors $\bar\varphi_j$ of the subspaces U_i:

$$U_1: \quad \bar\varphi_1 = \tfrac{1}{4}(\varphi_1 + \varphi_4 + \varphi_7 + \varphi_{10}), \quad \bar\varphi_2 = \tfrac{1}{4}(\varphi_2 + \varphi_5 + \varphi_8 + \varphi_{11}),$$
$$\bar\varphi_3 = \tfrac{1}{4}(\varphi_3 + \varphi_6 + \varphi_9 + \varphi_{12})$$

$$U_2: \quad \bar\varphi_4 = \tfrac{1}{4}(\varphi_1 + \varphi_4 - \varphi_7 - \varphi_{10}), \quad \bar\varphi_5 = \tfrac{1}{4}(\varphi_2 + \varphi_5 - \varphi_8 - \varphi_{11}),$$
$$\bar\varphi_6 = \tfrac{1}{4}(\varphi_3 + \varphi_6 - \varphi_9 - \varphi_{12})$$

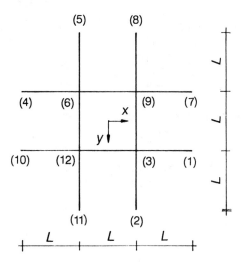

Fig. 6.7. Girder grillage with the nodal numbering in accordance with the group C_{2v}.

U_3: $\bar{\varphi}_7 = \frac{1}{4}(\varphi_1 - \varphi_4 + \varphi_7 - \varphi_{10})$, $\bar{\varphi}_8 = \frac{1}{4}(\varphi_2 - \varphi_5 + \varphi_8 - \varphi_{11})$,

$\bar{\varphi}_9 = \frac{1}{4}(\varphi_3 - \varphi_6 + \varphi_9 - \varphi_{12})$

U_4: $\bar{\varphi}_{10} = \frac{1}{4}(\varphi_1 - \varphi_4 - \varphi_7 + \varphi_{10})$, $\bar{\varphi}_{11} = \frac{1}{4}(\varphi_2 - \varphi_5 - \varphi_8 + \varphi_{11})$,

$\bar{\varphi}_{12} = \frac{1}{4}(\varphi_3 - \varphi_6 - \varphi_9 + \varphi_{12})$.

The twelve nodes of the girder grillage are grouped into the nonoverlapping nodal sets $S_1(1, 4, 7, 10)$, $S_2(2, 5, 8, 11)$, $S_3(3, 6, 9, 12)$, where within each set the nodes are permuted by action of symmetry operations of the group.

As shown in section 3.5, the group supermatrix in normal form, pertaining to the group C_{2v}, is

$$G = \begin{bmatrix} A & B & C & D \\ B & A & D & C \\ C & D & A & B \\ D & C & B & A \end{bmatrix},$$

and it can be transformed by the group supermatrix transformation TGT^{-1} into the group supermatrix in diagonal form

$$\bar{G} = \begin{bmatrix} A+B+C+D & & & \\ & A+B-C-D & & \\ & & A-B+C-D & \\ & & & A-B-C+D \end{bmatrix}$$

if the correspondence with the positive directions of generalized nodal displacements of the subspace U_1 is realized, and if positions of columns of matrices A, B, C, D are coordinated in such a way that in $A + B + C + D$, $A + B - C - D$, $A - B + C - D$, $A - B - C + D$ the variables $E\varphi_j, C_2\varphi_j, \sigma_1\varphi_j, \sigma_2\varphi_j$ are superposed for $j = 1, 2, \ldots, 12$. In agreement with these conditions are the numbering of the nodes and variables (displacements and rotations of the nodes), given in Fig. 6.8, as well as the positive directions of variables, so that this three-dimensional graphic diagram of the girder grillage possesses the symmetry type of the subspace U_1. This can be verified by application of each symmetry operation of the group C_{2v}, with positive sign, to the diagram, which must transform it into itself for all operations.

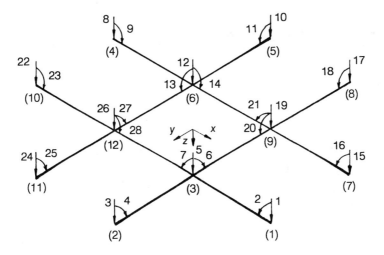

Fig. 6.8. Three-dimensional graphic diagram of the girder grillage with positive directions of displacements and rotations of the nodes corresponding to positive directions in the subspace U_1.

The stiffness matrix of the beam element is derived with respect to the coordinate system and positive directions of the displacements and rotations of the nodes given in Fig. 6.9. The member is loaded only at its end points, has uniform flexural rigidity and undergoes no shear deformation.

Consequently, the beam element stiffness matrix is

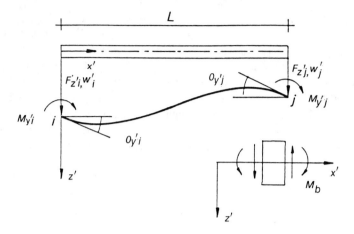

Fig. 6.9. Local coordinate system and directions of positive displacements and rotations of the nodes with respect to the axis of the beam element and the numbering of the nodes.

$$K'_{ij} = \frac{E'I_{y'}}{L^3} \begin{array}{c} w'_i \theta_{y'i} w'_j \theta_{y'j} \\ \begin{bmatrix} 12 & 6L & -12 & 6L \\ 6L & 4L^2 & -6L & 2L^2 \\ -12 & -6L & 12 & -6L \\ 6L & 2L^2 & -6L & 4L^2 \end{bmatrix} \end{array}$$

$$= E'I \begin{array}{c} w'_i \theta_{y'i} w'_j \theta_{y'j} \\ \begin{bmatrix} s & t & -s & t \\ t & 2v & -t & v \\ -s & -t & s & -t \\ t & v & -t & 2v \end{bmatrix} \end{array}$$

with $s = 12/L^3$, $t = 6/L^2$ and $v = 2/L$. As a reminder of the meaning of the elements of K'_{ij}, the displacements and rotations w'_i, $\theta_{y'i}$, w'_j, $\theta_{y'j}$ are denoted in a row above the matrix.

All beam elements in the direction x have the local coordinate systems with axes x', y', z' parallel to and in the same direction with respect to the axes x, y, z in the global coordinate system. They correspond to the beam elements of the types I and II, which were derived in the preceding case of the beam element assembly in section 6.2.

In the direction y the beam element stiffness matrices in the global coordinate system will be found by means of the direction cosines

Girder grillage

$$\lambda_{x'} = \cos\theta_{xx'} = 0, \quad \mu_{x'} = \cos\theta_{yx'} = 1, \quad \nu_{x'} = \cos\theta_{zx'} = 0,$$
$$\lambda_{y'} = \cos\theta_{xy'} = -1, \quad \mu_{y'} = \cos\theta_{yy'} = 0, \quad \nu_{y'} = \cos\theta_{zy'} = 0,$$
$$\lambda_{z'} = \cos\theta_{xz'} = 0, \quad \mu_{z'} = \cos\theta_{yz'} = 0, \quad \nu_{z'} = \cos\theta_{zz'} = 1,$$

which can be expressed by the matrix

$$\Lambda = \begin{bmatrix} 0 & 1 & 0 \\ -1 & 0 & 0 \\ 0 & 0 & 1 \end{bmatrix}$$

so that the 12 × 12 transformation matrix is

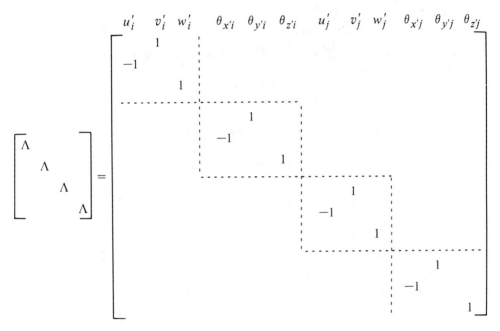

After deleting the nonexistent (zero) displacements and rotations $u'_i, v'_i, \theta_{x'i}, \theta_{z'i}, u'_j, v'_j, \theta_{z'j}$, the variables $w'_i, \theta_{y'i}, w'_j, \theta_{y'j}$ remain, so that the 4 × 4 transformation matrix is obtained:

$$T = \begin{bmatrix} 1 & & & \\ & 1 & & \\ & & 1 & \\ & & & 1 \end{bmatrix}.$$

Since this is an identity matrix, the basis transformation from the local into the global coordinate system does not change the beam element stiffness matrix formulated in the local coordinate system

344 Group supermatrices in formulation and assembly of stiffness equations [Ch. 6

$$K_{ij} = T^T K'_{ij} T = E^T K'_{ij} E = K'_{ij}.$$

It can be concluded that the beam element stiffness matrices of the elements in the direction y are identical with the corresponding beam element stiffness matrices of elements in the direction x.

As in the case of the linear beam assembly in section 6.2, according to Fig. 6.4 the beam elements of the girder grillage, given in the three-dimensional graphic display in Fig. 6.8, can be classified into

type I: (1)(3), (4)(6), (7)(9), (10)(12)   (x direction)

(2)(3), (5)(6), (8)(9), (11)(12)   (y direction)

type II: (3)(12), (6)(9)   (x direction)

(3)(9), (6)(12)   (y direction).

The beam element stiffness matrices of types I and II (from section 6.2) are associated with respective beam elements as follows:

type I

x direction

$$K'_{ij} = \begin{matrix} & \begin{matrix} 22 & 15 & 8 & 1 \\ (10) & (7) & (4) & (1) \\ 23 & 16 & 9 & 2 \\ 26 & 19 & 12 & 5 \\ (12) & (9) & (6) & (3) \\ 28 & 21 & 14 & 7 \end{matrix} \\ & \begin{bmatrix} s & t & -s & t \\ t & 2v & -t & v \\ -s & -t & s & -t \\ t & v & -t & 2v \end{bmatrix} E'I \end{matrix}$$

with column headings:
22 (10) 23   26 (12) 28
15 (7) 16   19 (9) 21
8 (4) 9   12 (6) 14
1 (1) 2   5 (3) 7

y direction

$$K'_{ij} = \begin{bmatrix} s & t & -s & t \\ t & 2v & -t & v \\ -s & -t & s & -t \\ t & v & -t & 2v \end{bmatrix} E'I$$

with row labels:
24 17 10 3
(11) (8) (5) (2)
25 18 11 4
26 19 12 5
(12) (9) (6) (3)
27 20 13 6

and column headings:
24 (11) 25   26 (12) 27
17 (8) 18   19 (9) 20
10 (5) 11   12 (6) 13
3 (2) 4   5 (3) 6

type II

$$K'_{ij} = \begin{array}{c} \\ \\ \begin{matrix} 12 & 5 & 12 & 5 \\ (6) & (3) & (6) & (3) \\ 13 & 6 & 14 & 7 \\ \\ 26 & 19 & 19 & 26 \\ (12) & (9) & (9) & (12) \\ 27 & 20 & 21 & 28 \end{matrix} \end{array} \begin{array}{c} \begin{matrix} 12 & (6) & 13 & 26 & (12) & 27 \\ 5 & (3) & 6 & 19 & (9) & 20 \\ 12 & (6) & 14 & 19 & (9) & 21 \\ 5 & (3) & 7 & 26 & (12) & 28 \end{matrix} \\ \begin{bmatrix} s & t & -s & -t \\ t & 2v & -t & -v \\ -s & -t & s & t \\ -t & -v & t & 2v \end{bmatrix} \end{array} E'I.$$

Assembling these beam element stiffness matrices produces the system stiffness group supermatrix K in normal form, which is given in Table 6.1 in the stiffness equation $K\Phi = P$, with positive directions of nodal displacements, rotations, forces and moments which correspond to the positive directions in the subspace U_1. This matrix can be written as

$$K = E'I \begin{bmatrix} A & B & C & D \\ B & A & D & C \\ C & D & A & B \\ D & C & B & A \end{bmatrix},$$

where B is a zero matrix, as evident in Table 6.1.

By applications of the group supermatrix transformation

$$TKT^{-1}T\Phi = TP,$$

as given in section 3.5, with

$$T = \frac{1}{4} \begin{bmatrix} E & E & E & E \\ E & E & -E & -E \\ E & -E & E & -E \\ E & -E & -E & E \end{bmatrix} \quad \text{and} \quad T^{-1} = 4T,$$

one obtains the stiffness equation $\bar{K}\bar{\Phi} = \bar{P}$ with the system stiffness group supermatrix $\bar{K}$ in diagonal form:

Table 6.1. The stiffness equation with the system stiffness group supermatrix in normal form with positive directions of displacements and rotations of the nodes that correspond to the positive directions in the subspace U_1

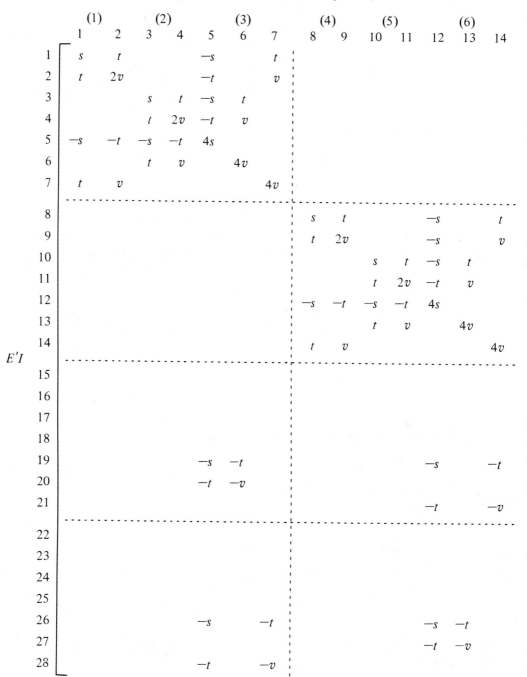

Sec. 6.3] Girder grillage 347

$$\begin{matrix}
(7) & (8) & (9) & (10) & (11) & (12) \\
15\ 16 & 17\ 18 & 19\ 20\ 21 & 22\ 23\ 24\ 25 & 26\ 27 & 28
\end{matrix}$$

$$\left[\begin{array}{cccccccccccccc}
 & & & & & & & & & & & & & \\
 & & & & -s & -t & & & & & -s & & -t & \\
 & & & & -t & -v & & & & & & & & \\
 & & & & & & & & & & -t & & -v & \\
 & & & & & & & & & & & & & \\
 & & & & & & & & & & & & & \\
 & & & & & & & & & & & & & \\
 & & & & & & & & & & & & & \\
 & & -s & & -t & & & & & & -s & -t & & \\
 & & & & & & & & & & -t & -v & & \\
 & & -t & & -v & & & & & & & & & \\
s & t & -s & & t & & & & & & & & & \\
t & 2v & -t & & v & & & & & & & & & \\
 & & s & t & -s & t & & & & & & & & \\
 & & t & 2v & -t & v & & & & & & & & \\
-s & -t & -s & -t & 4s & & & & & & & & & \\
 & & t & v & & 4v & & & & & & & & \\
t & v & & & 4v & & & & & & & & & \\
 & & & & & & s & t & & -s & & & t & \\
 & & & & & & t & 2v & & -t & & & v & \\
 & & & & & & & & s & t & -s & & t & \\
 & & & & & & & & t & 2v & -t & & v & \\
 & & & & & & -s & -t & -s & -t & 4s & & & \\
 & & & & & & & & t & v & & & 4v & \\
 & & & & & & t & v & & & & & & 4v
\end{array}\right]
\begin{bmatrix} w_1 \\ \theta_{y1} \\ w_2 \\ \theta_{x2} \\ w_3 \\ \theta_{x3} \\ \theta_{y3} \\ \hdashline w_4 \\ \theta_{y4} \\ w_5 \\ \theta_{x5} \\ w_6 \\ \theta_{x6} \\ \theta_{y6} \\ \hdashline w_7 \\ \theta_{y7} \\ w_8 \\ \theta_{x8} \\ w_9 \\ \theta_{x9} \\ \theta_{y9} \\ \hdashline w_{10} \\ \theta_{y10} \\ w_{11} \\ \theta_{x11} \\ w_{12} \\ \theta_{x12} \\ \theta_{y12} \end{bmatrix}
=
\begin{bmatrix} F_{z1} \\ M_{y1} \\ F_{z2} \\ M_{x2} \\ F_{z3} \\ M_{x3} \\ M_{y3} \\ \hdashline F_{z4} \\ M_{y4} \\ F_{z5} \\ M_{x5} \\ F_{z6} \\ M_{x6} \\ M_{y6} \\ \hdashline F_{z7} \\ M_{y7} \\ F_{z8} \\ M_{x8} \\ F_{z9} \\ M_{x9} \\ M_{y9} \\ \hdashline F_{z10} \\ M_{y10} \\ F_{z11} \\ M_{x11} \\ F_{z12} \\ M_{x12} \\ M_{y12} \end{bmatrix}$$

$$\begin{bmatrix} \bar{K}_1 & & & \\ & \bar{K}_2 & & \\ & & \bar{K}_3 & \\ & & & \bar{K}_4 \end{bmatrix} \begin{bmatrix} \bar{\Phi}^{(1)} \\ \bar{\Phi}^{(2)} \\ \bar{\Phi}^{(3)} \\ \bar{\Phi}^{(4)} \end{bmatrix} =$$

$$= E'I \begin{bmatrix} A+B+C+D & & & \\ & A+B-C-D & & \\ & & A-B+C-D & \\ & & & A-B-C+D \end{bmatrix} \begin{bmatrix} \bar{\Phi}^{(1)} \\ \bar{\Phi}^{(2)} \\ \bar{\Phi}^{(3)} \\ \bar{\Phi}^{(4)} \end{bmatrix}$$

$$= \begin{bmatrix} \bar{P}^{(1)} \\ \bar{P}^{(2)} \\ \bar{P}^{(3)} \\ \bar{P}^{(4)} \end{bmatrix}.$$

Thus, the system stiffness matrices $\bar{K}_i$ ($i = 1, 2, 3, 4$) in G-invariant subspaces U_i are

$$\bar{K}_1 = E'I(A+O+C+D) = E'I \begin{bmatrix} s & t & & & & -s & t \\ t & 2v & & & & -t & v \\ & & s & t & -s & t & \\ & & t & 2v & -t & v & \\ -s & -t & -s & -t & 2s & -t & -t \\ & & t & v & -t & 3v & \\ t & v & & & -t & & 3v \end{bmatrix}$$

$$\bar{K}_2 = E'I(A+O-C-D) = E'I \begin{bmatrix} s & t & & & & -s & t \\ t & 2v & & & & -t & v \\ & & s & t & -s & t & \\ & & t & 2v & -t & v & \\ -s & -t & -s & -t & 6s & t & t \\ & & t & v & t & 5v & \\ t & v & & & t & & 5v \end{bmatrix}$$

$$\bar{K}_3 = E'I(A - O + C - D) = E'I \begin{bmatrix} s & t & & & -s & t \\ t & 2v & & & -t & v \\ & & s & t & -s & t \\ & & t & 2v & -t & v \\ -s & -t & -s & -t & 4s & -t & t \\ & & t & v & -t & 3v \\ t & v & & & & t & 5v \end{bmatrix}$$

$$\bar{K}_4 = E'I(A - O - C + D) = E'I \begin{bmatrix} s & t & & & -s & t \\ t & 2v & & & -t & v \\ & & s & t & -s & t \\ & & t & 2v & -t & v \\ -s & -t & -s & -t & 4s & t & -t \\ & & t & v & t & 5v \\ t & v & & & -t & 3v \end{bmatrix}.$$

The symmetry-adapted generalized displacements are

$$\bar{\Phi} = T\Phi,$$

with

$$\bar{\Phi} = \begin{bmatrix} \bar{\Phi}^{(1)} \\ \bar{\Phi}^{(2)} \\ \bar{\Phi}^{(3)} \\ \bar{\Phi}^{(4)} \end{bmatrix} \qquad T = \frac{1}{4} \begin{bmatrix} E & E & E & E \\ E & E & -E & -E \\ E & -E & E & -E \\ E & -E & -E & E \end{bmatrix}$$

and

$$\begin{bmatrix} \bar{\Phi}^{(1)} \\ \bar{\Phi}^{(2)} \\ \bar{\Phi}^{(3)} \\ \bar{\Phi}^{(4)} \end{bmatrix} = \frac{1}{4} \begin{bmatrix} E & E & E & E \\ E & E & -E & -E \\ E & -E & E & -E \\ E & -E & -E & E \end{bmatrix} \begin{bmatrix} \Phi^{(1)} \\ \Phi^{(2)} \\ \Phi^{(3)} \\ \Phi^{(4)} \end{bmatrix}$$

with

$$[\Phi^{(1)} \ \Phi^{(2)} \ \Phi^{(3)} \ \Phi^{(4)}] = \begin{bmatrix} w_1 & w_4 & w_7 & w_{10} \\ \theta_{y1} & \theta_{y4} & \theta_{y7} & \theta_{y10} \\ w_2 & w_5 & w_8 & w_{11} \\ \theta_{x2} & \theta_{x5} & \theta_{x8} & \theta_{x11} \\ w_3 & w_6 & w_9 & w_{12} \\ \theta_{x3} & \theta_{x6} & \theta_{x9} & \theta_{x12} \\ \theta_{y3} & \theta_{y6} & \theta_{y9} & \theta_{y12} \end{bmatrix}.$$

Thus, the symmetry-adapted generalized displacements $\bar{\Phi}$ of G-invariant subspaces U_i are

$$\bar{\Phi}^{(1)} = \begin{bmatrix} \bar{\Psi}_1 \\ \bar{\Psi}_2 \\ \bar{\Psi}_3 \\ \bar{\Psi}_4 \\ \bar{\Psi}_5 \\ \bar{\Psi}_6 \\ \bar{\Psi}_7 \end{bmatrix} = \frac{1}{4}(\Phi^{(1)} + \Phi^{(2)} + \Phi^{(3)} + \Phi^{(4)}) = \frac{1}{4}\begin{bmatrix} w_1 + w_4 + w_7 + w_{10} \\ \theta_{y1} + \theta_{y4} + \theta_{y7} + \theta_{y10} \\ w_2 + w_5 + w_8 + w_{11} \\ \theta_{x2} + \theta_{x5} + \theta_{x8} + \theta_{x11} \\ w_3 + w_6 + w_9 + w_{12} \\ \theta_{x3} + \theta_{x6} + \theta_{x9} + \theta_{x12} \\ \theta_{y3} + \theta_{y6} + \theta_{y9} + \theta_{y12} \end{bmatrix}$$

$$\bar{\Phi}^{(2)} = \begin{bmatrix} \bar{\Psi}_8 \\ \bar{\Psi}_9 \\ \bar{\Psi}_{10} \\ \bar{\Psi}_{11} \\ \bar{\Psi}_{12} \\ \bar{\Psi}_{13} \\ \bar{\Psi}_{14} \end{bmatrix} = \frac{1}{4}(\Phi^{(1)} + \Phi^{(2)} - \Phi^{(3)} - \Phi^{(4)}) = \frac{1}{4}\begin{bmatrix} w_1 + w_4 - w_7 - w_{10} \\ \theta_{y1} + \theta_{y4} - \theta_{y7} - \theta_{y10} \\ w_2 + w_5 - w_8 - w_{11} \\ \theta_{x2} + \theta_{x5} - \theta_{x8} - \theta_{x11} \\ w_3 + w_6 - w_9 - w_{12} \\ \theta_{x3} + \theta_{x6} - \theta_{x9} - \theta_{x12} \\ \theta_{y3} + \theta_{y6} - \theta_{y9} - \theta_{y12} \end{bmatrix}$$

$$\bar{\Phi}^{(3)} = \begin{bmatrix} \bar{\Psi}_{15} \\ \bar{\Psi}_{16} \\ \bar{\Psi}_{17} \\ \bar{\Psi}_{18} \\ \bar{\Psi}_{19} \\ \bar{\Psi}_{20} \\ \bar{\Psi}_{21} \end{bmatrix} = \frac{1}{4}(\Phi^{(1)} - \Phi^{(2)} + \Phi^{(3)} - \Phi^{(4)}) = \frac{1}{4}\begin{bmatrix} w_1 - w_4 + w_7 - w_{10} \\ \theta_{y1} - \theta_{y4} + \theta_{y7} - \theta_{y10} \\ w_2 - w_5 + w_8 - w_{11} \\ \theta_{x2} - \theta_{x5} + \theta_{x8} - \theta_{x11} \\ w_3 - w_6 + w_9 - w_{12} \\ \theta_{x3} - \theta_{x6} + \theta_{x9} - \theta_{x12} \\ \theta_{y3} - \theta_{y6} + \theta_{y9} - \theta_{y12} \end{bmatrix}$$

Sec. 6.3]   Girder grillage   351

$$\bar{\Phi}^{(4)} = \begin{bmatrix} \bar{\Psi}_{22} \\ \bar{\Psi}_{23} \\ \bar{\Psi}_{24} \\ \bar{\Psi}_{25} \\ \bar{\Psi}_{26} \\ \bar{\Psi}_{27} \\ \bar{\Psi}_{28} \end{bmatrix} = \frac{1}{4}(\Phi^{(1)} - \Phi^{(2)} - \Phi^{(3)} + \Phi^{(4)}) = \frac{1}{4} \begin{bmatrix} w_1 - w_4 - w_7 + w_{10} \\ \theta_{y1} - \theta_{y4} - \theta_{y7} + \theta_{y10} \\ w_2 - w_5 - w_8 + w_{11} \\ \theta_{x2} - \theta_{x5} - \theta_{x8} + \theta_{x11} \\ w_3 - w_6 - w_9 + w_{12} \\ \theta_{x3} - \theta_{x6} - \theta_{x9} + \theta_{x12} \\ \theta_{y3} - \theta_{y6} - \theta_{y9} + \theta_{y12} \end{bmatrix}$$

In a similar way to the generalized displacements, the symmetry-adapted generalized nodal forces are formulated:

$$\bar{P} = TP,$$

with

$$\bar{P} = \begin{bmatrix} \bar{P}^{(1)} \\ \bar{P}^{(2)} \\ \bar{P}^{(3)} \\ \bar{P}^{(4)} \end{bmatrix} \qquad T = \frac{1}{4} \begin{bmatrix} E & E & E & E \\ E & E & -E & -E \\ E & -E & E & -E \\ E & -E & -E & E \end{bmatrix}$$

and

$$\begin{bmatrix} \bar{P}^{(1)} \\ \bar{P}^{(2)} \\ \bar{P}^{(3)} \\ \bar{P}^{(4)} \end{bmatrix} = \frac{1}{4} \begin{bmatrix} E & E & E & E \\ E & E & -E & -E \\ E & -E & E & -E \\ E & -E & -E & E \end{bmatrix} \begin{bmatrix} P^{(1)} \\ P^{(2)} \\ P^{(3)} \\ P^{(4)} \end{bmatrix}$$

with

$$[P^{(1)} \quad P^{(2)} \quad P^{(3)} \quad P^{(4)}] = \begin{bmatrix} F_{z1} & F_{z4} & F_{z7} & F_{z10} \\ M_{y1} & M_{y4} & M_{y7} & M_{y10} \\ F_{z2} & F_{z5} & F_{z8} & F_{z11} \\ M_{x2} & M_{x5} & M_{x8} & M_{x11} \\ F_{z3} & F_{z6} & F_{z9} & F_{z12} \\ M_{x3} & M_{x6} & M_{x9} & M_{x12} \\ M_{y3} & M_{y6} & M_{y9} & M_{y12} \end{bmatrix}.$$

Thus, the symmetry-adapted generalized forces $\bar{P}$ in G-invariant subspaces U_i are

$$\bar{P}^{(1)} = \begin{bmatrix} \bar{P}_1 \\ \bar{P}_2 \\ \bar{P}_3 \\ \bar{P}_4 \\ \bar{P}_5 \\ \bar{P}_6 \\ \bar{P}_7 \end{bmatrix} = \frac{1}{4}(P^{(1)} + P^{(2)} + P^{(3)} + P^{(4)}) = \frac{1}{4} \begin{bmatrix} F_{z1} + F_{z4} + F_{z7} + F_{z10} \\ M_{y1} + M_{y4} + M_{y7} + M_{y10} \\ F_{z2} + F_{z5} + F_{z8} + F_{z11} \\ M_{x2} + M_{x5} + M_{x8} + M_{x11} \\ F_{z3} + F_{z6} + F_{z9} + F_{z12} \\ M_{x3} + M_{x6} + M_{x9} + M_{x12} \\ M_{y3} + M_{y6} + M_{y9} + M_{y12} \end{bmatrix}$$

$$\bar{P}^{(2)} = \begin{bmatrix} \bar{P}_8 \\ \bar{P}_9 \\ \bar{P}_{10} \\ \bar{P}_{11} \\ \bar{P}_{12} \\ \bar{P}_{13} \\ \bar{P}_{14} \end{bmatrix} = \frac{1}{4}(P^{(1)} + P^{(2)} - P^{(3)} - P^{(4)}) = \frac{1}{4} \begin{bmatrix} F_{z1} + F_{z4} - F_{z7} - F_{z10} \\ M_{y1} + M_{y4} - M_{y7} - M_{y10} \\ F_{z2} + F_{z5} - F_{z8} - F_{z11} \\ M_{x2} + M_{x5} - M_{x8} - M_{x11} \\ F_{z3} + F_{z6} - F_{z9} - F_{z12} \\ M_{x3} + M_{x6} - M_{x9} - M_{x12} \\ M_{y3} + M_{y6} - M_{y9} - M_{y12} \end{bmatrix}$$

$$\bar{P}^{(3)} = \begin{bmatrix} \bar{P}_{15} \\ \bar{P}_{16} \\ \bar{P}_{17} \\ \bar{P}_{18} \\ \bar{P}_{19} \\ \bar{P}_{20} \\ \bar{P}_{21} \end{bmatrix} = \frac{1}{4}(P^{(1)} - P^{(2)} + P^{(3)} - P^{(4)}) = \frac{1}{4} \begin{bmatrix} F_{z1} - F_{z4} + F_{z7} - F_{z10} \\ M_{y1} - M_{y4} + M_{y7} - M_{y10} \\ F_{z2} - F_{z5} + F_{z8} - F_{z11} \\ M_{x2} - M_{x5} + M_{x8} - M_{x11} \\ F_{z3} - F_{z6} + F_{z9} - F_{z12} \\ M_{x3} - M_{x6} + M_{x9} - M_{x12} \\ M_{y3} - M_{y6} + M_{y9} - M_{y12} \end{bmatrix}$$

$$\bar{P}^{(4)} = \begin{bmatrix} \bar{P}_{22} \\ \bar{P}_{23} \\ \bar{P}_{24} \\ \bar{P}_{25} \\ \bar{P}_{26} \\ \bar{P}_{27} \\ \bar{P}_{28} \end{bmatrix} = \frac{1}{4}(P^{(1)} - P^{(2)} - P^{(3)} + P^{(4)}) = \frac{1}{4} \begin{bmatrix} F_{z1} - F_{z4} - F_{z7} + F_{z10} \\ M_{y1} - M_{y4} - M_{y7} + M_{y10} \\ F_{z2} - F_{z5} - F_{z8} + F_{z11} \\ M_{x2} - M_{x5} - M_{x8} + M_{x11} \\ F_{z3} - F_{z6} - F_{z9} + F_{z12} \\ M_{x3} - M_{x6} - M_{x9} + M_{x12} \\ M_{y3} - M_{y6} - M_{y9} + M_{y12} \end{bmatrix}$$

Sec. 6.3]   Girder grillage   353

Boundary conditions of the girder grillage
If the end points of supports of the girder grillage are the hinged ends at nodes (1), (2), (4), (5), (7), (8), (10), (11), the vertical displacements of these nodes are zero:

$$w_1 = w_2 = w_4 = w_5 = w_7 = w_8 = w_{10} = w_{11} = 0.$$

Since these displacements are designated as variables 1, 3, 8, 10, 15, 17, 22, 24 in Fig. 6.8 and in the system stiffness supermatrix in normal form in Table 6.1, the system stiffness matrix with the above boundary conditions will be obtained when columns and rows with numbers 1, 3, 8, 10, 15, 17,, 22, 24 are deleted. These boundary conditions preserve the symmetry types of the group C_{2v}, so that the system stiffness matrix with boundary conditions can be put to block diagonal form. Thus, in the subspace U_1 the symmetry-adapted displacements at the supports are

$$4\bar{\Psi}_1 = w_1 + w_4 + w_7 + w_{10} = 0 \qquad 4\bar{\Psi}_1 = \text{``1''} + \text{``8''} + \text{``15''} + \text{``22''} = 0$$
$$\text{or}$$
$$4\bar{\Psi}_3 = w_2 + w_5 + w_8 + w_{11} = 0 \qquad 4\bar{\Psi}_3 = \text{``3''} + \text{``10''} + \text{``17''} + \text{``24''} = 0$$

because all terms in these expressions are zero.

Since the symmetry-adapted displacements $\bar{\Psi}_8$, $\bar{\Psi}_{10}$ (of U_2), $\bar{\Psi}_{15}$, $\bar{\Psi}_{17}$ (of U_3) and $\bar{\Psi}_{22}$, $\bar{\Psi}_{24}$ (of $\bar{U}_4$) are linear combinations of the same terms in the above expressions, all these displacements must equal zero too. Thus, by deleting the first and the third column and row of each system stiffness matrix of the subspaces U_i ($i = 1, 2, 3, 4$), which were derived previously, the system stiffness matrices of the subspaces U_i with boundary conditions will be obtained. This is given in Table 6.2 in the stiffness equations for determination of the symmetry-adapted generalized displacements for the symmetry-adapted generalized loads in G-invariant subspaces U_1, U_2, U_3, U_4.

When the supports of the girder grillage are the rigidly fixed ends at nodes (2), (5), (8), (11) and the hinged ends at nodes (1), (4), (7), (10), as shown in Fig. 6.10, the boundary conditions are

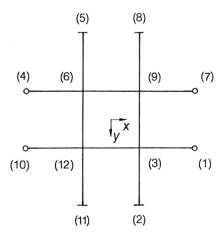

Fig. 6.10. Girder grillage with fixed end supports at nodes (2), (5), (8), (11) and hinged end supports at nodes (1), (4), (7), (10).

Table 6.2. The stiffness equation with the system stiffness group supermatrix in diagonal form for determination of generalized displacements in G-invariant subspaces

$$
\begin{bmatrix}
2v & -t &   & v &   &   &   &   &   &   &   &   &   &   &   &   &   &   &   &   &   &   \\
   & 2v & -t &   &   &   &   &   &   &   &   &   &   &   &   &   &   &   &   &   &   &   \\
-t & -t & 2s & -t & -t &   &   &   &   &   &   &   &   &   &   &   &   &   &   &   &   &   \\
   & v & -t & 3v &   &   &   &   &   &   &   &   &   &   &   &   &   &   &   &   &   &   \\
   &   & -t &   & 3v &   &   &   &   &   &   &   &   &   &   &   &   &   &   &   &   &   \\
   &   &   &   &   & 2v & -t &   & v &   &   &   &   &   &   &   &   &   &   &   &   &   \\
   &   &   &   &   &   & 2v & -t &   &   &   &   &   &   &   &   &   &   &   &   &   &   \\
   &   &   &   &   & -t & -t & 6s & -t & t &   &   &   &   &   &   &   &   &   &   &   &   \\
   &   &   &   &   &   & v & -t & 3v &   &   &   &   &   &   &   &   &   &   &   &   &   \\
   &   &   &   &   &   &   & t &   & 5v &   &   &   &   &   &   &   &   &   &   &   &   \\
   &   &   &   &   &   &   &   &   &   & 2v & -t &   & v &   &   &   &   &   &   &   &   \\
   &   &   &   &   &   &   &   &   &   &   & 2v & -t &   &   &   &   &   &   &   &   &   \\
   &   &   &   &   &   &   &   &   &   & -t & -t & 4s & -t & t &   &   &   &   &   &   &   \\
   &   &   &   &   &   &   &   &   &   &   & v & -t & 3v &   &   &   &   &   &   &   &   \\
   &   &   &   &   &   &   &   &   &   &   &   & t &   & 5v &   &   &   &   &   &   &   \\
   &   &   &   &   &   &   &   &   &   &   &   &   &   &   & 2v & -t &   & v &   &   &   \\
   &   &   &   &   &   &   &   &   &   &   &   &   &   &   &   & 2v & -t &   &   &   &   \\
   &   &   &   &   &   &   &   &   &   &   &   &   &   &   & -t & -t & 4s & t & -t &   &   \\
   &   &   &   &   &   &   &   &   &   &   &   &   &   &   &   & v & t & 5v &   &   &   \\
   &   &   &   &   &   &   &   &   &   &   &   &   &   &   & v &   & -t &   & 3v &   &   \\
\end{bmatrix}
\begin{bmatrix}
\bar{\Psi}_{2} \\ \bar{\Psi}_{4} \\ \bar{\Psi}_{5} \\ \bar{\Psi}_{6} \\ \bar{\Psi}_{7} \\ \bar{\Psi}_{9} \\ \bar{\Psi}_{11} \\ \bar{\Psi}_{12} \\ \bar{\Psi}_{13} \\ \bar{\Psi}_{14} \\ \bar{\Psi}_{16} \\ \bar{\Psi}_{18} \\ \bar{\Psi}_{19} \\ \bar{\Psi}_{20} \\ \bar{\Psi}_{21} \\ \bar{\Psi}_{23} \\ \bar{\Psi}_{25} \\ \bar{\Psi}_{26} \\ \bar{\Psi}_{27} \\ \bar{\Psi}_{28}
\end{bmatrix}
=
\begin{bmatrix}
\bar{P}_{2} \\ \bar{P}_{4} \\ \bar{P}_{5} \\ \bar{P}_{6} \\ \bar{P}_{7} \\ \bar{P}_{9} \\ \bar{P}_{11} \\ \bar{P}_{12} \\ \bar{P}_{13} \\ \bar{P}_{14} \\ \bar{P}_{16} \\ \bar{P}_{18} \\ \bar{P}_{19} \\ \bar{P}_{20} \\ \bar{P}_{21} \\ \bar{P}_{23} \\ \bar{P}_{25} \\ \bar{P}_{26} \\ \bar{P}_{27} \\ \bar{P}_{28}
\end{bmatrix}
$$

Sec. 6.3] Girder grillage 355

$$w_2 = w_5 = w_8 = w_{11} = 0$$
$$\theta_{x2} = \theta_{x5} = \theta_{x8} = \theta_{x11} = 0$$
and $w_1 = w_4 = w_7 = w_{10} = 0.$

These boundary conditions preserve the symmetry types of the group C_{2v}, because the symmetry operations (identity, rotation through 180° and reflections in xz and yz planes) transform a rectangle into itself, and this may be regarded as an analogy to the rigidly fixed and hinged ends of the girder grillage in Fig. 6.10.

Following the procedure from the preceding case, where all supports were hinged ends, the system stiffness matrices of the subspaces U_i, with boundary conditions given in Fig. 6.10, are obtained from the 7 × 7 matrices of U_i by deletion of the first, second and third column and row of each matrix. Thus, the system stiffness matrices of subspaces U_i are

U_1:

$$\bar{K}_1 = E'I \begin{bmatrix} \Psi_4 & \Psi_5 & \Psi_6 & \Psi_7 \\ 2v & -t & v & \\ -t & 2s & -t & -t \\ v & -t & 3v & \\ & -t & & 3v \end{bmatrix}$$

U_2:

$$\bar{K}_2 = E'I \begin{bmatrix} \Psi_{11} & \Psi_{12} & \Psi_{13} & \Psi_{14} \\ 2v & -t & v & \\ -t & 6s & t & t \\ v & t & 5v & \\ & t & & 5v \end{bmatrix}$$

U_3:

$$\bar{K}_3 = E'I \begin{bmatrix} \Psi_{18} & \Psi_{19} & \Psi_{20} & \Psi_{21} \\ 2v & -t & v & \\ -t & 4s & -t & t \\ v & -t & 3v & \\ & t & & 5v \end{bmatrix}$$

U_4:

$$\bar{K}_4 = E'I \begin{bmatrix} \Psi_{25} & \Psi_{26} & \Psi_{27} & \Psi_{28} \\ 2v & -t & v & \\ -t & 4s & t & -t \\ v & t & 5v & \\ & -t & & 3v \end{bmatrix}.$$

When the supports of the girder grillage are rigidly fixed ends at nodes (1), (2), (4), (5), (7), (8), (10), (11), the boundary conditions are

$$w_1 = w_4 = w_7 = w_{10} = 0$$

$$\theta_{y1} = \theta_{y4} = \theta_{y7} = \theta_{y10} = 0$$

and

$$w_2 = w_5 = w_8 = w_{11} = 0$$

$$\theta_{x2} = \theta_{x5} = \theta_{x8} = \theta_{x11} = 0.$$

According to previous considerations, with these boundary conditions the stiffness matrices of the subspaces U_i are obtained from the 7×7 matrices by deleting the first, second, third, and fourth column and row in each matrix, which results in

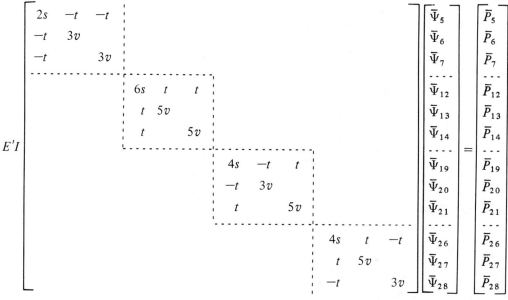

Appendix: Character tables

Symmetry groups and character tables were discussed in sections 1.2 and 1.4 respectively. Character tables contain characters of irreducible group representations, designated by Mullikan's symbols, which are given as follows.

Symbol	Dimension		Rotation C_n	$C_2 \perp C_n$ or $\sigma_v(\sigma_d)$	$\sigma \perp C_n$ σ_h	Inversion centre i
A, B	1	Symmetric	A	1	$'$	g
E	2	Antisymmetric	B	2	$''$	u
$T(F)$	3					

Character tables also include special basis sets of irreducible representations:

(1) Cartesian coordinates x, y, z;
(2) rotations about Cartesian axes R_x, R_y, R_z;
(3) binary Cartesian products $xy, xz, yz; x^2 + y^2, x^2 - y^2; x^2, y^2, z^2$.

Thus against a particular representation there are some common functions which form a basis of the representation. If one knows how x, y, z transform under operations of the group, it is not difficult to determine how any polynomial function of x, y, z transforms. For example, from the character table of the group C_{2v}, one can see that x transforms like B_1, y like B_2 and z like A_1.

If x, y, z belong to one-dimensional representations of the group, one can immediately assign the second-degree functions (x^2, y^2, xy, etc.) to the appropriate irreducible representation. When x, y, z belong to representations of dimensions greater than one, it is necessary to know the matrices of symmetry operations and not just their characters. In these cases, binary Cartesian products $x^2, y^2, z^2, xy, xz, yz$ are included in the tables.

Appendix: Character tables

The x, y, z axes referred to in the tables are a set of three mutually perpendicular axes with the following properties:

(1) in the group C_s the z axis is perpendicular to the reflection plane;
(2) in groups with one main axis of symmetry the z axis points along the main axis of symmetry, while the x axis, if applicable, lies in one of the σ_v planes or coincides with one of the C_2' axes; in groups D_2 and D_{2h} the x, y, z axes coincide with three equivalent twofold axes;
(3) in groups O_h, O, T_d the x, y, z axes lie along the three mutually perpendicular fourfold (O_h and O) or twofold (T_d) axes.

The following character tables with real characters are given in condensed form, i.e. groups having identical character tables are put into one common table.

Appendix: Character tables

C_s	E	σ_h		
A'	1	1	x, y, R_z	x^2, y^2, z^2, xy
A''	1	-1	z, R_x, R_y	yz, xz

C_i	E	i		
A_g	1	1	R_x, R_y, R_z	$x^2, y^2, z^2, xy, xz, yz$
A_u	1	-1	x, y, z	

C_2	E	C_2		
A	1	1	z, R_z	x^2, y^2, z^2, xy
B	1	-1	x, y, R_x, R_y	yz, xz

C_{3v}	E	$2C_3$	$3\sigma_v$		
D_3	E	$2C_3$	$3C_2$		
A_1	1	1	1	z	$x^2 + y^2, z^2$
A_2	1	1	-1	R_z	
E	2	-1	0	$(x, y); (R_x, R_y)$	$(x^2 - y^2, xy); (yz, xz)$

C_3	E	$2C_3$		
A	1	1	z, R_z	$x^2 + y^2, z^2$
E	2	-1	$(x, y); (R_x, R_y)$	$(x^2 - y^2, xy); (yz, xz)$

Appendix: Character tables

C_4 / S_4	E / E	$2C_4$ / $2S_4$	C_2 / C_2	C_4	S_4	C_4 / S_4
A	1	1	1	z, R_z	R_z	x^2+y^2, z^2
B	1	-1	1		z	x^2-y^2, xy
E	2	0	-2	$(x,y); (R_x, R_y)$	$(x,y); (R_x, R_y)$	(yz, xz)

C_5	E	$2C_5$	$2C_5^2$		
A	1	1	1	z, R_z	x^2+y^2, z^2
E_1	2	$\dfrac{1}{\varphi}$	$-\varphi$	$(x,y); (R_x, R_y)$	(yz, xz)
E_2	2	$-\varphi$	$\dfrac{1}{\varphi}$		(x^2-y^2, xy)

$$\dfrac{1}{\varphi} = 2\cos 72° = \dfrac{\sqrt{5}-1}{2}.$$

T	E	$8C_3$	$3C_2$		
A	1	1	1		$x^2+y^2+z^2$
E	2	-1	2		$(3x^2-r^2, x^2-y^2)$
T	3	0	-1	$(x,y,z); (xy, xz, yz)$	
				(R_x, R_y, R_z)	

Appendix: Character tables

C_{2v}	C_{2h}	D_2	E	C_2^z	σ_v^{xz}	σ_v^{yz}	C_{2v}		C_{2h}		D_2	
			E	C_2^z	i	σ_h^{xy}						
			E	C_2^z	C_2^y	C_2^x						
A_1	A_g	A	1	1	1	1	z	x^2, y^2, z^2	R_z	x^2, y^2, z^2, xy		x^2, y^2, z^2
A_2	A_u	B_1	1	1	-1	-1	R_z	xy	z		z, R_z	xy
B_1	B_g	B_2	1	-1	1	-1	x, R_y	xz	R_x, R_y	xz, yz	y, R_y	xz
B_2	B_u	B_3	1	-1	-1	1	y, R_x	yz	x, y		x, R_x	yz

C_{5v}	D_5	E	$2C_5$	$2C_5^2$	$5\sigma_v$	C_{5v}	D_5	C_{5v} D_5	
		E	$2C_5$	$2C_5^2$	$5C_2$				
A_1		1	1	1	1	z			$x^2 + y^2, z^2$
A_2		1	1	1	-1	R_z	z, R_z		
E_1		2	$\dfrac{1}{\varphi}$	$-\varphi$	0	$(x, y); (R_x, R_y)$	$(x, y); (R_x, R_y)$		(xz, yz)
E_2		2	$-\varphi$	$\dfrac{1}{\varphi}$	0				$(x^2 - y^2, xy)$

$$\varphi = -2\cos 144° = \dfrac{\sqrt{5} + 1}{2}.$$

$$\dfrac{1}{\varphi} = 2\cos 72° = \dfrac{\sqrt{5} - 1}{2}.$$

I	E	$12C_5$	$12C_5^2$	$20C_3$	$15C_2$		
A	1	1	1	1	1		$x^2+y^2+z^2$
T_1	3	φ	$-\dfrac{1}{\varphi}$	0	-1	$(R_x,R_y,R_z); (x,y,z)$	
T_2	3	$-\dfrac{1}{\varphi}$	φ	0	-1		
G	4	-1	-1	1	0		
H	5	0	0	-1	1		

C_{4h}	E	$2C_4$	C_2	σ_h	$2S_4$	i		
A_g	1	1	1	1	1	1	R_z	x^2+y^2, z^2
A_u	1	1	1	-1	-1	-1	z	
B_g	1	-1	1	1	-1	1		x^2-y^2, xy
B_u	1	-1	1	-1	1	-1		
E_g	2	0	-2	-2	0	2	(R_x,R_y)	(xz, yz)
E_u	2	0	-2	2	0	-2	(x,y)	

C_{5h}	E	$2C_5$	$2C_5^2$	σ_h	$2S_5^3$	$2S_5^2$		
A'	1	1	1	1	1	1	R_z	x^2+y^2, z^2
A''	1	1	1	-1	-1	-1	z	
E_1'	2	$\dfrac{1}{\varphi}$	$-\varphi$	2	$\dfrac{1}{\varphi}$	$-\varphi$	(x,y)	
E_1''	2	$\dfrac{1}{\varphi}$	$-\varphi$	-2	$-\dfrac{1}{\varphi}$	φ	(R_x,R_y)	(xz, yz)
E_2'	2	$-\varphi$	$\dfrac{1}{\varphi}$	2	$-\varphi$	$\dfrac{1}{\varphi}$		(x^2-y^2, xy)
E_2''	2	$-\varphi$	$\dfrac{1}{\varphi}$	-2	φ	$-\dfrac{1}{\varphi}$		

$$\varphi = -2\cos 144° = \frac{\sqrt{5}+1}{2},$$

$$\frac{1}{\varphi} = 2\cos 72° = \frac{\sqrt{5}-1}{2}.$$

Appendix: Character tables

C_{6v}	E	$2C_6$	$2C_3$	C_2	$3\sigma_v$	$3\sigma_d$					
D_6	E	$2C_6$	$2C_3$	C_2	$3C_2'$	$3C_2''$	C_{6v}	D_6	D_{3h}	$\begin{array}{c}C_{6v}\\D_6\\D_{3h}\end{array}$	D_{3d}
D_{3h}	E	$2S_3$	$2C_3$	σ_h	$3C_2'$	$3\sigma_v$					
D_{3d}	E	$2S_6$	$2C_3$	i	$3C_2'$	$3\sigma_d$					
$A_1\ A_1'\ A_{1g}$	1	1	1	1	1	1	z			$x^2+y^2,$ z^2	x^2+y^2, z^2
$A_2\ A_2'\ A_{2g}$	1	1	1	1	-1	-1	R_z	z, R_z	R_z		R_z
$B_1\ A_1''\ A_{1u}$	1	-1	1	-1	1	-1					
$B_2\ A_2''\ A_{2u}$	1	-1	1	-1	-1	1			z		z
$E_1\ E''\ E_u$	2	-2	-1	1	0	0	$(x,y);$ (R_x,R_y)	(x,y) (R_x,R_y)	(R_x,R_y)	(xy, yz) (x,y)	(R_x, R_y)
$E_2\ E'\ E_g$	2	2	-1	-1	0	0	(x^2-y^2, xy)		(x,y)	$(x^2-y^2, (R_x,R_y)$ (x^2-y^2, xy)	$(x^2-y^2, xy); (xz, yz)$

C_{3h}	E	$2C_3$	σ_h	$2S_3$			
C_6	E	$2C_6^2$	C_6	$2C_6$	C_{3h}	C_6	S_6
S_6	E	$2C_3$	i	$2S_6$			
$A'\ A\ A_g$	1	1	1	1	R_z	x^2+y^2, z^2	x^2+y^2, z^2
$A''\ B\ A_u$	1	1	-1	-1	z	z, R_z	z
$E'\ E_2\ E_g$	2	-1	2	-1	(x,y) (R_x, R_y)	(x^2-y^2, xy) (xz, yz)	R_z (R_x, R_y) (x,y)	$(x^2-y^2, xy);$ (xz, yz)
$E''\ E_1\ E_u$	2	-1	-2	1		(x,y) (R_x, R_y)	

| C_{4v} | E | $2C_4$ | C_2 | $2\sigma_v$ | $2\sigma_d$ | | | |
| D_4 | E | $2C_4$ | C_2 | $2C_2'$ | $2C_2''$ | C_{4v} | D_4 | D_{2d} |
D_{2d}	E	$2S_4$	C_2	$2C_2'$	$2\sigma_d$			
A_1	1	1	1	1	1	z		x^2+y^2, z^2
A_2	1	1	1	-1	-1	R_z	z, R_z	R_z
B_1	1	-1	1	1	-1			x^2-y^2
B_2	1	-1	1	-1	1			xy; z
E	2	0	-2	0	0	(x,y); (R_x,R_y)	(x,y); (R_x,R_y)	(x,y); (R_z,R_y) (xz, yz)

| T_d | E | $8C_3$ | $3C_2$ | $6S_4$ | $6\sigma_d$ | | | |
O	E	$8C_3$	$3C_2$	$6C_4$	$6C_2'$	T_d	O	
A_1	1	1	1	1	1			x^2+y^2, z^2
A_2	1	1	1	-1	-1			
E	2	-1	2	0	0			$(2z^2-x^2-y^2, x^2-y^2)$
T_1	3	0	-1	1	-1	(R_x, R_y, R_z)	(x,y,z); (R_x, R_y, R_z)	
T_2	3	0	-1	-1	1	(x,y,z)		(xy, xz, yz)

Appendix: Character tables

D_{4d}	E	$2S_8$	$2C_4$	$2S_8^3$	C_2	$4C_2'$	$4\sigma_d$		
A_1	1	1	1	1	1	1	1		x^2+y^2, z^2
A_2	1	1	1	1	1	−1	−1	R_z	
B_1	1	−1	1	−1	1	1	−1		
B_2	1	−1	1	−1	1	−1	1	z	
E_1	2	$\sqrt{2}$	0	$-\sqrt{2}$	−2	0	0	(x,y)	
E_2	2	0	−2	0	2	0	0		(x^2-y^2, xy)
E_3	2	$-\sqrt{2}$	0	$\sqrt{2}$	−2	0	0	(R_x, R_y)	(xz, yz)

C_{6h}	E	$2C_6$	$2C_3$	C_2	σ_h	$2S_6$	$2S_3$	i		
A_g	1	1	1	1	1	1	1	1	R_z	x^2+y^2, z^2
A_u	1	1	1	1	−1	−1	−1	−1	z	
B_g	1	−1	1	−1	−1	1	−1	1		
B_u	1	−1	1	−1	1	−1	1	−1		
E_{1g}	2	1	−1	−2	−2	−1	1	2	(R_x, R_y)	(xz, yz)
E_{1u}	2	1	−1	−2	2	1	−1	−2	(x,y)	
E_{2g}	2	−1	−1	2	2	−1	−1	2		(x^2-x^2, xy)
E_{2u}	2	−1	−1	2	−2	1	1	−2		

D_{2h}	E	C_2^z	C_2^y	C_2^x	i	σ^{xy}	σ^{xz}	σ^{yz}		
A_g	1	1	1	1	1	1	1	1		x^2, y^2, z^2
B_{1g}	1	1	−1	−1	1	1	−1	−1	R_z	xy
B_{2g}	1	−1	1	−1	1	−1	1	−1	R_y	xz
B_{3g}	1	−1	−1	1	1	−1	−1	1	R_x	yz
A_u	1	1	1	1	−1	−1	−1	−1		
B_{1u}	1	1	−1	−1	−1	−1	1	1	z	
B_{2u}	1	−1	1	−1	−1	1	−1	1	y	
B_{3u}	1	−1	−1	1	−1	1	1	−1	x	

D_{5h}	D_{5d}	E	$2C_5$	$2C_5^2$	σ_h	$5C_2$	$5\sigma_v$	$2S_5$	$2S_5^3$	D_{5h}		D_{5d}	
		E	$2C_5$	$2C_5^2$	i	$5C_2$	$5\sigma_d$	$2S_{10}^3$	$2S_{10}$				
A_1'	A_{1g}	1	1	1	1	1	1	1	1		x^2+y^2, z^2		x^2+y^2, z^2
A_1''	A_{1u}	1	1	1	-1	1	-1	-1	-1				
A_2'	A_{2g}	1	1	1	1	-1	-1	1	1	R_z		R_z	
A_2''	A_{2u}	1	1	1	-1	-1	1	-1	-1	z		z	
E_1'	E_{1g}	2	$1/\varphi$	$-\varphi$	2	0	0	$1/\varphi$	$-\varphi$	(x,y)		(R_x, R_y)	(xz, yz)
E_1''	E_{1u}	2	$1/\varphi$	$-\varphi$	-2	0	0	$-1/\varphi$	φ	(R_x, R_y)	(xy, yz)	(x, y)	
E_2'	E_{2g}	2	$-\varphi$	$1/\varphi$	2	0	0	$-\varphi$	$1/\varphi$		(x^2-y^2, xy)		(x^2-y^2, xy)
E_2''	E_{2u}	2	$-\varphi$	$1/\varphi$	-2	0	0	φ	$-1/\varphi$				

$$\varphi = -2\cos 144° = \frac{\sqrt{5}+1}{2}.$$

$$\frac{1}{\varphi} = 2\cos 72° = \frac{\sqrt{5}-1}{2}.$$

Appendix: Character tables   367

D_{6d}	E	$2S_{12}$	$2C_6$	$2S_4$	$2C_3$	$2S_{12}^5$	C_2	$6C_2'$	$6\sigma_d$		
A_1	1	1	1	1	1	1	1	1	1		x^2+y^2, z^2
A_2	1	1	1	1	1	1	1	−1	−1	R_z	
B_1	1	−1	1	−1	1	−1	1	1	−1		
B_2	1	−1	1	−1	1	−1	1	−1	1	z	
E_1	2	$\sqrt{3}$	1	0	−1	−$\sqrt{3}$	−2	0	0	(x,y)	
E_2	2	1	−1	−2	−1	1	2	0	0		(x^2-y^2, xy)
E_3	2	0	−2	0	2	0	−2	0	0		
E_4	2	−1	−1	2	−1	−1	2	0	0		(xz, yz)
E_5	2	−$\sqrt{3}$	1	0	−1	$\sqrt{3}$	−2	0	0	(R_x, R_y)	

D_{4h}	E	$2C_4$	C_2	$2C_2'$	$2C_2''$	i	$2S_4$	σ_h	$2\sigma_v$	$2\sigma_d$		
A_{1g}	1	1	1	1	1	1	1	1	1	1		x^2+y^2, z^2
A_{2g}	1	1	1	−1	−1	1	1	1	−1	−1	R_z	
B_{1g}	1	−1	1	1	−1	1	−1	1	1	−1		x^2-y^2
B_{2g}	1	−1	1	−1	1	1	−1	1	−1	1		xy
E_g	2	0	−2	0	0	2	0	−2	0	0	(R_x, R_y)	(xz, yz)
A_{1u}	1	1	1	1	1	−1	−1	−1	−1	−1		
A_{2u}	1	1	1	−1	−1	−1	−1	−1	1	1	z	
B_{1u}	1	−1	1	1	−1	−1	1	−1	−1	1		
B_{2u}	1	−1	1	−1	1	−1	1	−1	1	−1		
E_u	2	0	−2	0	0	−2	0	2	0	0	(x, y)	

O_h	E	$8C_3$	$6C_2$	$6C_4$	$3C_4^2$	i	$6S_4$	$8S_6$	$3\sigma_h$	$6\sigma_d$		
A_{1g}	1	1	1	1	1	1	1	1	1	1		x^2+y^2, z^2
A_{1u}	1	1	1	1	1	−1	−1	−1	−1	−1		
A_{2g}	1	1	−1	−1	1	1	−1	1	1	−1		
A_{2u}	1	1	−1	−1	1	−1	1	−1	−1	1		
E_g	2	−1	0	0	2	2	0	−1	2	0		$(2z^2-x^2-y^2, x^2-y^2)$
E_u	2	−1	0	0	2	−2	0	1	−2	0		
T_{1g}	3	0	−1	1	−1	3	1	0	−1	−1	(R_x, R_y, R_z)	
T_{1u}	3	0	−1	1	−1	−3	−1	0	1	1	(x, y, z)	
T_{2g}	3	0	1	−1	−1	3	−1	0	−1	1		(xz, yz, xy)
T_{2u}	3	0	1	−1	−1	−3	1	0	1	−1		

I_h	E	$12C_5$	$12C_5^2$	$20C_3$	$15C_2$	i	$12S_{10}$	$12S_{10}^3$	$20S_6$	15σ		
A_g	1	1	1	1	1	1	1	1	1	1		$x^2+y^2+z^2$
A_u	1	1	1	1	1	-1	-1	-1	-1	-1		
T_{1g}	3	φ	$-1/\varphi$	0	-1	3	$-1/\varphi$	φ	0	-1	(R_x, R_y, R_z)	
T_{1u}	3	φ	$-1/\varphi$	0	-1	-3	$1/\varphi$	$-\varphi$	0	1	(x, y, z)	
T_{2g}	3	$-1/\varphi$	φ	0	-1	3	φ	$-1/\varphi$	0	-1		
T_{2u}	3	$-1/\varphi$	φ	0	-1	-3	$-\varphi$	$1/\varphi$	0	1		
G_g	4	-1	-1	1	0	4	-1	-1	1	0		
G_u	4	-1	-1	1	0	-4	1	1	-1	0		
H_g	5	0	0	-1	1	5	0	0	-1	1		$(2z^2-x^2-y^2,$ $x^2-y^2,$ $xy, yz, zx)$
H_u	5	0	0	-1	1	-5	0	0	1	-1		

$$\varphi = -2\cos 144° = \frac{\sqrt{5}+1}{2},$$

$$\frac{1}{\varphi} = 2\cos 72° = \frac{\sqrt{5}-1}{2}.$$

Appendix: Character tables

D_{6h}	E	$2C_6$	$2C_3$	C_2	$3C_2'$	$3C_2''$	σ_h	$3\sigma_v$	$3\sigma_d$	$2S_6$	$2S_3$	i		
A_{1g}	1	1	1	1	1	1	1	1	1	1	1	1		x^2+y^2, z^2
A_{1u}	1	1	1	1	1	1	−1	−1	−1	−1	−1	−1		
A_{2g}	1	1	1	1	−1	−1	1	−1	−1	1	1	1	R_z	
A_{2u}	1	1	1	1	−1	−1	−1	1	1	−1	−1	−1	z	
B_{1g}	1	−1	1	−1	1	−1	1	1	−1	1	−1	1		
B_{1u}	1	−1	1	−1	1	−1	−1	−1	1	−1	1	−1		
B_{2g}	1	−1	1	−1	−1	1	1	−1	1	1	−1	1		
B_{2u}	1	−1	1	−1	−1	1	−1	1	−1	−1	1	−1		
E_{1g}	2	1	−1	−2	0	0	−2	0	0	−1	1	2	(R_x, R_y)	(xz, yz)
E_{1u}	2	1	−1	−2	0	0	2	0	0	1	−1	−2	(x, y)	
E_{2g}	2	−1	−1	2	0	0	2	0	0	−1	−1	2		(x^2-y^2, xy)
E_{2u}	2	−1	−1	2	0	0	−2	0	0	1	1	−2		

Bibliography

Adini, A. & Clough, R. W. (1961) Analysis of plate bending by the finite element method. *Grant G 7337*, National Science foundation, USA.

Argyris, J. H., & Fried, I. (1968) The LUMINA element for the matrix displacement method. *The Aeronautical Journal of the Royal Aeronautical Society*, 72 No. 690, 514–17.

Argyris, J. H. & Kelsey, S. (1960) *Energy theorems and structural analysis*. Butterworths, London.

Au, T. & Christiano, P. (1987) *Structural analysis*. Prentice-Hall, Englewood Cliffs, New Jersey.

Bogner, F. K., Fox, R. L. & Schmit, L. A., Jr. (1966) The generation of interelement-compatible stiffness and mass matrices by the use of interpolation formulas. *Proceedings, Conference on Matrix Methods in Structural Mechanics*, AFFDL TR-66-80. Fairborn, Ohio, 397–444.

Clough, R. W. (1969) Comparison of three-dimensional finite elements. *Proc. Symp. on Application of Finite Element Methods in Civil Engineering*. Nashville, Tennessee.

Dawe, D. J. (1965) A finite element approach to plate vibration problems. *J. Mech. Eng. Sci.*, 7, 28–32.

Dawe, D. J. (1984) *Matrix and finite element displacement analysis of structures*. Clarendon Press, Oxford.

Ergatoudis, J. G., Irons, B. M. & Zienkiewicz, O. C. (1968a) Three-dimensional analysis of arch dams and their foundation. *Symposium on Arch Dams at the Institution of Civil Engineers*, London.

Ergatoudis, J. G., Irons, B. M. & Zienkiewicz, O. C. (1968b) Curved isoparametric quadrilateral elements for finite element analysis. *International Journal of Solids and Structures*, 4 No. 1, 31–42.

Falicov, L. M. (1966) *Group theory and its physical applications*. University of Chicago Press, Chicago.

Hall, L. H. (1969) *Group theory and symmetry in chemistry*. McGraw-Hill, New York.

Hamermesh, M. (1962) *Group theory and its applications to physical problems*. McGraw-Hill, New York.

Kolář, V., Kratochvil, L., Leitner, F. & Ženišek, A. (1972) *Berechnung von Flächen- und Raumtragwerken.* Springer-Verlag, Vienna, New York. (in German).

Kurosh, A. G. (1967) *Group theory*, 3rd (expanded) edn. Nauka, Moscow (in Russian).

Mathiak, K. & Stingl, P. (1968) *Gruppentheorie.* Vieweg, Braunschweig (in German).

Melosh, R. J. (1961) A stiffness matrix for the analysis of thin plates in bending. *Journal of Aerospace Sciences*, **28** No. 1, 34–42.

Melosh, R. J. (1963a) Basis for derivation of matrices for the direct stiffness method. *Journal of the American Institute of Aeronautics and Astronautics*, **1**, 1631–7.

Melosh, R. J. (1963b) Structural analysis of solids. *Journal of the Structural Division ASCE*, **89** No. ST4, 205–23.

Pawsey, S. F. & Clough, R. W. (1971) Improved numerical integration of thick shell finite elements. *International Journal for Numerical Methods in Engineering*, **3**, 375–580.

Rigby, G. L. & McNeice, G. N. (1972) A strain energy basis of element stiffness matrices. *Journal of the American Institute of Aeronautics and Astronautics*, **11**, 1490–3.

Rockey, K. C., Evans, H. R. Griffiths, D. W. & Nethercot, D. A. (1975) *The finite element method.* Crosby Lockwood Staples, London.

Schonland, D. (1965) *Molecular symmetry.* Van Nostrand, London.

Sekulović, M. (1984) *The finite element method.* Gradjevinska knjiga, Beograd (in Serbo-Croatian).

Szilard, R. (1974) *Theory and analysis of plates.* Prentice-Hall, Englewood Cliffs, New Jersey.

Taylor, R. L. (1972) On completeness of shape functions for finite element analysis. *International Journal for Numerical Methods in Engineering*, **4**, 17–22.

Yang, T. Y. (1986) *Finite element structural analysis.* Prentice-Hall, Englewood Cliffs, New Jersey.

Zienkiewicz, O. C., The finite element method for analysis of elastic isotropic and orthotropic slabs. *Proc. Inst. Civ. Eng.*, **28**, 471–88.

Zienkiewicz, O. C., Taylor, R. L. & Too, J. M. (1971) Reduced integration technique in general analysis of plates and shells. *International Journal for Numerical Methods in Engineering*, **3**, 275–90.

Zloković, G. (1958) Moments in the centre of elastic plates subjected to uniform load. *Tehnika* XIII, 2; Naše gradjevinarstvo **XII** No. 2, 34–40 (in Serbo-Croatian; German summary).

Zloković, G. (1969a) *The co-ordinated system of constructions.* Gradjevinska knjiga, Beograd (in English and Serbo-Croatian).

Zloković, G. (1969b) *Space structures.* Gradjevinska knjiga, Beograd (in English and Serbo-Croatian).

Zloković, G. (1973) *Group theory and G-vector spaces in vibration, stability and statics of structures.* ICS, Beograd (in English and Serbo-Croatian).

Zloković, G. (1974) Vibration of cable networks and the G-invariant subspaces method. *Proceedings of the V Congress of the Yugoslav society of structural engineers*, Budva, A-30, 429–440 (in Serbo-Croatian; English summary).

Zloković, G. (1976) Vibration of cable networks and the G-invariant subspaces method.

Stroitelnaya mehanika, raschot i konstruirovanie sooruzeniy, Institute of Architecture, Moscow, **5**, 149–159 (in Russian, translated from Serbo-Croatian).

Zloković, G. (1977) *Group theory and G-vector spaces in vibrations, stability and statics of structures*. Stroizdat, Moscow (in Russian, translated from Serbo-Croatian).

Zloković, G. (1989) *Group theory and G-vector spaces in structural analysis: vibration, stability and statics*. Ellis Horwood, Chichester.

Index

Abelian group, 14
abstract group theory, 14
Adini, 257, 371
Argyris, 153, 257, 371
assembly of system stiffness matrices, 329, 335, 345
Au, 371

bases of G-invariant subspaces, 46
basis
    of a function space, 32
    of a vector space, 21
    of the centre of the group algebra, 33, 40
    transformation, 27
    vectors, 21
basis transformation supermatrix, 93
basis vectors of G-invariant subspaces, 46, 48, 93, 147, 254, 328
Bogner, 257, 371
boundary conditions
    of the linear beam element assembly, 334, 336
    of the girder grillage, 353

Cartesian sets and products
    in addition to the character table of the group, 150, 357
    in addition to C_2, 257
    in addition to C_{2v}, 154, 273
    in addition to D_{2h}, 199
centre
    of a group, 16
    of symmetry, 17
    of the group algebra, 33, 40
character, 34, 35
character tables, 34, 38, 357
    in supermatrix form, 91
Christiano, 371
class
    of conjugate group elements, 16, 33, 40
    sum, 33, 40

classification
    of irreducible group representations, 43
    of symmetry groups, 17
Clough, 153, 257, 371, 372
conjugate element in a group, 16
conjugate transpose, 31
correspondence of operations, 25
cyclic group, 15

Dawe, 257, 371
decomposition of a G-vector space, 46
derivation of basis vectors of G-invariant subspaces with respect to the nodal numbering, 147, 254
determination pf the symmetry group
    of a space object, 20
    of a configuration analysed by the direct stiffness method, 327
dimension
    of a character, 43
    of a vector space, 21
    of the centre of the group algebra, 33
dimensions of G-invariant subspaces
    equation for analytical determination of, 46
    in special form, 71
    for configuration compatible with the icosahedral group, 80
    in relation to nodal positions with respect to reflection planes, 327
displacement fields in G-invariant subspaces, 151, 255
    beam element, 269
    rectangular element
        with four nodes, 156
        with eight nodes, 164
        with twelve nodes, 173
        with sixteen nodes, 189
    rectangular element for planar analysis, 281
    rectangular element for plate flexure (12 d.-o.-f.), 301

Index   375

rectangular element for plate flexure
    (16 d.-o.-f.), 324
rectangular hexahedral element
    with eight nodes, 201
    with twenty nodes, 215
    with thirty-two nodes, 235
    with sixty-four nodes, 247
displacement function decomposed into G-invariant subspaces, 149, 255
    beam element, 268
    rectangular element
        with four nodes, 156
        with eight nodes, 163
        with twelve nodes, 172
        with sixteen nodes, 189
    rectangular element for planar analysis, 280
    rectangular element for plate flexure
        (12-d.-o.-f.), 299
    rectangular element for plate flexure
        (16-d.-o.-f.), 324
    rectangular hexahedral element
        with eight nodes, 200
        with twenty nodes, 212
        with thirty-two nodes, 229
        with sixty-four nodes, 247
dodecahedral–icosahedral configuration, 67
    general, 75
    general expression for the total number of nodes, 75
    node classification, 73, 80
    with 120 nodes, 81
    with 182 nodes, 72
    with 752 nodes, 75

Ergatoudis, 153, 371
Evans, 372

Falicov, 371
Fox, 371
Fried, 153, 371
function spaces, 32

generators of a group, 15
generator system of a vector space, 21
geometrical framework, 14
G-invariant subspaces, 34
girder grillage, 339
global coordinate system in the group supermatrix procedure, 327, 332, 342
Griffiths, 372
group, 14
    axioms, 14
    definition, 14
    element, 14
group algebra, 33, 39
    centre of, 33, 40
group representation, 34
    equivalent, 34
    irreducible, 33, 35
    reducible, 35
group supermatrices
    group C_2
        for a two-node line element, 94
        for a four-node line element, 98, 99
        for a three-node line element, 102
    group C_3, for the three-node equilateral triangle element, 104
    group C_4, for the four-node square element, 107
    group C_{2v}
        for the four-node rectangular element, 111
        for the four-node rhombic element, 119
    group C_{3v}, for the six-node hexagonal element, 121
    group C_{4v}, for the eight-node octogonal element, 126
    group D_{2h}
        for the eight-node rectangular hexahedral element, 130
        for the twelve-node truncated rectangular hexahedral element, 141
group supermatrix
    in normal form, 90
    in diagonal form, 93
group supermatrix nodal numbering, 90, 146, 254
    position of the first node, 90
    use of the first nodes in the nodal sets, 147, 254
group supermatrix procedure
    for derivation of shape functions in G-invariant subspaces, 146
    for derivation of stiffness equations in G-invariant subspaces, 254
    in the direct stiffness method, 326
group supermatrix transformation, 93, 256
    inverse, 93
group table, 14
G-vector analysis, 47
    algorithmic scheme of, 84, 89
    in the direct stiffness method, 327
G-vector analysis of node patterns
    based on the cube, 58
    based on the right parallelepiped, 62
    based on the square, 48
        with four nodes, 48
        with twelve nodes, 53
    based on the regular icosahedron, 67
G-vector space, 34, 44
    decomposition of a, 46
    group representation as a, 44
    subspace of a, 44

Hall, 371
Hammermesh, 13, 371

idempotents of the centre of group algebra, 34, 41

linear independence of, 41
maximum system of orthogonal, 41
orthogonal, 34, 41
in supermatrix form, 92
identity, 14
operation, 17
invariant subspace, 35
inverse symmetry operation, 17
inversion, 17
Irons, 371
isomorphic groups, 15
set of, 84, 85

Kelsey, 153, 257, 371
Kolář, 372
Kratochvíl, 372
Kurosh, 13, 372

Leitner, 372
length
of a vector, 30
of a function, 33
linear beam element assembly, 330
linear group of V, 24
linear independence of vectors, 21, 31
linear operators, 23
regular, 24, 25
local coordinate system in the group supermatrix procedure, 329, 331, 342

Mathiak, 13, 372
matrix, 24
adjoint of a, 31
conjugate matrix, 27
transpose of a, 31
inverse, 26
invertible, 26
multiplication rule, 26
metric, 31
nonsingular, 25
regular, 25
self-adjoint, 31
unitary, 32
matrix representation, 44
McNeice, 153, 372
Melosh, 153, 372
Mulliken's notation for irreducible representations, 44, 357

n-dimensional representation of the group, 33, 34
n-dimensional vector space, 20
Nethercot, 372
nodal framework, 14
nodal numbering in the group supermatrix procedure, 90, 145, 254, 328
group C_2, 94, 98, 99, 102, 257, 259, 266, 330
group C_3, 105
group C_4, 108

group C_{2v}, 112, 119, 153, 159, 170, 186, 273, 290, 339
group C_{3v}, 121
group C_{4v}, 127
group D_{2h}, 130, 141, 199, 206, 224, 242
nodal sets, 146, 254, 328
normalization
of a vector, 30
of a function, 33
numbering of nodal functions (variables), 104, 328, 330, 341

operator of a vector space, 23
order
of a group, 14, 16
of a rotation axis, 17
orthogonal
functions, 33
subspaces, 30
vectors, 30, 31
orthogonality relation
of vectors, 30
of functions, 33
orthogonality relation in representation theory
first, 37
second, 38
orthonormal bases, 31

Pawsey, 153, 372
permutation group, 15
point groups, 17
positive directions of generalized displacements and forces suiting the first symmetry type of the group, 50, 254, 266, 273, 290, 326, 329, 330, 333, 341
procedure of the G-vector analysis, 48
projection operator, 34, 43
properties of irreducible representations, 37

reflection, 16
relation between characters and idempotents, 42
relation between two bases of an n-dimensional vector space, 27
relation of basis vectors and nodal functions as systems of equations with supermatrices in diagonal form, 147
in derivations of group supermatrices
group C_2, 96, 99, 102
group C_{2v}, 116, 120
group D_{2h}, 137, 144
in derivations of shape functions
group C_{2v}, 161, 172, 187
group D_{2h}, 208, 227, 245
in derivations of stiffness equations, 255
group C_2, 260, 267
group C_{2v}, 279, 298, 317
relations of displacements to coefficients of the polynomial in G-invariant subspaces, 150, 255
beam element, 269

Index   377

rectangular element
　with four nodes, 156
　with eight nodes, 163
　with twelve nodes, 173
　with sixteen nodes, 189
rectangular element for planar analysis, 280
rectangular element for plate flexure
　(12 d.-o.-f.), 299
rectangular element for plate flexure
　(16 d.-o.-f.), 324
rectangular hexahedral element
　with eight nodes, 200
　with twenty nodes, 212
　with thirty-two nodes, 230
　with sixty-four nodes, 248
representation theory, 33
　first orthogonality relation, 37
　second orthogonality relation, 38
　theorems of, (1.1) 37, (1.2) 37, (1.3) 38,
　　(1.4) 38, (1.5) 40, (1.6) 41, (1.7) 41,
　　(1.8) 41, (1.9) 46, (1.10) 46
Rigby, 153, 372
Rockey, 257, 372
rotary reflection, 16
rotation, 16

scalar product, 30, 32, 33
Schmit, 371
Schonland, 13, 372
Sekulović, 372
shape functions at the nodes in G-invariant
　subspaces, 151
　rectangular element
　　with four nodes, 157
　　with eight nodes, 167
　　with twelve nodes, 180
　　with sixteen nodes, 196
　rectangular hexahedral element
　　with eight nodes, 204
　　with thirty-two nodes, 241
shape functions in G-invariant subspaces, 146,
　151, 255
　beam element, 270
　rectangular element
　　with four nodes, 153, 157
　　with eight nodes, 159, 165
　　with twelve nodes, 170, 173
　　with sixteen nodes, 185, 196
　rectangular element for planar analysis, 282
　rectangular element for plate flexure
　　(12 d.-o.-f.), 301
　rectangular element for plate flexure
　　(16 d.-o.-f.), 324
　rectangular hexahedral element
　　with eight nodes, 199, 203
　　with twenty nodes, 205, 215
　　with thirty-two nodes, 224, 237
　　with sixty-four nodes, 241, 252

shape functions of elements in G-invariant sub-
　spaces derived from shape functions of
　serendipity elements, 152, 158, 168, 180, 205
stiffness equations in G-invariant subspaces, 254
　comparison with derivations of stiffness
　　equations and matrices by conventional
　　methods, 273, 290, 315
　group supermatrix transformations of, 256,
　　259, 265, 266, 272, 286, 307
　transformed into conventional stiffness
　　equations, 265, 272, 287, 311
　with the stiffness group supermatrix in
　　diagonal form, 256, 259, 264, 271, 286,
　　307
　with the stiffness group supermatrix in
　　normal form, 256, 259, 265, 271, 286,
　　307
stiffness equations of finite elements in
　G-invariant subspaces
　beam element, 266
　one-dimensional elements,
　　two-node bar element, 257, 259
　　four-node bar element, 259
　rectangular element for planar analysis, 273
　rectangular element for plate flexure
　　with 12 d.-o.-f., 290
　　with 16 d.-o.-f., 315
stiffness group supermatrix
　in diagonal form, 255, 264, 271, 286, 304
　in normal form, 256, 259, 265, 271, 286, 311
stiffness matrices of finite elements in G-invariant
　subspaces, 255, 329
　beam element, 270
　　in girder grillage, 342
　　in linear beam element assembly, 331
　one-dimensional elements
　　two-node bar element, 259
　　four-node bar element, 263
　rectangular element for planar analysis, 283
　rectangular element for plate flexure
　　with 12 d.-o.-f., 301
　　with 16 d.-o.-f., 324
Stingl, 13, 372
subgroup, 16
subspace of a vector space, 22
symmetry-adapted generalized displacements,
　330, 337, 349
symmetry-adapted generalized nodal forces, 351
symmetry
　element, 17
　operation, 16
　operator, 35
symmetry group, 16, 34
　data on
　　C_n, D_n, C_{nv}, C_{nh}, 86
　　D_{nh}, D_{nd}, S_n, 87, 88
symmetry groups applied in derivations of group
　supermatrices
　　C_2, 94, 98, 99, 102

C_3, 104
C_4, 107
C_{2v}, 111, 119
C_{3v}, 121
C_{4v}, 126
D_{2h}, 130, 141
symmetry groups applied in G-vector analysis
   C_{2v}, 52
   C_{4v}, 48, 53
   D_{2h}, 63
   I_h, 67
   O, 59
symmetry groups applied in the group supermatrix procedure in the direct stiffness method
   C_2, 330
   C_{2v}, 339
symmetry type of the first irreducible representation of the group, 50
symmetry types of the group, 34, 46, 149, 254, 329
system of generators, 15
system of orthogonal idempotents, 41
system stiffness equations with group supermatrices in normal and diagonal forms, 329, 345
Szilard, 272

Taylor, 153, 372
Too, 372
trace of a matrix, 27, 34
traces of matrices of elements of the icosahedral group, 71
transformation
   of a matrix by change of the basis, 29
   of a vector by a linear operator, 24
   of vector coordinates by change of the basis, 29
transformation supermatrix for formulation of the idempotents, 92

unit
   column 22
   vector, 22
unitary spaces, 30

vector spaces, 20

Yang, 257, 372

Ženišek, 372
Zienkiewicz, 153, 371, 372
Zloković, 7, 8, 9, 13, 47, 327, 372, 373